Deciphering The Role of Succinate Dehydrogenase in Drug Discovery

Vaishali Manikrao Patil
Dileep Kumar • Satya P. Gupta
Editors

Deciphering The Role of Succinate Dehydrogenase in Drug Discovery

SDH in Drug Discovery

Springer

Editors
Vaishali Manikrao Patil
Charak School of Pharmacy
Chaudhary Charan Singh
University (CCSU)
Meerut, Uttar Pradesh, India

Satya P. Gupta
Ex-Professor
Birla Institute of Technology and
Science (BITS)
Pilani, Rajasthan, India

Meerut Institute of Engineering
and Technology
Meerut, Uttar Pradesh, India

Dileep Kumar
Department of Pharmaceutical Chemistry
Manipal College of Pharmaceutical
Sciences
Manipal Academy of Higher Education
Manipal, Karnataka, India

ISBN 978-981-96-6830-4 ISBN 978-981-96-6831-1 (eBook)
https://doi.org/10.1007/978-981-96-6831-1

This Springer imprint is published by the registered company Springer Nature Singapore Pte Ltd.
The registered company address is: 152 Beach Road, #21-01/04 Gateway East, Singapore 189721, Singapore

Preface

In the evolving landscape of biochemistry and medical research, succinate dehydrogenase (SDH) has moved from a relatively niche area to become a focal point of inquiry. Once primarily recognized as a key enzyme within the citric acid cycle and the electron transport chain, SDH is now understood to play critical roles across a range of biological processes, with implications for human health, agriculture, and for the development of therapeutics. This book explores expanding role of SDH across various scientific domains, offering readers a comprehensive view of its functions, pathological relevance, and applications.

The journey begins in Chapter 1, which provides a structural and functional overview of SDH, highlighting its role as Complex II in the electron transport chain and its involvement in the citric acid cycle. Through this, SDH could facilitate crucial electron transfer steps via its subunits, namely, SDHA, SDHB, SDHC, and SDHD. These processes are essential for cellular energy production, and the chapter details how disruptions in SDH activity can lead to metabolic imbalances and disease. Recent research has shed light on SDH's assembly and regulation, showing it as an enzyme whose mutations can lead to various disorders.

In Chapter 2, SDH's involvement in cellular respiration is examined in greater depth, with a focus on how it influences hypoxic response and gene expression. Dysfunction in SDH, caused by either genetic mutations or environmental stressors, disrupts cellular energy balance, leading to conditions that range from benign metabolic disorders to severe pathologies.

Chapter 3 explores SDH's role in maintaining cellular homeostasis, emphasizing how it prevents oxidative stress and supports normal physiological function. Importance of proper SDH activity is essential for energy production. Any type of disruptions in this enzyme can lead to cellular damage and further impacts overall health.

The significance of SDH as a therapeutic target in neglected diseases is examined in Chapter 4. Enzymes that participate in energy metabolism, like SDH, are pivotal for parasite survival, making them attractive drug targets. This chapter explores the potential of SDH inhibitors for treating diseases such as trypanosomiasis, schistosomiasis, and leishmaniasis, which disproportionately affect low-income populations. Current research on SDH-targeted therapies highlights its potential as a treatment target in diseases often overlooked in mainstream drug development.

In Chapter 5, the focus shifts to mitochondrial disorders linked to SDH dysfunction. The chapter provides an analysis of how disruptions in SDH activity contribute to mitochondrial pathologies, which in turn affect cellular respiration and energy production. As mitochondrial diseases are often complex, this chapter delves into how SDH dysfunction interacts with other mitochondrial components, leading to multifactorial pathophysiological effects.

Chapter 6 examines SDH's influence on cardiovascular health, detailing how defects in SDH contribute to conditions like ischemia and Barth syndrome. Using animal models, the chapter investigates how inhibiting SDH during reperfusion can have protective effects on the heart by preventing production of reactive oxygen species and mitochondrial permeability transitions. SDH inhibitors like diazoxide and 3-nitropropionic acid are explored for their cardioprotective potential, alongside discussion on how SDH dysfunction can exacerbate ischemic damage, suggesting new therapeutic strategies for cardiovascular conditions.

In Chapter 7, SDH's connection to cancer is explored, where it functions as a metabolic checkpoint with implications for tumorigenesis. Beyond its enzymatic activity, SDH acts as a tumor suppressor, with its substrate succinate classified as an oncometabolite. SDH mutations lead to succinate accumulation, affecting processes like epigenetic regulation, angiogenesis, and cancer progression. SDH's role in cancer diagnosis is also discussed, presenting it as a biomarker for various cancers.

Chapter 8 focuses on neurodegenerative diseases, explaining how SDH dysfunction can impair mitochondrial function and contribute to conditions like Alzheimer's, Parkinson's, and Huntington's diseases. The chapter explores the enzyme's critical role in the brain's high-energy demand and its potential in neuroprotection, with SDH emerging as a regulator in managing neurodegenerative disease progression.

Chapter 9 further explores SDH's potential in treating neurodegenerative disorders. Given SDH's role in energy homeostasis, the chapter investigates how SDH dysfunction contributes to lipid synthesis abnormalities and excitotoxicity, worsening neurodegenerative conditions. The chapter underscores SDH's significance in developing neuroprotective therapies.

Chapter 10 examines the use of SDH inhibitors in agriculture, particularly as fungicides for crop protection. The chapter provides an overview of SDH inhibitors, their classification, and mechanism of action. These fungicides play a critical role in combating fungal diseases in crops. However, the chapter also addresses resistance in fungal pathogens, environmental concerns, and health risks from long-term SDH inhibitor use. Sustainable approaches are discussed to ensure the efficacy and safety of these fungicides in agriculture.

Chapter 11 highlights importance of SDH in oxidative phosphorylation and the tricarboxylic acid cycle, emphasizing its relevance in metabolic diseases. Known also as electron transport chain complex II, SDH has garnered attention due to its role in regulating mitochondrial metabolism. This chapter explores SDH's involvement in metabolic reprogramming and energy balance, as well as its effects on glucose-induced oxidative phosphorylation. The role of SDH as a potential generator of reactive oxygen species, especially in obesity and inflammatory bowel diseases, is also discussed, alongside its impact on immune system inflammation. By examining SDH's biological and pathological functions, this chapter emphasizes the enzyme's importance in understanding metabolic disorders.

Throughout these chapters, this book offers a deep understanding of SDH's complex biology and its broad impact on human and environmental health. By bridging research across cellular biology, medicine, and agriculture, it highlights SDH's centrality in essential processes and its therapeutic potential. This volume will serve as a resource for scientists, clinicians, and researchers interested in unraveling the role of SDH across disease treatment, drug discovery, and agriculture. Through this exploration, we hope to inspire continued research into SDH and its roles in maintaining cellular health and addressing pressing health and environmental challenges.

Meerut, Uttar Pradesh, India — Vaishali Manikrao Patil
Manipal, Karnataka, India — Dileep Kumar
Meerut, Uttar Pradesh, India — Satya P. Gupta

Contents

Editors and Contributors

About the Editors

Vaishali Manikrao Patil is Associate Professor and Principal at Charak School of Pharmacy, Chaudhary Charan Singh University, Meerut, India; Visiting Professor, Faculty of Pharmaceutical Sciences, UCSI University, Kuala Lumpur, Malaysia; and editorial board member of indexed journals, *viz. Anti-Infective Agents, Coronaviruses*. She has earlier served as Associate Professor and Head-D.Pharm at KIET Group of Institutions (KIET School of Pharmacy), India; Associate Professor and head in the Department of Pharmaceutical Chemistry at SV Subharti University, India; and Assistant Professor at BIT, Meerut, India. Her research interest is in medicinal chemistry and computer-aided drug design with specialization in quantitative structure-activity relationship studies of biologically active molecules, drug-receptor interactions, and synthesis of small heterocyclic molecules. She has been conferred with various prestigious awards notably, Dr. A. P. J. Abdul Kalam Technical University Best Teacher Award. She has served as a referee for several international journals. She has more than 18 years of experience in medicinal chemistry and computer-aided drug design. She has also published more than 80 research articles in the peer-reviewed journals and authored numerous books and book chapters. She has delivered several lectures on application of rational drug design across the world and is a community member of American Chemical Society and a life member of Association of Pharmaceutical Teachers (APTI), India.

Dileep Kumar is an Assistant Professor in Manipal College of Pharmaceutical Sciences, India and was a visiting faculty at the University of California, USA. He received his B.Pharm from Manipal College of Pharmaceutical Sciences, Manipal University, India, and M.Pharm from Shri Govindram Seksaria Institute of Technology and Science, Indore, Madhya Pradesh, India. He was awarded junior research fellowship by National Medicinal Plant Board (NMPB), New Delhi, senior research fellowship from University Grants Commission (UGC), and teaching assistantship and institute post-doctoral fellowship from IIT (BHU) Varanasi, India. He worked as an assistant professor in Poona College of Pharmacy, Bharati Vidyapeeth University, Pune, India. His current research interests include design and development of adamantyl analogues as GluN2B selective N-methyl-D-aspartate (NMDA) receptors antagonist and amyloid-β protein aggregation inhibitors for treatment of Alzheimer's disease. He was also involved in synthesis of soluble epoxide hydrolase (sEH), inhibitors as an anticancer agent at UC Davis Comprehensive Cancer Center, University of California Davis, USA.

Satya P. Gupta is presently a Professor Emeritus at Meerut Institute of Engineering and Technology (MIET), Meerut, India, after retiring as Professor in the Department of Applied Sciences at National Institute of Technical Teachers' Training and Research (NITTTR), Bhopal, India. Earlier he served at Tata Institute of Fundamental Research (TIFR), Mumbai, Birla Institute of Technology and Science (BITS), Pilani, and then Meerut Institute of Engieering and Technology as its Director-cum Professor of Eminence in the Department of Pharmacy. Professor Gupta has a very long standing of teaching quantum chemistry, pharmaceutical chemistry, biophysics, and drug design. He had obtained his M.Sc. and D.Phil degrees from University of Allahabad, Allahabad, India, in 1967 and 1971, respectively. For his work in drug design, he has bagged several honors and awards to his credit, Professor Gupta has more than 200 research publications in highly reputed national and international journals and several reviews in prestigious periodicals as *Chemical Reviews* (American Chemical Society), *Progress in Drug Research* (Birkhäuser Verlag Basel, Switzerland), and *Current Medicinal Chemistry* (Bentham Science, Netherlands), and has been on editorial board of several international journals.

Contributors

Diksha Bharti Laboratory of Drug Discovery and Ecotoxicology, Department of Pharmacy, Guru Ghasidas Vishwavidyalaya (A Central University), Bilaspur, Chhattisgarh, India

Jaishree Biswas Laboratory of Drug Discovery and Ecotoxicology, Department of Pharmacy, Guru Ghasidas Vishwavidyalaya (A Central University), Bilaspur, Chhattisgarh, India

Pablo Rayff da Silva Postgraduate Program of Natural and Synthetic Bioactive Products (PgPNSB), Health Sciences Center, Federal University of Paraíba, João Pessoa, PB, Brazil

Natália Ferreira de Sousa Postgraduate Program of Natural and Synthetic Bioactive Products (PgPNSB), Health Sciences Center, Federal University of Paraíba, João Pessoa, PB, Brazil

Aditya Ganeshpurkar Shri Ram Institute of Technology-Pharmacy, Jabalpur, Madhya Pradesh, India

Ankit Ganeshpurkar Department of Pharmaceutical Sciences, Dr. Harisingh Gour Vishwavidyalaya (A Central University), Sagar, Madhya Pradesh, India

Pragya Gawande Laboratory of Drug Discovery and Ecotoxicology, Department of Pharmacy, Guru Ghasidas Vishwavidyalaya (A Central University), Bilaspur, Chhattisgarh, India

Archit Gupta SAV Pharmacy, Ghaziabad, Uttar Pradesh, India

Satya P. Gupta Ex-Professor, Birla Institute of Technology and Science (BITS), Pilani, Rajasthan, India

Meerut Institute of Engineering and Technology, Meerut, Uttar Pradesh, India

Sukesh Kumar Gupta School of Medicine, Wayne State University, Detroit, MI, USA

Souranava Jana Laboratory of Drug Discovery and Ecotoxicology, Department of Pharmacy, Guru Ghasidas Vishwavidyalaya (A Central University), Bilaspur, Chhattisgarh, India

Haldia Institute of Pharmacy, Haldia, West Bengal, India

Surbhi Jyoti Laboratory of Drug Discovery and Ecotoxicology, Department of Pharmacy, Guru Ghasidas Vishwavidyalaya (A Central University), Bilaspur, Chhattisgarh, India

Dhruv Kansal Central Drugs Standard Control Organization, New Delhi, India

Praveen Thaggikuppe Krishnamurthy Department of Pharmacology, JSS College of Pharmacy, JSS Academy of Technical Education, Noida, Uttar Pradesh, India

Ajay Kumar Sardar Patel College of Pharmacy, Gorakhpur, Uttar Pradesh, India

Akshay Kumar Department of Pharmacy, Guru Ghasidas Vishwavidyalaya (A Central University), Bilaspur, Chhattisgarh, India

Deepak Kumar Sardar Patel College of Pharmacy, Gorakhpur, Uttar Pradesh, India

Devendra Kumar Department of Pharmaceutical Chemistry, School of Pharmacy and Technology Management, Shirpur Campus, Shirpur, Dhule, Maharashtra, India

Dileep Kumar Department of Pharmaceutical Chemistry, Manipal College of Pharmaceutical Sciences, Manipal Academy of Higher Education, Manipal, Karnataka, India

Department of Entomology and Nematology, UC Davis Comprehensive Cancer Center, University of California Davis, Davis, CA, USA

Mohd Qasid Lari Sardar Patel College of Pharmacy, Gorakhpur, Uttar Pradesh, India

Deepa S. Mandlik Department of Pharmacology, Bharti Vidyapeeth (Deemed to be University), Poona College of Pharmacy, Pune, Maharashtra, India

Satish K. Mandlik Department of Pharmacology, Bharti Vidyapeeth (Deemed to be University), Poona College of Pharmacy, Pune, Maharashtra, India

Neeraj Masand Department of Pharmacy, Lala Lajpat Rai Memorial Medical College, Meerut, Uttar Pradesh, India

Balaji Wamanrao Matore Laboratory of Drug Discovery and Ecotoxicology, Department of Pharmacy, Guru Ghasidas Vishwavidyalaya (A Central University), Bilaspur, Chhattisgarh, India

Anjali Murmu Laboratory of Drug Discovery and Ecotoxicology, Department of Pharmacy, Guru Ghasidas Vishwavidyalaya (A Central University), Bilaspur, Chhattisgarh, India

Vaishali Manikrao Patil Charak School of Pharmacy, Chaudhary Charan Singh University (CCSU), Meerut, Uttar Pradesh, India

Partha Pratim Roy Laboratory of Drug Discovery and Ecotoxicology, Department of Pharmacy, Guru Ghasidas Vishwavidyalaya (A Central University), Bilaspur, Chhattisgarh, India

Luciana Scotti Postgraduate Program of Natural and Synthetic Bioactive Products (PgPNSB), Health Sciences Center, Federal University of Paraíba, João Pessoa, PB, Brazil

Marcus Tullius Scotti Postgraduate Program of Natural and Synthetic Bioactive Products (PgPNSB), Health Sciences Center, Federal University of Paraíba, João Pessoa, PB, Brazil

Yashika Sharma Poona College of Pharmacy, Bharati Vidyapeeth (Deemed to be) University, Pune, Maharashtra, India

Jagadish Singh Laboratory of Drug Discovery and Ecotoxicology, Department of Pharmacy, Guru Ghasidas Vishwavidyalaya (A Central University), Bilaspur, Chhattisgarh, India

Siddhant Tripathi Poona College of Pharmacy, Bharati Vidyapeeth (Deemed to be) University, Pune, Maharashtra, India

Viswapriya Viswalingam Department of Pharmaceutical Chemistry, Poona College of Pharmacy, Bharati Vidyapeeth (Deemed to be University), Pune, Maharashtra, India

Succinate Dehydrogenase: Structure, Function and Significance

1

Dileep Kumar, Aditya Ganeshpurkar, and Ankit Ganeshpurkar

Abstract

The chapter underscores succinate dehydrogenase (SDH) as a pivotal enzyme fundamental to cellular life, tracing its historical evolution and structural intricacies. Its central role in the Krebs cycle, efficiently converting succinate to fumarate, is pivotal for electron transport chain function and ATP production. Regulation of SDH activity is crucial for cellular health maintenance. Beyond its metabolic role, SDH dysfunction is linked to various diseases like hereditary paraganglioma–pheochromocytoma syndrome, leiomyomatosis and neurodegeneration, opening avenues for further investigation. Understanding how mutations or altered SDH activity contribute to these conditions can inform novel therapeutic approaches. In summary, SDH transcends its metabolic function, offering rich potential for future research and therapeutic development. Further exploration promises insights into healthcare pathways and a deeper comprehension of cellular dynamics. SDH's complexity and significance ensure ongoing intrigue and inspiration in scientific pursuits.

Keywords

Succinate dehydrogenase · Electron transport chain · ATP · Succinate · Ubiquonoe

D. Kumar
Department of Pharmaceutical Chemistry, Manipal College of Pharmaceutical Sciences, Manipal Academy of Higher Education, Manipal, Karnataka, India

A. Ganeshpurkar
Shri Ram Institute of Technology-Pharmacy, Jabalpur, Madhya Pradesh, India

A. Ganeshpurkar (✉)
Department of Pharmaceutical Sciences, Dr. Harisingh Gour Vishwavidyalaya (A Central University), Sagar, Madhya Pradesh, India

V. M. Patil et al. (eds.), *Deciphering The Role of Succinate Dehydrogenase in Drug Discovery*, https://doi.org/10.1007/978-981-96-6831-1_1

Abbreviations

AMS	Acute mountain sickness
ATP	Adenosine triphosphate
CoQ	Coenzyme Q
Cyt c	Cytochrome c
DCPIP	2,6-Ddichlorophenolindophenol
ETC	Electron transport chain
FAD	Flavin adenine dinucleotide
FADH2	Flavin adenine dinucleotide
GIST	Gastrointestinal stromal tumour
HLRCC	Hereditary leiomyomatosis and renal cell cancer
IMS	Intermembrane space
NADH	Nicotinamidc adenine dinucleotide
NPA	3-Nitro propionic acid
SDH	Succinate dehydrogenase
TCA	Tricarboxylic acid

1.1 Introduction

Mitochondria, the powerhouse of cells, plays a crucial role in cellular metabolism, particularly in the 'tricarboxylic acid' (TCA) cycle, also known as the 'citric acid cycle or Krebs cycle'. The TCA cycle takes place within the mitochondrial matrix. The cycle serves as a primary metabolic pathway for the oxidation of acetyl-CoA obtained from metabolic processes, including fatty acid oxidation, glycolysis and amino acid catabolism (Martínez-Reyes and Chandel 2020).

In the TCA cycle, acetyl-CoA is chemically added to oxaloacetate forming citric acid, which further undergoes a series of enzymatic reactions, ultimately regenerating oxaloacetate. This process generates reducing equivalents viz. reduced form of 'nicotinamide adenine dinucleotide' (NADH) and 'flavin adenine dinucleotide' ($FADH_2$) (Fernie et al. 2004). These reducing equivalents are then utilized by the 'electron transport chain' (ETC) to generate 'adenosine triphosphate' (ATP) through 'oxidative phosphorylation'. Additionally, the intermediates formed in the TCA cycle are utilized for various biosynthetic anabolic pathways, including the synthesis of nucleotides, amino acids and the production of heme rings (Chandel 2021). Furthermore, the TCA cycle is tightly regulated by allosteric regulators, substrate availability and post-translational modifications to ensure proper metabolic flux and energy production in response to cellular demands (Mayes and Bender 2003).

Succinate dehydrogenase (SDH), a pivotal enzyme in cellular respiration, plays a crucial role in energy metabolism and redox balance within biological systems. It is a key component of both the TCA cycle and ETC and is anchored in the inner mitochondrial membrane (Moosavi et al. 2019). The unique structural composition of SDH allows its biochemical catalytic function and also facilitates the electron transfer from 'succinate to ubiquinone', contributing to ATP generation. It is of paramount importance in various biological processes, serving as a critical enzyme with diverse functions. It has significance spanning across cellular respiration, energy production, redox balance and the regulation of key metabolic pathways (Kregiel 2012).

The primary function of SDH lies in its involvement in the TCA cycle, where it catalyses the oxidation of 'succinate to fumarate'. Simultaneously, SDH transfers electrons to ubiquinone facilitating the flow of electrons in the ETC. The enzyme's catalytic mechanism involves succinate oxidation and the subsequent reduction of flavin adenine dinucleotide (FAD) to $FADH_2$. The electrons are then transferred to ubiquinone, contributing to the proton motive force and ATP synthesis (Marín-García and Marín-García 2013).

Aberrations in SDH have clinical implications, particularly in hereditary cancers and neurodegenerative diseases. Mutations in subunits of the SDH enzyme are linked to certain cancers, such as paragangliomas and pheochromocytomas, emphasizing its role as a tumour suppressor. Understanding the clinical relevance of SDH provides insights into potential therapeutic targets for diseases associated with mitochondrial dysfunction (Rasheed and Tarjan 2018). SDH stands at the junctions of vital biological processes, influencing energy metabolism, redox balance and overall cellular health. Its complex involvement in both the TCA cycle and the ETC emphasizes its indispensability for sustaining life and maintaining cellular functionality. The multifaceted importance of SDH extends from fundamental bioenergetics to clinical implications, making it a crucial player in the intricate web of cellular processes.

1.2 Historical Perspectives

1.2.1 Early Discoveries

The origins of SDH's discovery date back to the late nineteenth century when researchers such as Eduard Buchner were pioneering the study of cellular respiration. Buchner's experiments with cell-free extracts revealed fermentation, showcasing the role of enzymes in sugar breakdown (Buchner 1897). This breakthrough encouraged deeper investigations into the specific enzymes involved in cellular metabolic processes. During a similar period, researchers such as Otto Warburg were focusing on isolating and studying individual enzymes implicated in respiration. Warburg's investigations with minced muscle tissue were pivotal in setting the stage for identifying SDH (Otto 2016). However, the formal discovery and characterization of SDH are commonly credited to scientists like Hans Thunberg in the early twentieth century. Thunberg's studies on tissue dehydrogenase activity offered compelling evidence for the presence of an enzyme targeting succinate, a critical component in cellular respiration (Holmes 1991).

In the early 1900s, Krebs and his contemporaries identified the series of chemical reactions involved in the breakdown of nutrients to produce energy, providing a conceptual framework for subsequent discoveries (Gasmi et al. 2021). This provided the groundwork for understanding cellular respiration and the TCA cycle. In the 1930s and 1940s, researchers focused on isolating and characterizing the enzymes associated with the TCA cycle. The initial investigations identified enzymes like citrate synthase and aconitase, shedding light on the sequential steps of the cycle. The observation of SDH, the conversion of 'succinate to fumarate', became a crucial clue in understanding the TCA cycle. Scientists noticed this enzymatic activity and recognized the significance of a specific enzyme in facilitating this process. Subsequent experiments and assays specifically aimed at SDH identified the presence of an enzyme responsible for catalysing this reaction. The isolation and characterization of this enzyme marked a pivotal point in the discovery of SDH. This discovery cemented SDH's place as a core component of cellular energy production.

As research progressed in the 1950s and 1960s, the integration of TCA cycle components with the ETC became apparent. SDH was identified not only as a participant in the TCA cycle but also as a crucial component linking it to the ETC. Advances in biochemical techniques allowed researchers to purify and characterize SDH. Biochemical assays, chromatography and biophysical methods provided insights into the enzyme's structure, function and cofactor involvement.

1.3 Structural Insights

1.3.1 Molecular Components of SDH

The structural insights and molecular composition of SDH have been extensively studied. It is a 'heterotetrameric protein complex' composed of four nuclear-encoded subunits that link the TCA with the ETC. In humans, SDH is a multi-subunit enzyme present in the inner mitochondrial membrane. It consists of four subunits: 'SDHA, SDHB, SDHC and SDHD'. SDHA and SDHB form the catalytic domain, while SDHC and SDHD constitute the membrane-anchor domain.

1.3.2 SDHA (Flavoprotein Subunit)

SdhA is the largest subunit and contains a FAD cofactor covalently linked to the protein subunit. It hosts a catalytic site that binds to organic molecules such as succinate or fumarate. Numerous effector molecules also bind to these subunits *viz.* ATP or oxaloacetate. FAD is essential for the initial dehydrogenation reaction in which succinate is oxidized to fumarate (Gutman and Silman 1975). The electrons derived from succinate are transferred to FAD during the catalytic process. SDHA is situated on the inner membrane, positioning itself to interface and extend into the space of the mitochondrial matrix (Van Vranken et al. 2015). SDHA adopts a 'Rossmann-type fold' comprising four distinct domains: an 'FAD-binding domain (residues 52-267 and 355-439)', a 'capping domain (residues 268-354)', a 'helical domain (residues 440-537)' and a 'C-terminal domain'. A covalent bond is present between FAD and 'His99' with additional coordination of FAD by several residues. These residues establish a structured hydrogen bonding network (Sun et al. 2005). 3-Nitro propionic acid (NPA) was used as a probe to identify the putative succinate binding site. The study indicated that NPA binds with the FMN portion of FAD, along with Glu261, Glu246, Thr260, His248, Arg403 and His359. FAD serves as a vital cofactor for SDH1. It originates from dietary riboflavin, also known as vitamin B_2, and its conversion involves 'two ATP-dependent enzymes', a 'riboflavin kinase' (Fmn1) and 'FAD synthetase' (Fad1). Fmn1 phosphorylates tricyclic isoalloxazine ring to produce FMN.

Putative mitochondrial FAD transporter (Flx1) plays a vital role in the insertion of FAS in enzymes. Flx1-deficient cells display decreased levels of FAD in the matrix, resulting in diminished activity of 'two matrix flavoproteins': 'SDH' and 'lipoamide dehydrogenase'. This impairment stems from a defect in flavinylation, the process of attaching FAD to these enzymes (Kim et al. 2012; Tzagoloff et al. 1996). SDH1 migrates into the mitochondrial matrix as an apoprotein and this process is allosterically affected by succinate along with several other intermediates of the TCA cycle (Brandsch and Bichler 1989; Robinson et al. 1994). SDH5 is another crucial protein that is responsible for the flavinylation of SDH1. The deletion of SDH5 gene in yeast inhibited growth dependent on respiration and led to a significant decrease in oxygen consumption (Hao et al. 2009; Kim et al. 2012). The binding of FAD is encouraged by, although not dependent upon, the presence of the 'iron–sulphur subunit' and intermediates of the TCA cycle, such as malate, succinate or fumarate (Robinson and Lemire 1996).

1.4 SDHB (Iron–Sulphur Subunit)

The 'N-terminal domain (residues 37-142)' comprises a small 'α-helix' along with a five-strand 'β-sheet'. This domain contains the FAD-proximal 2Fe-2S centre, coordinated by 'Cys93, Cys98, Cys101 and Cys113'. On the other hand, the 'C-terminal domain' (residues 142–280) consists of six 'α-helices', 4Fe-4S and 3Fe-4S centres. These centres are coordinated by 'Cys186, Cys189 and Cys192', 'Cys253 and Cys196' and 'Cys243 and Cys249', respectively. SDHB is a critical subunit for the enzyme activity that contains 'iron–sulphur clusters', including a [2Fe-2S] and a [4Fe-4S] cluster. Adjacent to the FAD site is a 2Fe-2S centre, followed by a neighbouring 4Fe-4S centre, and subsequently, a 3Fe-4S centre (Read et al. 2021). Consequently, the terminal 3Fe-4S centre is strategically positioned in proximity to ubiquinone, facilitating the transfer of ultimate electrons (Sun et al. 2005). It also abridges the membrane anchoring and catalytic domains of the enzyme (Hägerhäll 1997). These clusters act as electron carriers, aiding in the electron transfer to ubiquinone from succinate. The Fe-S clusters also play a role in the electron transport chain, assisting in the creation of a proton gradient across the inner mitochondrial membrane (Rouault and Maio 2017). These clusters are synthesized within the matrix of mitochondria utilizing a complex scaffold (ISU), composed of four proteins: Isd11, Yfh1, Nfs1 and Isu1 (Lill et al. 2012; Schmucker et al.

2011; Tsai and Barondeau 2010). Sulphide ions, essential for cluster formation, are supplied by the 'Nfs1 cysteine desulfurase', in conjunction with its associated proteins 'Isd11 and Yfh1' (Pandey et al. 2012; Pandey et al. 2013). Additionally, the Yah1 ferredoxin reductant also supports this process (Sheftel et al. 2010). Before being transferred to client proteins, Fe(II) and sulphide ions combine to form '2Fe-2S clusters' on the scaffold proteins *viz.* Isu1 (or Isu2). The preformed cluster then binds to the monothiol Grx5, a Fe-S shuttle protein. This binding is facilitated by the DnaJ protein Jac1, which interacts with Isu1, leading to the dissociation of the ISU complex (Majewska et al. 2013). Additionally, the Hsp70 enzyme Ssq1 is recruited during this process. Mutations in mitochondrial protein BolA3 have been found to cause malfunctions in respiratory complexes as well as 2-oxoacid dehydrogenases, akin to Nfu1 mutations (Baker et al. 2014; Cameron et al. 2011). This suggests that BolA3 might be involved in the later phases of mitochondrial Fe-S biogenesis or transfer (Li and Outten 2012).

1.5 SDHC (Membrane Anchor Subunit)

SDHC is a transmembrane protein composed of four helices, serving to anchor the SDH complex firmly to the inner mitochondrial membrane. It contains two 'heme b' groups, aiding in the coordination of electrons during the electron transfer. The 'heme b' groups in SDHC also contribute to the overall function of SDH in the mitochondrial ETC (Rutter et al. 2010; Van Vranken et al. 2015).

1.6 SDHD (Membrane Anchor Subunit)

SDHD is another transmembrane subunit containing five helices that assist in the attachment of the complex to the inner mitochondrial membrane. It interacts with SDHC to ensure proper localization and is crucial for the integrity and stability of the SDH complex within the membrane of the mitochondria. 'N-terminal helix' of SDHD resides within the matrix of mitochondria and facilitates the hydrophilic dimer localization through interaction with SDHB.

SDHC and SDHD interact with the porphyrin ring of the 'heme B cofactor' situated at the core four-helix bundle interface. Within this arrangement, conserved histidine residues of both subunits coordinate with the iron of the heme group. Additionally, two arginine residues, one from each subunit, along with SDHC another histidine residue, interact with the propionate groups of heme moiety (Sun et al. 2005).

1.7 Active Sites and Catalytic Mechanisms

The hydrophilic head comprises two subunits, SDHA and SDHB, which together form the catalytic core in mammals. The catalytic core hosts the redox cofactors essential for the transfer of electrons to ubiquinone. SDHA houses the FAD cofactor and hosts the succinate binding site. SDHB encompasses the three Fe-S centres responsible for facilitating electron transfer to ubiquinone (Hägerhäll 1997; Sun et al. 2005). SDHB connects the catalytic and membrane anchor domains within the complex. The interaction between SDHB and both SDHA and SDHC involves a similar surface area, indicating that the catalytic core does not function as an independent dimer without a membrane anchor. SDHC and SDHD contribute two transmembrane helices each to form a 'four-helix bundle', constituting the membrane anchor core (Hao et al. 2009). Each subunit contributes a transmembrane helix that encircles the core (Van Vranken et al. 2015). Within the membrane domain, a heme b moiety is bound at the interface of the subunits SDHC and SDHD, with each subunit providing one of the two axial His ligands. Additionally, two separate ubiquinone binding sites have been detected in mammalian SDH complexes (Sun et al. 2005; Yankovskaya et al. 2003). The residues Thr254, His354 and Arg399 within subunit A provide stabilization to the molecule. Meanwhile, FAD undergoes oxidation and transports the electrons to the initial iron–sulphur cluster, [2Fe-2S] (Kenney 1975).

The 'succinate-binding site' and 'ubiquinone-binding site' are connected through a sequence of redox centres, including FAD and iron–sulphur clusters. This chain extends over 40 Å across the enzyme monomer (Yankovskaya et al. 2003). The high-affinity ubiquinone site (QP-proximal) is located on the matrix side of the inner mitochondrial membrane (IM) and involves residues from SDHB, SDHC and SDHD. Within proximity of 7 Å to the '3Fe–4S redox centre', the QP site serves as the primary ubiquinone binding site (Oyedotun and Lemire 2001; Silkin et al. 2007). In contrast, the second, less tightly bound ubiquinone site (QD-distal) is located closer to the intermembrane space (IMS) side of the inner membrane. Ubiquinone reduction occurs through two individual electron reactions, differing from the two-electron reduction of FAD. The QP site notably stabilizes the partially reduced semiquinone, aiding in its full reduction to ubiquinol. The QP site notably enhances the stability of the partially reduced semiquinone, enabling its complete reduction to ubiquinol (Yankovskaya et al. 2003). The role of the conserved heme group in eukaryotic SDHs and the distant QD site remains unclear. While the heme is not essential for ubiquinone reduction at the QP site, it may aid in electron transfer to the distant QD site. SDH complexes capable of succinate reduction at heme might also produce ubiquinol at the QD site, but definitive proof is lacking. Unlike the bc1 Complex III, SDH lacks a Q cycle and does not actively transport protons, despite having two Q sites.

1.8 Functional Significance

1.8.1 Role in Cellular Respiration

SDH is an important membrane dehydrogenase allied with the respiratory chain and is part of the TCA cycle. Its primary role is to catalyse the oxidation of succinate to fumarate. This reaction involves the removal of two hydrogen atoms from succinate, which are transferred to FAD, a prosthetic group, which is tightly attached to the SDH enzyme. Various structural analyses displayed that arginine residue, for instance, Arg 286 of *E. coli*, serves as a proton shuttle for the reaction. It is followed by two possible eliminations, *i.e.* E_2 or E_{1cb}. In the E_2 reaction, a basic amino acid or cofactor removes a proton from the α-carbon, while FAD accepts a hydride from the β-carbon, leading to the oxidation of succinate. On the other hand, an enolate is formed before accepting the proton by FAD in the E_{1cb} reaction (Fig. 1.1). The product formed loses affinity for the active site and the depart forms the enzyme cavity (Dorota 2012).

1.8.2 Contribution to the Electron Transport Chain

Electron transport through SDH involves a series of redox reactions within the enzyme complex. Initially, electrons are extracted from succinate during its oxidation to fumarate. These electrons are transferred to the FAD cofactor, leading to its reduction to $FADH_2$. Subsequently, electrons are shuttled through a series of 'iron–sulphur clusters', i.e. traversing the [Fe-S] relay until reaching the [3Fe-4S] cluster present within SDH. The clusters serve as electron carriers, facilitating the transfer of electrons along the enzyme complex. Ultimately, the electrons are transferred to a mobile electron carrier named ubiquinone present in the mitochondrial membrane. It is postulated that electrons received by ubiquinone through the 3Fe-4S cluster oscillate between the 'heme moiety' and 'ubiquinone' intermediate resulting in the development of an electron sink. The orientation and reduction of ubiquinone within SDH involve specific interactions with amino acid residues within the enzyme complex. Initially, hydrogen bonds between ubiquinone and specific residues guide its positioning at the active site. Upon the arrival of electrons from the 'iron–sulphur cluster', ubiquinone undergoes a 'conformational change', leading to further interactions with the enzyme. This enables stepwise reduction of ubiquinone to ubiquinol, involving the formation of a semiquinone radical species followed by the addition of a 'second electron' to complete the reduction process. The electron transfer process within SDH links succinate oxidation in the TCA cycle to the ETC, aiding in the formation of a proton gradient across the inner mitochondrial membrane (Fig. 1.2) (Moser et al. 2006).

1.8.3 Impact on ATP Production

In contrast to other respiratory complexes, SDH does not actively pump protons across the inner mitochondrial membrane. Instead, it indirectly contributes to proton gradient generation by reducing ubiquinone to ubiquinol. Ubiquinol then transfers electrons to complex III consisting of 'cytochrome c reductase' and complex IV made up of 'cytochrome c oxidase', which directly transfers protons across the inner membrane (Tretter et al. 2016). This results in the build-up of the necessary

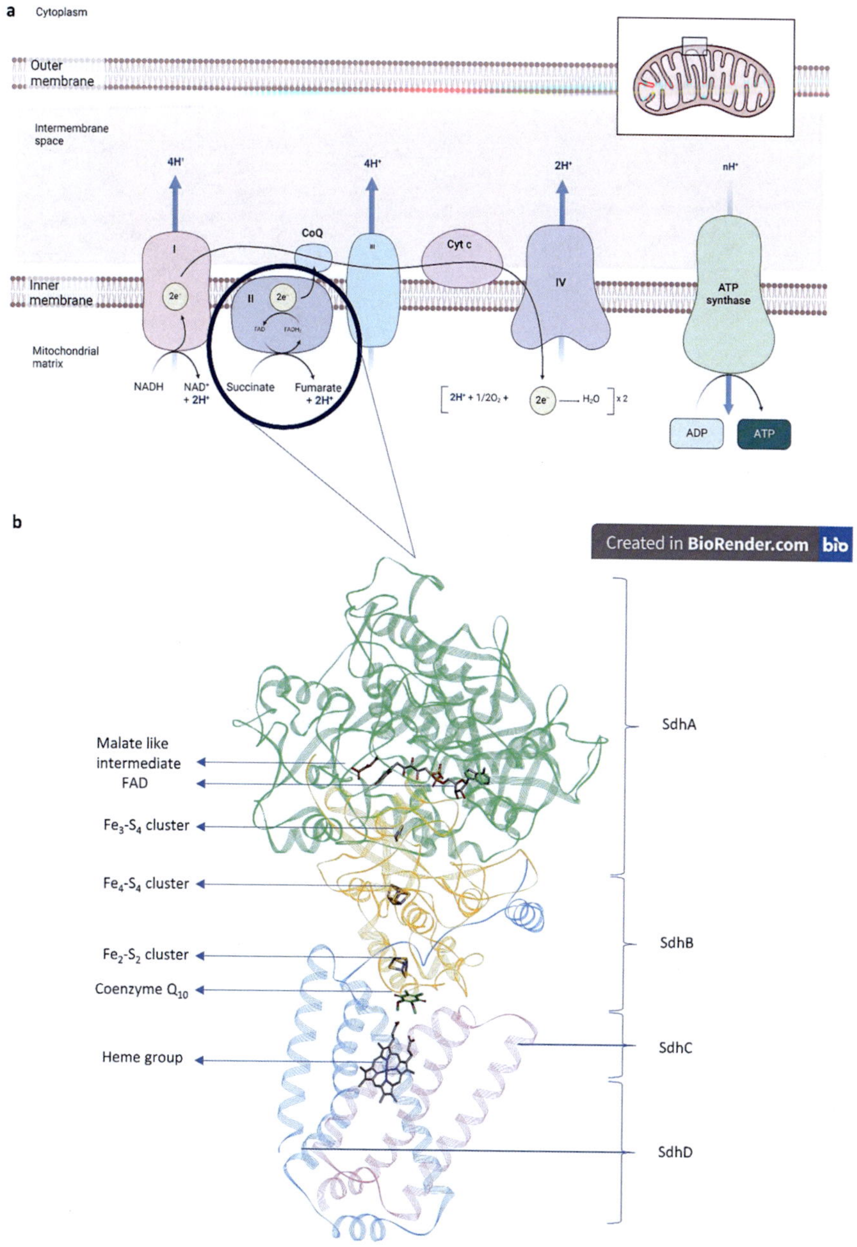

Fig. 1.1 (**a**) Electron transport chain, within the mitochondria, showcasing the sequential flow of electrons through respiratory complexes, leading to the generation of ATP and the establishment of a proton gradient across the inner mitochondrial membrane. (**b**) Representation of the subunit arrangement within the SDH enzyme complex, highlighting the interaction between various subunits and cofactors essential for its catalytic activity. (Created from BioRender.com)

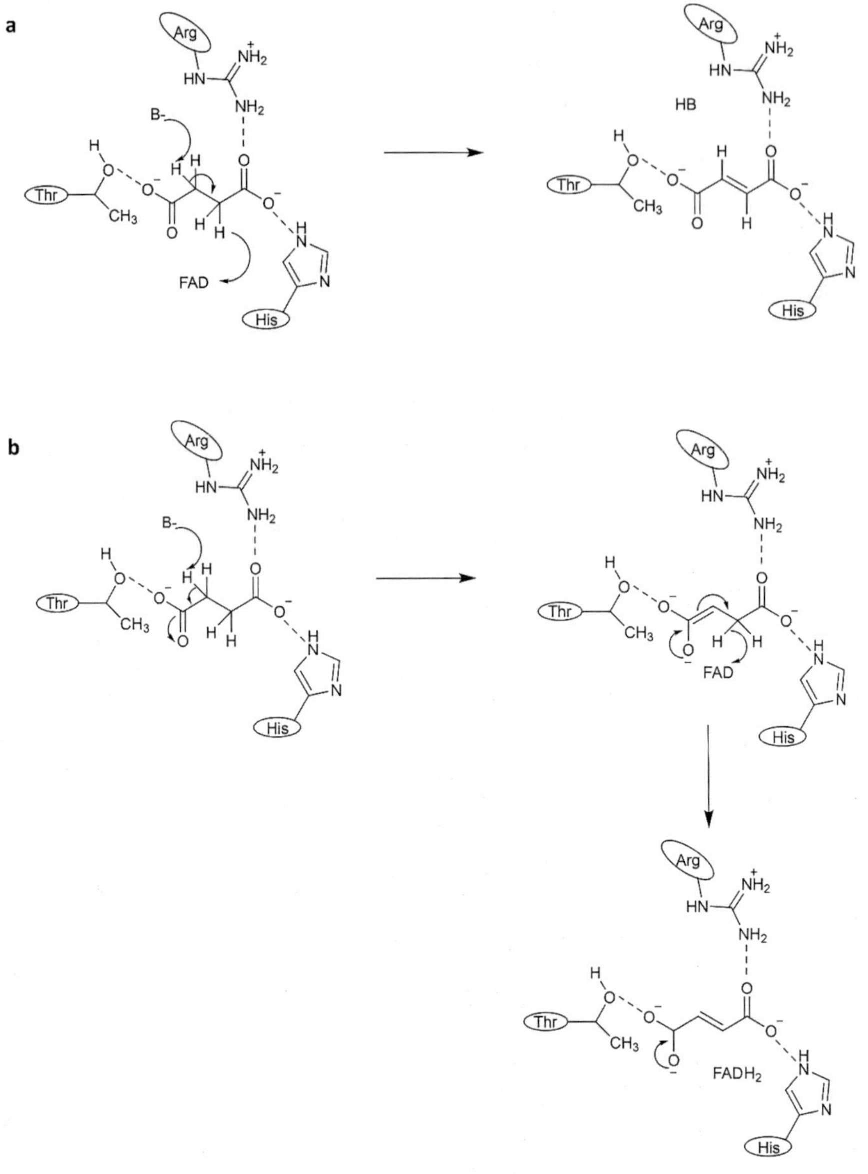

Fig. 1.2 (**a**) E_2 and (**b**) E_{1cb} mechanisms involved in the conversion of fumarate to malate, depicting the stepwise enzymatic reactions and the intermediates formed during the process

proton gradient for ATP synthesis. The gradient hence produced powers the F_1F_0 ATP synthase, enabling the phosphorylation of ADP for the synthesis of ATP. This enzyme is driven by protons, harnessing the energy stored within the proton gradient, to facilitate the phosphorylation of ADP to ATP (Rustin et al. 2002).

1.9 Regulation of Enzyme

Post-translational modifications play a crucial role in regulating SDH activity. The acetylation and phosphorylation of the active site cause inhibition of activity. Studies demonstrated that reversible acetylation at numerous lysine residues within mouse SDH1 leads to the reduction of its catalytic activity (Cimen et al. 2010). On the other hand, sirtuins, notably SIRT3, facilitates the deacetylation of SDH1's active site, thereby regulating the acetylation level and activity of SDH. Further, the phosphorylation of the SDH1 subunit also deactivates the enzyme similar to acetylation (Tomitsuka et al. 2009). In vitro experiments have demonstrated that Fgr tyrosine kinase can phosphorylate tyrosine residues of SDH1 in two instances. However, the biological relevance of this modification by Fgr remains unexplored (Salvi et al. 2007). The enzymatic function of SDH is affected by Krebs cycle intermediates, with oxaloacetate serving as a strong inhibitor. Conversely, succinate triggers the release of oxaloacetate from SDH, activating the enzyme. This oxaloacetate-mediated inhibition may contribute to the modulation of SDH activity depending on the mitochondrial metabolic state (Gutman and Silman 1975).

'Necrosis Factor Receptor-Associated Protein 1' (TRAP1) is also an inhibitor of SDH activity of the complex II through 'activity down-modulation'. It triggers pseudohypoxia by stabilizing HIF1α and also inhibits the 'ROS-dependent' opening of the 'mitochondrial permeability transition pores', reducing oxidative damage in tumour cells (Guzzo et al. 2014). Malonate, a substrate analogue of succinate, competitively inhibits the SDH complex. It interacts with positively charged amino acid residues within the active site, akin to succinate. However, the absence of ethylenic linkage of malonate, necessary for dehydration, prevents it from oxidation (Hajjawi 2010; Hajjawi and Hider 2009).

1.10 Determination of SDH Activity

Succinate is an efficient energy substrate making the SDH activity assay a vital method for assessing yeast and other cells' vitality, particularly monitoring various fermentation processes. SDH activity can be assessed both in vitro, using cell lysates or mitochondrial fractions, and in situ, within individual cells (Kregiel et al. 2008). Its localization on the inner mitochondrial membrane makes it readily isolable alongside mitochondria using various techniques such as 'sucrose density gradient ultracentrifugation', 'free-flow electrophoresis' or commercially available 'kit-based methods'. The mitochondrial fraction obtained by the above methods represents the primary source of this enzyme (Hartwig et al. 2009).

The feasibility of using fumarate or succinate for the estimation of the enzyme activity is not very favourable; hence, synthetic electron donors are employed. Additionally, the normal flow of electrons should be blocked in ETC for such estimation. Poisons such as sodium azide and potassium cyanide block electron transfer from cytochrome a_3 to oxygen (electron acceptor). Consequently, the preceding cytochromes and 'coenzyme Q' are unable to facilitate the transfer of electrons. 2,6-Dichlorophenolindophenol (DCPIP), an artificial electron acceptor, captures electrons instead of $FADH_2$. The reduced product of DCPIP can be measured spectrometrically to colourless from the blue product.

$$SDH\text{-FADH}_2 + \text{DCPIP}_{\text{oxid}} \rightarrow SDH\text{-FAD} + \text{DCPIP}_{\text{red}} + 2\text{H}^+$$

The alteration in absorbance, quantified at '600 nm', is employed to track the progression of the reaction throughout its duration. The rate at which the blue colour diminishes correlates directly with the enzyme concentration. By monitoring the change in absorbance of the mixture over time, the enzyme concentration can be accurately determined from the resulting data (Mahesha 2014).

Various tetrazolium salts also act as substrates for SDH. The metabolically active cells reduce different tetrazolium salts by dehydrogenases, resulting in the formation of intensely coloured end products known as formazans, thus making in situ estimation of SDH feasible. In the case of in situ estimation of SDH enzymatic activity, the plasma membrane acts as a barrier and reduces the cellular penetration of these tetrazolium salts (Kregiel et al. 2008). Hence, cell permeabilization, such as with digitonin, is suggested as an alternative approach to investigate intracellular enzyme activities. Berlowska et al. demonstrated the efficacy of digitonin in 'membrane permeabilization' without adverse effects on cell morphology. After treatment with digitonin, formazan crystals were visible inside the yeast cells but not outside (Berlowska et al. 2006). The formazan products, while insoluble in water, readily diffuse out of cells once solubilized in dimethyl sulfoxide.

1.11 Role of SDH in Diseases

SDH plays a significant role in mitochondrial diseases by contributing to mitochondrial dysfunction. Dysregulation of SDH activity can lead to disruptions in energy production, increased oxidative stress and impaired cellular function, contributing to the pathogenesis of various 'mitochondrial disorders'. These disorders encompass a wide range of conditions, including Leigh syndrome, mitochondrial encephalomyopathy and mitochondrial myopathy, among others. SDH dysfunction may result from mutations in genes encoding its subunits or regulatory proteins, leading to a cascade of biochemical abnormalities that compromise mitochondrial function and cellular health (Table 1.1).

1.12 Hereditary Paraganglioma–Pheochromocytoma Syndrome

'Hereditary paraganglioma–pheochromocytoma syndrome' is an inherited disorder marked by tumour growth in paraganglia, clusters of cells near nerve ganglia. These tumours, known as paragangliomas, may include pheochromocytomas, which arise in the adrenal glands responsible for stress hormone production (Else et al. 1993). Paragangliomas primarily occur in the 'head and neck', particularly at the 'carotid

Table 1.1 Diseases associated with succinate dehydrogenase deficiency

Disease	Associated genes	Signs and symptoms	Effects	References
Leigh syndrome	SDHA, SDHB, SDHC, SDHD	Early-onset encephalopathy, developmental delay, exercise intolerance, muscle weakness, seizures, vision problems, lactic acidosis.	Severe brain damage, intellectual disability, movement problems, heart failure.	Karimzadeh et al. (2020)
Hereditary paraglioma/ pheochromocytoma syndrome (HAPP)	SDHB, SDHC, SDHD	Increased risk of paragangliomas (tumours of the nervous system), pheochromocytomas (tumours of the adrenal gland), headaches, sweating, palpitations, high blood pressure.	Cancers, hormonal imbalances, severe hypertension.	Nazar et al. (2019)
Gastrointestinal stromal tumours (GISTs)	SDHB	Abdominal pain, bleeding, nausea, early satiety.	Cancers of the digestive tract.	Nazar et al. (2019)
Renal cell carcinoma	SDHB (less common)	Blood in urine, flank pain, weight loss, fatigue.	Kidney cancer.	Nazar et al. (2019)
Leukodystrophy	SDHB (limited studies)	Progressive loss of coordination, muscle weakness, vision problems, seizures, dementia.	Degeneration of white matter in the brain, leading to severe neurological disability.	Moreno et al. (2020)
Ataxia	SDHB (limited studies)	Difficulty walking, speaking and coordinating movements, tremors.	Loss of balance and coordination.	Moreno et al. (2020)
Optic atrophy	SDHB (limited studies)	Vision loss, decreased colour vision, night blindness.	Degeneration of the optic nerve, leading to permanent vision impairment.	Moreno et al. (2020)
Acute mountain sickness (AMS)	Likely epigenetic modifications of SDHA/SDHB	Headache, nausea, vomiting, dizziness, fatigue, shortness of breath at high altitudes.	Difficulty acclimatizing to high altitude, potentially life-threatening.	Moreno et al. (2020)
Carney-Stratakis syndrome	SDHB, SDHC	Gastrointestinal stromal tumours (GISTs), paragangliomas, pheochromocytomas, skin pigmentation changes, myxomas (benign tumours).	Increased risk of multiple tumours in various organs.	Negro et al. (2017)
Neuroblastoma	SDHB (rare)	Abdominal mass, pain, fatigue, weight loss, bone pain.	Aggressive childhood cancer.	Rapizzi et al. (2014)

bifurcation (80%)', 'jugular bulb or tympanic nerve (17.5%)' and 'vagal nerve (4.5%)' (Lack 1997). In recent times, genetic mutations linked to the mitochondrial SDH complex have been demonstrated to lead 'head and neck paragangliomas', 'extra-adrenal paragangliomas' and 'pheochromocytomas'. Gene SDHD, situated at 11q23, encodes the transmembrane subunit of Complex II in the ETC and displays 110 distinct mutations related to the disease. These tumours predominantly manifest in the 'head and neck' region, specifically within the parasympathetic trunk (Baysal et al. 2000).

SDHB, situated at 1p36.1-p35, is responsible for encoding the catalytic subunit of iron–sulphur protein within Complex II. With 175 distinct mutations identified, SDHB mutations are linked to tumours found in abdominal/thoracic paraganglia, adrenal glands and the sympathetic trunk (Astuti et al. 2001). SDHC, located at 1q23.3, is responsible for encoding another transmembrane subunit of 'Complex II'. It has been associated with 34 distinct mutations. These tumours primarily emerge in the 'head and neck' region, particularly in the parasympathetic trunk (Mannelli et al. 2007). SDHAF2, at 11q12.2, is linked to hereditary paraganglioma/pheochromocytoma syndrome, with one unique mutation reported. Tumours associated with SDHAF2 mutations are located in the 'head and neck', particularly in the parasympathetic trunk (Bayley et al. 2010). SDHA, at 5p15, contributes to hereditary paraganglioma/pheochromocytoma syndrome, with one unique mutation identified. Tumours associated with SDHA mutations predominantly occur in the abdominal paraganglia and the sympathetic trunk (Burnichon et al. 2010).

1.13 Hereditary Leiomyomatosis and Renal Cell Cancer

'Hereditary leiomyomatosis and renal cell cancer' (HLRCC) is an 'autosomal dominant' tumour susceptibility syndrome. It is characterized by a predisposition to benign leiomyomas of the skin and uterus (Kiuru and Launonen 2004). Genetic mutations associated with SDHB and SDHD genes, which encode subunits of SDH, are linked to HLRCC. These mutations interfere with normal cellular functions, promoting tumourigenesis through mechanisms such as pseudohypoxia (Kiuru and Launonen 2004; Nazar et al. 2019). Individuals with HLRCC who harbour mutations in SDHB and SDHD genes are at an increased risk of developing aggressive types of kidney cancer. Regular screening is crucial for early detection and effective management of renal cell carcinoma in these individuals.

1.14 Gastrointestinal Stromal Tumours

SDH-deficient gastrointestinal stromal tumours (GISTs) exhibit a scarcity of SDH, resulting in a distinct multinodular/plexiform architecture and epithelioid cell composition. Typically found in the stomach, these tumours display unique morphological traits like multinodular growth patterns and lymphovascular involvement

(Ibrahim and Chopra 2019; Lv et al. 2021). Patients with SDH-deficient GISTs often present at a young age, with symptoms like gastrointestinal bleeding, abdominal pain and fatigue.

1.15 Renal Oncocytomas

Renal oncocytomas present as well-defined tumours characterized by typical gross features, including a 'tan or mahogany coloured mass' with a 'central stellate scar'. Microscopically, they demonstrate a nested architecture with 'bland cytology', regular nuclei featuring 'prominent central nucleoli' and 'eosinophilic cytoplasm'. These tumours primarily consist of oncocytes, characterized by an abnormal abundance of mitochondria (Lieber et al. 1981). Studies have identified cases where renal oncocytomas initially diagnosed based on histological features showed characteristic histologic features of 'SDH-deficient renal cell carcinomas'. Loss of SDHB gene expression by immunohistochemistry was observed in these cases, indicating a potential link between SDH alterations and the development of oncocytomas (Gupta et al. 2019).

1.16 Cardiomyopathy

Cardiomyopathy encompasses a group of heart muscle diseases that impair the heart's ability to pump blood effectively. 'Dilated cardiomyopathy' involves the enlargement of one of the ventricles and is more prevalent in males. 'Hypertrophic cardiomyopathy' is characterized by thickening of the heart muscle posing a risk of sudden death, particularly in young athletes. 'Arrhythmogenic cardiomyopathy' causes irregular heartbeats or rhythms and can lead to sudden cardiac death, especially in children and young adults (Wexler et al. 2009). SDH plays a vital role in myocardial homeostasis by overseeing fatty acid oxidation and glycolysis. Mice deficient in SDH show disturbances in critical metabolic pathways for cardiac function, underscoring SDH's involvement in epigenetic and metabolic regulation in the heart (Li et al. 2023). Research has revealed a specific shortage of SDH in the heart, notably in conditions such as Barth syndrome, a genetic ailment linked to cardiomyopathy. This insufficiency prompts structural alterations in respiratory chain complexes, diminishing enzymatic capabilities and elevating reactive oxygen species production, which exacerbates cardiac issues (Dudek et al. 2016).

1.17 Neurodegeneration

Research highlights that oxidative stress and inflammation are major contributors to neurodegeneration. SDH deficiency can exacerbate oxidative stress within neurons, leading to cellular damage and dysfunction. Disruption in SDH function may result in mitochondrial dysfunction, reduced energy generation and heightened oxidative stress, thereby playing a role in neuronal injury and neurodegeneration seen in conditions such as 'Alzheimer's, Parkinson's and Huntington's disease' (Farshbaf and Kiani-Esfahani 2018). Neurodegenerative disorders are marked by aberrant protein aggregation and subsequent cell demise. Changes in SDH function might impact protein misfolding occurrences, thus playing a role in the neuronal decline seen in diseases such as 'amyotrophic lateral sclerosis', multiple system atrophy and tauopathies.

1.18 Conclusion

SDH, a multi-subunit enzyme nestled within the inner mitochondrial membrane, stands as a pivotal player in cellular respiration. It bridges the gap between the TCA cycle and the ETC, performing the seemingly simple yet vital task of oxidizing succinate to fumarate. However, this seemingly straightforward reaction lies at the core of a complex task of influencing energy metabolism, redox balance and overall cellular health. SDH's intricate structure, with its unique blend of subunits and cofactors, facilitates its diverse functions. The FAD-bound SDHA subunit catalyses the initial dehydrogenation reaction, while the iron–sulphur clusters housed within SDHB act as electron carriers, shuttling electrons to ubiquinone. SDHC and SDHD, the membrane anchor subunits, ensure proper localization within the mitochondrial membrane.

Beyond its role in the TCA cycle, SDH contributes significantly to the ETC. Electrons extracted from succinate are transferred through a series of redox centres, ultimately reaching ubiquinone, a mobile electron carrier. This electron transfer helps establish the proton gradient across the inner mitochondrial membrane, a crucial step in ATP synthesis. The activity of SDH is tightly regulated by various mechanisms, including post-translational modifications and metabolite availability. Understanding these regulatory mechanisms sheds light on how cellular demands are met and on potential therapeutic targets for diseases associated with mitochondrial dysfunction. The importance of SDH extends far beyond basic cellular processes. Mutations in genes encoding SDH subunits have been linked to various cancers, including paragangliomas, pheochromocytomas and certain kidney cancers. These findings highlight the role of SDH as a tumour suppressor and underscore the potential for SDH-based therapeutic strategies.

In conclusion, SDH stands as a remarkable example of how a seemingly simple enzyme can orchestrate a multitude of complex cellular functions. Its role in energy production, redox balance and disease development makes it a captivating target for further research and a potential key to unlocking new avenues in medicine.

Conflict of Interest The authors declare that they have no conflict of interest.

References

Astuti D, Latif F, Dallol A, Dahia PL, Douglas F, George E, Sköldberg F, Husebye ES, Eng C, Maher ER (2001) Gene mutations in the succinate dehydrogenase subunit SDHB cause susceptibility to familial pheochromocytoma and to familial paraganglioma. Am J Hum Genet 69(1):49–54. https://doi.org/10.1086/321282

Baker PR, Friederich MW, Swanson MA, Shaikh T, Bhattacharya K, Scharer GH, Aicher J, Creadon-Swindell G, Geiger E, MacLean KN (2014) Variant non ketotic hyperglycinemia is caused by mutations in LIAS, BOLA3 and the novel gene GLRX5. Brain 137(2):366–379. https://doi.org/10.1093/brain/awt328

Bayley J-P, Kunst HP, Cascon A, Sampietro ML, Gaal J, Korpershoek E, Hinojar-Gutierrez A, Timmers HJ, Hoefsloot LH, Hermsen MA (2010) SDHAF2 mutations in familial and sporadic paraganglioma and phaeochromocytoma. Lancet Oncol 11(4):366–372. https://doi.org/10.1016/S1470-2045(10)70007-3

Baysal BE, Ferrell RE, Willett-Brozick JE, Lawrence EC, Myssiorek D, Bosch A, Mey AVD, Taschner PE, Rubinstein WS, Myers EN (2000) Mutations in SDHD, a mitochondrial complex II gene, in hereditary paraganglioma. Science 287(5454):848–851. https://doi.org/10.1126/science.287.5454.848

Berlowska J, Kregiel D, Klimek L, Orzeszyna B, Ambroziak W (2006) Novel yeast cell dehydrogenase activity assay in situ. Pol J Microbiol 55(2):127–131. https://www.ncbi.nlm.nih.gov/pubmed/17419290

Brandsch R, Bichler V (1989) Covalent cofactor binding to flavoenzymes requires specific effectors. Eur J Biochem 182(1):125–128. https://doi.org/10.1111/j.1432-1033.1989.tb14808.x

Buchner E (1897) Alkoholische gärung ohne hefezellen. Ber Dtsch Chem Ges 30(1):1110–1113

Burnichon N, Brière J-J, Libé R, Vescovo L, Riviere J, Tissier F, Jouanno E, Jeunemaitre X, Bénit P, Tzagoloff A (2010) SDHA is a tumor suppressor gene causing paraganglioma. Hum Mol Genet 19(15):3011–3020. https://doi.org/10.1093/hmg/ddq206

Cameron JM, Janer A, Levandovskiy V, MacKay N, Rouault TA, Tong W-H, Ogilvie I, Shoubridge EA, Robinson BH (2011) Mutations in iron-sulfur cluster scaffold genes NFU1 and BOLA3 cause a fatal deficiency of multiple respiratory chain and 2-oxoacid dehydrogenase enzymes. Am J Hum Genet 89(4):486–495. https://doi.org/10.1016/j.ajhg.2011.08.011

Chandel NS (2021) Amino Acid Metabolism. Cold Spring Harb Perspect Biol 13(4). https://doi.org/10.1101/cshperspect.a040584

Cimen H, Han M-J, Yang Y, Tong Q, Koc H, Koc EC (2010) Regulation of succinate dehydrogenase activity by SIRT3 in mammalian mitochondria. Biochemistry 49(2):304–311. https://doi.org/10.1021/bi901627u

Dorota K (2012) Succinate dehydrogenase of saccharomyces cerevisiae – the unique enzyme of TCA cycle – current knowledge and new perspectives. In: Rosa Angela C (ed) Dehydrogenases. IntechOpen, p 9. https://doi.org/10.5772/48413

Dudek J, Cheng IF, Chowdhury A, Wozny K, Balleininger M, Reinhold R, Grunau S, Callegari S, Toischer K, Wanders RJ, Hasenfuß G, Brügger B, Guan K, Rehling P (2016) Cardiac-specific succinate dehydrogenase deficiency in Barth syndrome. EMBO Mol Med 8(2):139–154. https://doi.org/10.15252/emmm.201505644

Else T, Greenberg S, Fishbein L (1993) Hereditary paraganglioma-pheochromocytoma syndromes. University of Washington, Seattle. http://europepmc.org/abstract/MED/20301715

Farshbaf MJ, Kiani-Esfahani A (2018) Succinate dehydrogenase: prospect for neurodegenerative diseases. Mitochondrion 42:77–83. https://doi.org/10.1016/j.mito.2017.12.002

Fernie AR, Carrari F, Sweetlove LJ (2004) Respiratory metabolism: glycolysis, the TCA cycle and mitochondrial electron transport. Curr Opin Plant Biol 7(3):254–261. https://doi.org/10.1016/j.pbi.2004.03.007

Gasmi A, Peana M, Arshad M, Butnariu M, Menzel A, Bjørklund G (2021) Krebs cycle: activators, inhibitors and their roles in the modulation of carcinogenesis. Arch Toxicol 95(4):1161–1178. https://doi.org/10.1007/s00204-021-02974-9

Gupta S, Swanson AA, Chen YB, Lopez T, Milosevic D, Kipp BR, Leibovich BC, Thompson RH, Herrera-Hernandez L, Cheville JC, Jimenez RE (2019) Incidence of succinate dehydrogenase and fumarate hydratase-deficient renal cell carcinoma based on immunohistochemical screening with SDHA/SDHB and FH/2SC. Hum Pathol 91:114–122. https://doi.org/10.1016/j.humpath.2019.07.004

Gutman M, Silman N (1975) The steady state activity of succinate dehydrogenase in the presence of opposing effectors: II. Reductive activation of succinate dehydrogenase in presence of oxaloacetate. Mol Cell Biochem 7(3):177–185

Guzzo G, Sciacovelli M, Bernardi P, Rasola A (2014) Inhibition of succinate dehydrogenase by the mitochondrial chaperone TRAP1 has anti-oxidant and anti-apoptotic effects on tumor cells. Oncotarget 5:23. https://doi.org/10.18632/oncotarget.2472

Hägerhäll C (1997) Succinate: quinone oxidoreductases: variations on a conserved theme. Biochim Biophys Acta 1320(2):107–141. https://doi.org/10.1016/s0005-2728(97)00019-4

Hajjawi OS (2010) Sequential mechanism of the malonate transport system in human red blood cells. Eur J Sci Res 46:18–27

Hajjawi OS, Hider RC (2009) Asymmetry of the malonate transport system in human red blood cells. Eur J Sci Res 31(4):534–545

Hao H-X, Khalimonchuk O, Schraders M, Dephoure N, Bayley J-P, Kunst H, Devilee P, Cremers CW, Schiffman JD, Bentz BG (2009) SDH5, a gene required for flavination of succinate dehydrogenase, is mutated in paraganglioma. Science 325(5944):1139–1142. https://doi.org/10.1126/science.1175689

Hartwig S, Feckler C, Lehr S, Wallbrecht K, Wolgast H, Müller-Wieland D, Kotzka J (2009) A critical comparison between two classical and a kit-based method for mitochondria isolation. Proteomics 9(11):3209–3214. https://doi.org/10.1002/pmic.200800344

Holmes FL (1991) Hans Krebs. Oxford University Press

Ibrahim A, Chopra S (2019) Succinate dehydrogenase–deficient gastrointestinal stromal tumors. Arch Pathol Lab Med 144(5):655–660. https://doi.org/10.5858/arpa.2018-0370-RS

Karimzadeh P, Keramatipour M, Karamzade A, Pourbakhtyaran E (2020) Succinate Dehydrogenase Deficiency: A Treatable Neurometabolic Disorder. Iran. J Child Neurol 14(4):111–116. https://doi.org/10.22037/ijcn.v14i4.28633

Kenney WC (1975) The reaction of N-ethylmaleimide at the active site of succinate dehydrogenase. J Biol Chem 250(8):3089–3094. https://www.ncbi.nlm.nih.gov/pubmed/235539

Kim HJ, Jeong M-Y, Na U, Winge DR (2012) Flavinylation and assembly of succinate dehydrogenase are dependent on the C-terminal tail of the flavoprotein subunit. J Biol Chem 287(48):40670–40679. https://doi.org/10.1074/jbc.M112.405704

Kiuru M, Launonen V (2004) Hereditary leiomyomatosis and renal cell cancer (HLRCC). Curr Mol Med 4(8):869–875. https://doi.org/10.2174/1566524043359638

Kregiel D (2012) Succinate dehydrogenase of Saccharomyces cerevisiae–the unique enzyme of TCA cycle–current knowledge and new perspectives. In: Canuto RA (ed) Dehydrogenases, pp 211–234. https://doi.org/10.5772/48413

Kregiel D, Berlowska J, Ambroziak W (2008) Succinate dehydrogenase activity assay in situ with blue tetrazolium salt in crabtree-positive Saccharomyces cerevisiae strain. Food Technol Biotechnol 46(4):376–380

Lack EE (1997) Tumors of the adrenal gland and extra-adrenal paraganglia. American Registry of Pathology

Lefebvre M, Foulkes WD (2014) Pheochromocytoma and paraganglioma syndromes: genetics and management update. Curr Oncol (Toronto, Ont.) 21(1):e8-e17 http://europepmc.org/books/NBK1548

Li H, Outten CE (2012) Monothiol CGFS glutaredoxins and BolA-like proteins:[2Fe-2S] binding partners in iron homeostasis. Biochemistry 51(22):4377–4389. https://doi.org/10.1021/bi300393z

Li W, Quan L, Peng K, Wang Y, Wang X, Chen Q, Cheng H, Ma Q (2023) Succinate dehydrogenase is essential for epigenetic and metabolic homeostasis in hearts. Basic Res Cardiol 118(1):45. https://doi.org/10.1007/s00395-023-01015-z

Lieber MM, Tomera KM, Farrow GM (1981) Renal oncocytoma. J Urol 125(4):481–485. https://doi.org/10.1016/s0022-5347(17)55079-x

Lill R, Hoffmann B, Molik S, Pierik AJ, Rietzschel N, Stehling O, Uzarska MA, Webert H, Wilbrecht C, Mühlenhoff U (2012) The role of mitochondria in cellular iron–sulfur protein biogenesis and iron metabolism. Biochim Biophys Acta 1823(9):1491–1508. https://doi.org/10.1016/j.bbamcr.2012.05.009

Lv B-B, Li J-M, Yao Z-G, Cheng X-K, Ren F-X, Su W-J, Qin Y-J, Wang Z, Cao ZX (2021) Succinate dehydrogenase deficient gastrointestinal stromal tumor in a three month old boy with a fatal clinical course: a case report and review of literature. Diagn Pathol 16(1):14. https://doi.org/10.1186/s13000-021-01077-4

Mahesha HB (2014) Estimation of succinate dehydrogenase activity. https://doi.org/10.13140/RG.2.2.30654.43849

Majewska J, Ciesielski SJ, Schilke B, Kominek J, Blenska A, Delewski W, Song J-Y, Marszalek J, Craig EA, Dutkiewicz R (2013) Binding of the chaperone Jac1 protein and cysteine desulfurase Nfs1 to the iron-sulfur cluster scaffold Isu protein is mutually exclusive. J Biol Chem 288(40):29134–29142. https://doi.org/10.1074/jbc.M113.503524

Mannelli M, Ercolino T, Giachè V, Simi L, Cirami C, Parenti G (2007) Genetic screening for pheochromocytoma: should SDHC gene analysis be included? J Med Genet 44(9):586–587. https://doi.org/10.1136/jmg.2007.051045

Marín-García J, Marín-García J (2013) Mechanisms of bioenergy production in mitochondria. In: Mitochondria and their role in cardiovascular disease, pp 99–121. https://doi.org/10.1007/978-1-4614-4599-9_5

Martínez-Reyes I, Chandel NS (2020) Mitochondrial TCA cycle metabolites control physiology and disease. Nat Commun 11(1):102. https://doi.org/10.1038/s41467-019-13668-3

Mayes PA, Bender DA (2003) The citric acid cycle: the catabolism of AcetylCoA. In: Murray RK, Granner DK, Mayes PA, Rodwell V (eds) Harper's illustrated biochemistry, pp 130–135

Moosavi B, Berry EA, Zhu X-L, Yang W-C, Yang G-F (2019) The assembly of succinate dehydrogenase: a key enzyme in bioenergetics. Cell Mol Life Sci 76(20):4023–4042. https://doi.org/10.1007/s00018-019-03200-7

Moreno C, Santos RM, Burns R, Zhang WC (2020) Succinate dehydrogenase and ribonucleic acid networks in cancer and other diseases. Cancers (Basel) 12(11). https://doi.org/10.3390/cancers12113237

Moser CC, Farid TA, Chobot SE, Dutton PL (2006) Electron tunneling chains of mitochondria. Biochim Biophys Acta 1757(9):1096–1109. https://doi.org/10.1016/j.bbabio.2006.04.015

Nazar E, Khatami F, Saffar H, Tavangar SM (2019) The emerging role of succinate dehyrogenase genes (SDHx) in tumorigenesis. Int J Hematol Oncol Stem Cell Res 13(2):72–82. https://www.ncbi.nlm.nih.gov/pubmed/31372201

Negro A, Nicoli D, Cavazza A, Santi R, Bonilauri S, Farnetti E, Panebianco M (2017) A rare case of carney-stratakis syndrome in a patient with SDHA mutation [Carney-Stratakis syndrome]. https://doi.org/10.14740/jmc2831w

Otto AM (2016) Warburg effect (s) – a biographical sketch of Otto Warburg and his impacts on tumor metabolism. Cancer Metab 4:1–8. https://doi.org/10.1186/s40170-016-0145-9

Oyedotun KS, Lemire BD (2001) The quinone-binding sites of the Saccharomyces cerevisiae succinate-ubiquinone oxidoreductase. J Biol Chem 276(20):16936–16943. https://doi.org/10.1074/jbc.M100184200

Pandey A, Golla R, Yoon H, Dancis A, Pain D (2012) Persulfide formation on mitochondrial cysteine desulfurase: enzyme activation by a eukaryote-specific interacting protein and Fe–S cluster synthesis. Biochem J 448(2):171–187. https://doi.org/10.1042/BJ20120951

Pandey A, Gordon DM, Pain J, Stemmler TL, Dancis A, Pain D (2013) Frataxin directly stimulates mitochondrial cysteine desulfurase by exposing substrate-binding sites, and a mutant Fe-S cluster scaffold protein with frataxin-bypassing ability acts similarly. J Biol Chem 288(52):36773–36786. https://doi.org/10.1074/jbc.M113.525857

Rapizzi E, Ercolino T, Fucci R, Zampetti B, Felici R, Guasti D, Morandi A, Giannoni E, Giaché V, Bani D, Chiarugi A, Mannelli M (2014) Succinate dehydrogenase subunit B mutations modify

human neuroblastoma cell metabolism and proliferation. Horm Cancer 5(3):174–184. https://doi.org/10.1007/s12672-014-0172-3
Rasheed MRHA, Tarjan G (2018) Succinate dehydrogenase complex: an updated review. Arch Pathol Lab Med 142(12):1564–1570. https://doi.org/10.5858/arpa.2017-0285-RS
Read AD, Bentley RET, Archer SL, Dunham-Snary KJ (2021) Mitochondrial iron–sulfur clusters: structure, function, and an emerging role in vascular biology. Redox Biol 47:102164. https://doi.org/10.1016/j.redox.2021.102164
Robinson KM, Lemire BD (1996) Covalent attachment of FAD to the yeast succinate dehydrogenase flavoprotein requires import into mitochondria, presequence removal, and folding. J Biol Chem 271(8):4055–4060. https://doi.org/10.1074/jbc.271.8.4055
Robinson KM, Rothery RA, Weiner JH, Lemire BD (1994) The covalent attachment of FAD to the flavoprotein of Saccharomyces cerevisiae succinate dehydrogenase is not necessary for import and assembly into mitochondria. Eur J Biochem 222(3):983–990. https://doi.org/10.1074/jbc.271.8.4055
Rouault TA, Maio N (2017) Biogenesis and functions of mammalian iron-sulfur proteins in the regulation of iron homeostasis and pivotal metabolic pathways. J Biol Chem 292(31):12744–12753. https://doi.org/10.1074/jbc.R117.789537
Rustin P, Munnich A, Rötig A (2002) Succinate dehydrogenase and human diseases: new insights into a well-known enzyme. Eur J Hum Genet 10(5):289–291. https://doi.org/10.1038/sj.ejhg.5200793
Rutter J, Winge DR, Schiffman JD (2010) Succinate dehydrogenase – assembly, regulation and role in human disease. Mitochondrion 10(4):393–401. https://doi.org/10.1016/j.mito.2010.03.001
Salvi M, Morrice NA, Brunati AM, Toninello A (2007) Identification of the flavoprotein of succinate dehydrogenase and aconitase as *in vitro* mitochondrial substrates of Fgr tyrosine kinase. FEBS Lett 581(29):5579–5585. https://doi.org/10.1016/j.febslet.2007.11.005
Schmucker S, Martelli A, Colin F, Page A, Wattenhofer-Donzé M, Reutenauer L, Puccio H (2011) Mammalian frataxin: an essential function for cellular viability through an interaction with a preformed ISCU/NFS1/ISD11 iron-sulfur assembly complex. PLoS One 6(1):e16199. https://doi.org/10.1371/journal.pone.0016199
Sheftel AD, Stehling O, Pierik AJ, Elsässer H-P, Mühlenhoff U, Webert H, Hobler A, Hannemann F, Bernhardt R, Lill R (2010) Humans possess two mitochondrial ferredoxins, Fdx1 and Fdx2, with distinct roles in steroidogenesis, heme, and Fe/S cluster biosynthesis. Proc Natl Acad Sci 107(26):11775–11780. https://doi.org/10.1073/pnas.1004250107
Silkin Y, Oyedotun KS, Lemire BD (2007) The role of Sdh4p Tyr-89 in ubiquinone reduction by the Saccharomyces cerevisiae succinate dehydrogenase. Biochim Biophys Acta 1767(2):143–150. https://doi.org/10.1016/j.bbabio.2006.11.017
Sun F, Huo X, Zhai Y, Wang A, Xu J, Su D, Bartlam M, Rao Z (2005) Crystal structure of mitochondrial respiratory membrane protein complex II. Cell 121(7):1043–1057. https://doi.org/10.1016/j.cell.2005.05.025
Tomitsuka E, Kita K, Esumi H (2009) Regulation of succinate-ubiquinone reductase and fumarate reductase activities in human complex II by phosphorylation of its flavoprotein subunit. Proc Jpn Acad Ser B 85(7):258–265. https://doi.org/10.2183/pjab.85.258
Tretter L, Patocs A, Chinopoulos C (2016) Succinate, an intermediate in metabolism, signal transduction, ROS, hypoxia, and tumorigenesis. Biochim Biophys Acta 1857(8):1086–1101. https://doi.org/10.1016/j.bbabio.2016.03.012
Tsai C-L, Barondeau DP (2010) Human frataxin is an allosteric switch that activates the Fe–S cluster biosynthetic complex. Biochemistry 49(43):9132–9139. https://doi.org/10.1021/bi1013062
Tzagoloff A, Jang J, Glerum DM, Wu M (1996) FLX1 codes for a carrier protein involved in maintaining a proper balance of flavin nucleotides in yeast mitochondria. J Biol Chem 271(13):7392–7397. https://doi.org/10.1074/jbc.271.13.7392

Van Vranken JG, Na U, Winge DR, Rutter J (2015) Protein-mediated assembly of succinate dehydrogenase and its cofactors. Crit Rev Biochem Mol Biol 50(2):168–180. https://doi.org/10.3109/10409238.2014.990556
Wexler R, Elton T, Pleister A, Feldman D (2009) Cardiomyopathy: an overview. Am Fam Physician 79(9):778. https://doi.org/10.3390/ijms22147722
Yankovskaya V, Horsefield R, Tornroth S, Luna-Chavez C, Miyoshi H, Léger C, Byrne B, Cecchini G, Iwata S (2003) Architecture of succinate dehydrogenase and reactive oxygen species generation. Science 299(5607):700–704. https://doi.org/10.1126/science.1079605

2 The Function of Succinate Dehydrogenase in Maintaining Normal Health

Mohd Qasid Lari, Deepak Kumar, Dileep Kumar, and Ajay Kumar

Abstract

Succinic dehydrogenase (SDH) is a vital enzyme involved in cellular respiration and metabolism of energy. This enzyme is much essential to maintain the normal health as it takes part in tricarboxylic acid cycle (TCA) and the electron transport chain. SDH acts as a catalyst in the conversion of succinate to fumarate by the oxidation reaction, and electron is produced in this procedure which is transported along the electron transport chain to form adenosine triphosphate (ATP). SDH must function properly for a number of physiological functions, including ATP synthesis, control of cellular metabolism and upkeep of mitochondrial homeostasis. Cancer, hereditary mitochondrial problems, neurological diseases and other diseases have all been linked to mutations or dysregulation of SDH function. The development of neurodegenerative disorders such as Parkinson's disease, Alzheimer's disease and Huntington's disease has also been linked to dysfunctional SDH activity. The rare genetic illness like hereditary paraganglioma and pheochromocytoma syndromes, which are characterised by the development of tumours in the paraganglia and adrenal glands, can also occur due to the inherited mutations in SDH genes (SDHD, SDHAF2, SDHB, SDHA, TMEM127 and MAX). Thus, it is necessary to learn more about SDH functions in order to improve our understanding of cellular metabolism, mitochondrial

M. Q. Lari · D. Kumar · A. Kumar (✉)
Sardar Patel College of Pharmacy, Gorakhpur, Uttar Pradesh, India

D. Kumar (✉)
Department of Pharmaceutical Chemistry, Manipal College of Pharmaceutical Sciences, Manipal Academy of Higher Education, Manipal, Karnataka, India

Department of Entomology and Nematology, UC Davis Comprehensive Cancer Center, University of California Davis, Davis, CA, USA

V. M. Patil et al. (eds.), *Deciphering The Role of Succinate Dehydrogenase in Drug Discovery*, https://doi.org/10.1007/978-981-96-6831-1_2

function and disease pathogenesis. So, in this chapter we will explore the relation between the SDH with our physiological phenomena and also about its importance for maintaining the proper health.

Keywords

Succinic dehydrogenase · TCA cycle · Homeostasis · Mitochondria · Neurodegenerative disorder · Cancer

Abbreviations

ATP	Adenosine triphosphate
CS	Citric synthase
ETC	Electron transport chain
FAD	Flavin-adenine dinucleotide
HIF1	Hypoxia-inducible factor 1
IDH	Isocitrate dehydrogenase
IGF1	Insulin-like growth factor-1
JMJC	Jumanji domain-containing
OGDH	Alpha-ketoglutarate dehydrogenase
OXPHOS	Oxidative phosphorylation
PE	Phosphatidylethanolamine
PGL	Paraganglioma
PHEO	Pheochromocytoma
RCC	Renal cell carcinoma
SDH	Succinic dehydrogenase
SDHAF2	Succinate dehydrogenase complex assembly factor 2
SDHB	Succinate dehydrogenase complex iron sulphur subunit B
SDHD	Succinate dehydrogenase complex subunit D
SQR	Succinate ubiquinone oxidoreductase
TCA	Tricarboxylic acid cycle
TET	Ten-eleven translocation
TMEM127	Transmembrane protein 127
VEGF	Vascular endothelial growth factors

2.1 Introduction

The only enzyme complex that participates in the electron transport chain in the citric acid cycle or tricarboxylic acid (TCA) cycle is succinate dehydrogenase (SDH) (Rasheed and Tarjan 2018). A crucial metabolic centre of the cell is the TCA cycle, sometimes referred to as the citric acid cycle or the Krebs cycle. With the exception of succinate dehydrogenase, an anomaly linked to the respiratory chain on the inner mitochondrial membrane, all eight of the enzymes that make up this complex are found within the mitochondrial matrix. Molecules that can transform into an acetyl group or dicarboxylic acid can enter aerobic metabolism through this cycle. Three different enzymes—citrate synthase, isocitrate dehydrogenase and alpha-ketoglutarate dehydrogenase—are involved in the regulation of the TCA cycle. Precursors for the storage form of fuels, like cholesterol and amino acids, are also replenished through the cycle (Alabduladhem and Bordoni 2022). SDH is the only one of the five oxidative phosphorylation (OXPHOS) complexes that participates in the TCA cycle, by catalysing the oxidation of succinic acid to fumaric acid in the TCA cycle, which is connected to the reduction of flavin-adenine dinucleotide (FAD) to produce $FADH_2$ and transfer electrons to ubiquinone (UQ) in the electron transport chain (ETC), SDH links the essential metabolic OXPHOS and TCA pathways in eukaryotic cells (Fullerton et al. 2020).

A vital metabolic route in cells is the TCA cycle, sometimes referred to as the citric acid cycle or the Krebs cycle. It occurs in the matrix of the mitochondria and is essential to the oxidation of carbon skeletons made of fatty acids, amino acids and carbohydrates. The majority of the enzymes involved in the eight enzymatic processes that make up the TCA cycle are found in the mitochondrial matrix. One exception exists, though: SDH. SDH is an enzyme complex that is a component of the electron transport chain and is located in the inner mitochondrial membrane. In the TCA cycle, succinate is converted to fumarate by SDH. To create $FADH_2$, this process is combined with the reduction of FAD. After that, in the ETC, the electrons from $FADH_2$ are converted to ubiquinone (UQ), aiding in the production of ATP by oxidative phosphorylation (Haddad and Mohiuddin 2022; Xiao et al. 2023).

2.2 Regulation and Functions of the TCA Cycle

The regulation of the TCA cycle maintains metabolic balance and effective energy generation. Citric synthase (CS), isocitrate dehydrogenase (IDH) and alpha-ketoglutarate dehydrogenase (OGDH) are the three major enzymes in the TCA cycle that are involved in its regulation.

In order to create citrate, the first step in the TCA cycle, CS catalyses the condensation of acetyl-CoA and oxaloacetate. Alpha-ketoglutarate is transformed from isocitrate to alpha-ketoglutarate by IDH, while succinyl-CoA is catalysed by OGDH. According to TCA cycle theory, these three enzymes are rate-limiting phases that are governed by allostery (Arnold and Finley 2023).

The TCA cycle is the central hub for the latter stages of the oxidative breakdown of fatty acids, amino acids and carbohydrates. Reduced coenzymes like NADH and $FADH_2$, which are crucial electron carriers for oxidative phosphorylation in the ETC, are produced by each oxidative step in the cycle. Precursors for the production of other compounds, such as cholesterol and amino acids, are also supplied by the TCA cycle. It is essential for preserving the cell's metabolic equilibrium and refills the fuel storage forms (Xiao et al. 2023) (Fig. 2.1).

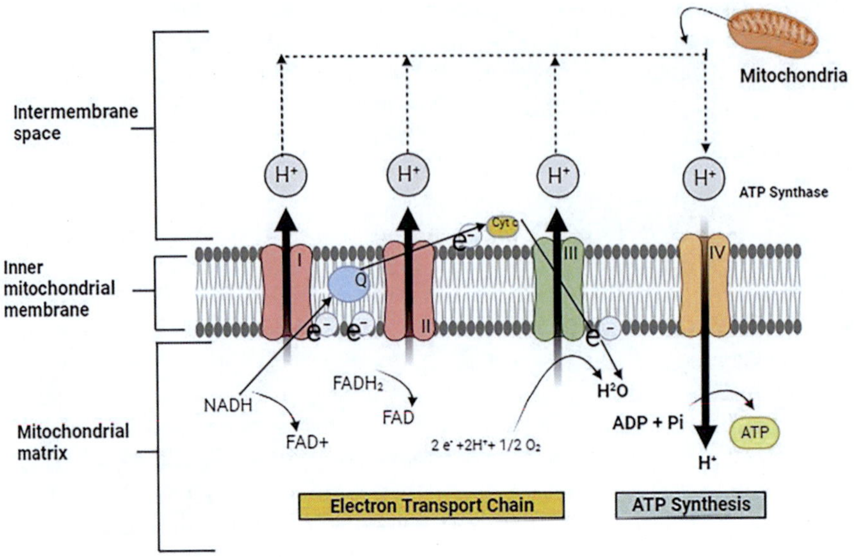

Fig. 2.1 Diagrammatic representation of oxidative phosphorylation for ATP synthesis and electron transport chain

2.3 Structure and Function of SDH

SDH is made up of only four subunits, called SDHA (~70 kDa), SDHB (~30 kDa), SDHC (~18 kDa) and SDHD (~17 kDa) (SDH1, SDH2, SDH3 and SDH4 in all *S. cerevisiae* orthologs), which are all encoded by the nuclear genome. This is in contrast to other mitochondrial complexes with many subunits and dual genetic origin sites (Mukherjee and Ghosh 2020). SDH is defined as succinate: ubiquinone oxidoreductase (SQR), which is made up of two transmembrane proteins, two hydrophilic proteins and prosthetic groups that are involved in electron transport. The large cytochrome b-binding protein, or CybL, and the small cytochrome b-binding protein, or CybS, are found in the hydrophobic region, whereas the flavoprotein, or Fp, and the iron–sulphur protein, or Ip, are found in the hydrophilic section (Protasoni and Zeviani 2021). The biggest subunit of SDH, SDHA oxidises succinate to produce fumarate and is covalently bonded by FAD. The FAD-binding domain of the four unique domains that make up the structure of human SDHA is essential for electron transport and catalytic activities (Zhu et al. 2019). The SDHB subunit is situated in between the SDHC–SDHD dimer and the SDHA subunit. Three Fe–S clusters arranged into two separate domains that resemble butterflies make up SDHB .The 3Fe–4S and 4Fe–4S clusters are located in the C-terminal domain of SDHB, while the 2Fe–2S cluster is located in the N-terminal domain .The electrons produced by succinate oxidation must be transferred to ubiquinone via these Fe–S clusters (Fullerton et al. 2020; Zhu et al. 2019). Three Fe–S clusters and FAD bind SDHA and SDHB covalently to generate the soluble subunits, which together create a catalytic heterodimer (Sun et al. 2005). The SDHC–SDHD dimer, which binds the whole complex to the IMM, makes up the hydrophobic domain. Three transmembrane helices are present in both SDHC and SDHD, and these helices exhibit a comparable folding pattern. One heme B group and a phosphatidylethanolamine (PE) unit, which is situated right below the heme B group, are present in each SDHC–SDHD dimer. Furthermore, the C-terminal portion of SDHB and the transmembrane helices of SDHC and SDHD form a pocket that has a ubiquinone-binding site at its entrance (Du et al. 2023).

Complex II does not pump protons across the inner membrane like other complexes do. On the other hand, it is proposed that complex II's major job is to preserve the integral membrane Q pool's decreased state. Normally, SDH catalyses the orderly transfer of electrons to ubiquinone via electron transport groups, such as FAD, Fe–S centres and a heme group, to oxidise succinate and produce fumarate (Cecchini 2003; Horsefield et al. 2006). It's interesting to note that increased succinate levels have reportedly been observed in response to hypoxia. Reverse electron transport can use accumulated succinate to provide a redox signal that increases the formation of reactive oxygen species (ROS). Additionally, succinate can be released into the cytosol to stabilise the expression of HIF-1α, a key player in the hypoxia response of tumour cells (Murphy and O'Neill 2018). In addition to more recent studies that demonstrate the connection between SDH and the TCA cycle and OXPHOS, SDH is now known to have a key role in controlling mitochondrial activity and metabolism in a number of clinical disorders.

2.4 Role of SDH in Cancer

A complex of mitochondrial enzymes called succinate dehydrogenase, or SDH for short, is involved in the electron transport chain (by directing the transfer of electrons to the terminal acceptor ubiquinone) and the citric acid cycle (by catalysing the oxidation of succinate to fumarate) (Bezawork-Geleta et al. 2017). The four subunits of the formed SDH complex—SDHA, SDHB, SDHC and SDHD, commonly referred to as SDHx—are encoded by nuclear genes and are found in the inner membrane of the mitochondria. The carcinogenesis of several malignancies, such as GIST, paraganglioma, pheochromocytoma, chronic lymphocytic leukaemia, renal cell carcinoma, thyroid cancer, Hodgkin lymphoma, pituitary adenomas and neuroendocrine tumours of the pancreas, is associated with all of the SDH genes (Barletta and Hornick 2012; Janeway et al. 2011; Niemeijer et al. 2015).

When one SDHx subunit completely stops functioning, the SDH complex becomes unstable and loses its enzymatic function, which is what causes tumour formation in cases of SDH deficiency. When SDH is disrupted, either by biallelic mutations or epigenetic silencing, succinate builds up and becomes an oncometabolite that triggers many pathways that aid in the development of tumours. Increased expression of the transcriptional factors vascular endothelial growth factor (VEGF) and insulin-like growth factor-1 (IGF1), which in turn cause increased angiogenic and cell growth signalling, is the result of succinate accumulation inhibiting prolyl-hydroxylase domain proteins (PHD) activity and stabilising hypoxia-inducible factor 1 (HIF1) (Pantaleo et al. 2009).

The SDH complex loses its enzymatic activity and becomes unstable when one of the SDHx subunits is totally dysfunctional. Succinate, an oncometabolite, may build up as a result of this disturbance in SDH function. Tumour formation is facilitated by a number of mechanisms that are activated by succinate build-up. HIF1 is stabilised as a result of the suppression of PHD, one of the effects of succinate build-up. Subsequently, stabilised HIF1 can promote the production of transcription factors including IGF1 and VEGF. Angiogenesis and cell growth signalling are facilitated by these substances and are linked to the advancement of tumours. The citric acid cycle and the electron transport chain are both facilitated by the complex of mitochondrial enzymes known as SDH. Mutations or epigenetic silencing that impair SDH function can accumulate succinate and act as a catalyst for many pathways involved in the formation of tumours, potentially contributing to tumourigenesis (Hao et al. 2009; Fujita et al. 2010).

2.5 SDH Dysfunction and Illness

2.5.1 Gastrointestinal Stromal Tumour (GIST)

The gastrointestinal system is the primary site of this disease. GIST-prone loci include BRAF, SDHx, PDGFRA, PDGFRB and KIT. GIST, also known as 'wild-type GIST' or dSDH-GIST, is mostly caused by mutations in genes that encode SDH subunits. Future research should determine if genes encoding these assembly factors contribute to the aetiology of GIST since they are involved in the maturation and assembly of functional SDH. These assembly factors include SDHAF1, SDHAF2, SDHAF3 and SDHAF4. Supporting evidence includes the observation that paediatric patients with metastatic GIST have negative staining for SDHB without observable mutations in any of the SDH subunit genes (Schipani et al. 2023).

2.5.2 Renal Cell Carcinoma (RCC)

This tumour originates from the kidney, which is the urinary tract. RCC has been linked to germline and somatic mutations of SDHx in addition to translocations of chromosome 3 and mutations in the FLCN, c-Met and VHL genes. SDH-deficient RCC resembles GIST and PGL linked to SDHx mutations morphologically (Paik et al. 2014; Ozluk et al. 2015). Unlike GIST, RCCs associated with mutations in SDH subunits other than SDHB are uncommon; around 80% of SDH-dysfunctional RCCs are caused by mutant SDHB. Unknown is the molecular process that underlies SDHx's involvement in RCC (Rogala and Zhou 2023).

2.5.3 Paraganglioma (PGL) and Pheochromocytoma (PHEO)

Paraganglia, which sense oxygen levels and are crucial for hypoxic adaption, are the source of these tumours (Welander et al. 2011). The paravertebral axis runs from the base of the skull and neck to the pelvis, where paraganglia are found. PGLs and PHEOs are classed as clinically separate tumours; PHEOs are defined as tumours generated from the adrenal medulla, whereas PGLs are defined as tumours developed from paraganglia extra-adrenally. PGLs and PHEOs are caused by mutations in approximately 17 genes, including RET, VHL, NF1, TMEM127, MAX, KIF1B, EGLN1 and SDHx (Gill 2012). Rarely, PGL can be formed from organs devoid of chromaffin cells. This could be because of developmental disorders that leave chromaffin cells in certain organs or regions (Martucci et al. 2015).

2.5.4 Leigh Syndrome

There is currently no treatment for this progressive neurological disease, which affects babies but is also occasionally seen in adults and teenagers (Finsterer 2008). A wide range of genetic abnormalities, including mutations in nuclear and mitochondrial genes, including SDHA, can cause Leigh syndrome. Leigh syndrome has been identified in patients with the SDHA gene variants R451C, A524V, R554Y or G555Q. However, determining which mutations in SDHA cause PGL and which result in Leigh syndrome requires a thorough biochemical and structural investigation (Van Coster et al. 2003) (Table 2.1).

Table 2.1 Neoplastic pathologies associated with mutations in SDH

Problems with complex II	Pathology	Manifestation	References
SDHA-D	PHEO/PGL	Hypertension, palpitation, sweating	Kantorovich et al. (2010)
SDHA-D	GIST	Abdominal pain, gastrointestinal bleeding	Boikos et al. (2016a, b)
SDHA-D	Renal cell carcinoma	Haematuria, abdominal pain	Udager and Mehra (2016), Gill et al. (2014)
SDHA-D	Carney triad	Similar to PHEO/PGL and GIST	Boikos et al. (2016a, 2016b)
SDHA-D	Carney dyad (Carney–Stratakis syndrome)	Similar to PHEO/PGL and GIST	Pasini et al. (2008)
SDAF2	PHEO/PLG, preferentially head and neck; PGL mainly in young adults	Similar to PHEO/PGL	Casey et al. (2014)
SDHB	Pancreatic NET	Abdominal pain	Niemeijer et al. (2015)
SDHB	Ganglioneuroma	Abdominal pain	
SDHAD	PHEO/PGL and pituitary adenoma (3PAs)	Similar to PHEO/PGL; headache, visual problems, clinical characteristics related to hypersecretion of GH and PRL	Xekouki et al. (2015)

2.6 Human Health and Disease: Assembly of Mitochondrial SDH

Despite having only four subunits, the SDH complex needs certain assembly elements for its formation, just like other OXPHOS complexes. Four assembly factors—namely SDHAF1, SDHAF2, SDHAF3 and SDHAF4 in humans (SDH6, SDH5, SDH7 and SDH8 in Saccharomyces cerevisiae)—fully coordinate the major steps of SDH biogenesis, including the maturation of individual subunits and assembly of the entire complex, as recently attributed to them by Dr. Jared Rutter's team (Bricker et al. 2012; Fullerton et al. 2020). While SDHAF4 reduces oxidative stress damage to the FAD–SDHA dimer, promoting SDHA interaction with SDHB and aiding in SDHA maturation, SDHAF2 mediates the flavin acylation of SDHA (Hao et al. 2009; Bricker et al. 2012). Fe–S clusters must be inserted for SDHB maturation, and HSPA9, HSC20 and ISCU catalyse their creation. It has been demonstrated that SDHAF1 and SDHAF3 encourage the insertion of Fe–S clusters and shield those that are exposed to the matrix, which makes it easier for SDHA and SDHB to connect. In the IMM, SDHC and SDHD dimerize after SDHA and SDHB mature. Following the release of the assembly factors from the subunits by an unidentified process, SDHA–SDHB attaches to SDHC–SDHD to form the whole complex (Na et al. 2014; Moosavi et al. 2019) (Fig. 2.2).

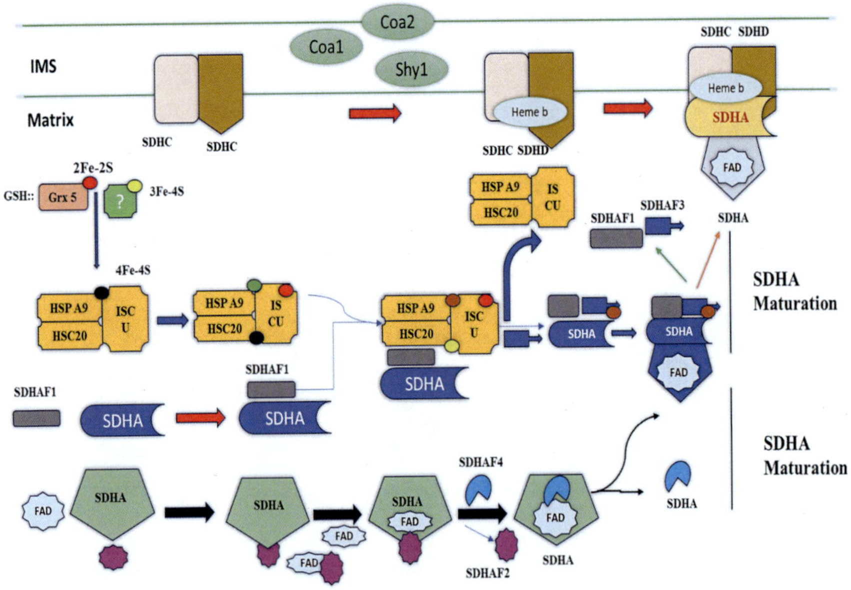

Fig. 2.2 Predicted assembly of SDH in humans

2.7 SDHA Maturation

SDH

SDH-dependent respiration and SDHA flavination are dependent on SDHAF2 (c11orf79, PGL2), which is also known as SDHE in bacteria (Hao et al. 2009). The catalytic phase produces oxidative electrons from succinate, which the FAD cofactor can absorb. However, the exact method by which FAD binds to SDHA is mostly unclear (Zafreen et al. 2016). A yeast strain with SDHAF2 loss had poor ETC assembly and stability, which resulted in a significant decrease in oxygen consumption and a shorter lifetime. These abnormalities were restored by SDHAF2 overexpression (Hao et al. 2009; Tivendale et al. 2021). These results suggested that SDHAF2 plays a crucial role in preserving SDH activity. This conclusion is supported by a recent work using an in vitro flavination system, which showed that FAD was unable to establish covalent connections with SDHA in the absence of SDHAF2, indicating that SDHAF2 serves a similar essential role in humans (Na et al. 2014).

There are now two different kinds of SDHAF2 insertion mechanisms. Studies have suggested that SDHAF2 probably serves as a chaperone or catalysing agent to modify the structure of SDHA, promoting the binding of FAD to SDHA. This is because, in situations where SDHAF2 levels are low or non-existent, FAD binds SDHA via an ineffective autocatalytic process (McNeil et al. 2014; McNeil and Fineran 2013). Contrary to this theory, some researchers think that SDHAF2 directly binds to FAD through covalent or noncovalent bonds, acting as a transporter of FAD when necessary. This suggests that SDHAF2 may be produced frequently in a form that is free of FAD. Since precise hypotheses have not been able to be proposed due to variances among different species and cellular circumstances, these hypotheses are generally generalised (Zafreen et al. 2016; Bezawork-Geleta et al. 2017).

SDHAF4 (c6orf57), another subunit-specific chaperone, engages in complex formation after SDHAF2-dependent covalent flavin acylation. SDHAF4 is a tiny protein that is roughly 15 kDa and a member of a family of eukaryotic proteins that is extensively expressed (Bricker et al. 2012). There is a suggestion that the main role of SDHAF4 is to safeguard the SDHA subunit in the mitochondrial matrix when it is unbound by SDHB. Earlier, SDHA lowers FAD to produce FADH2 and oxidises succinate to produce fumarate. Extra electrons during these events encourage the generation of superoxide when they are inefficiently transported to SDHB. As a result, SDHAF4 acts as a chaperone by binding to FAD–SDHA directly, preventing FAD-mediated oxidative stress and obstructing the production of excessive ROS (Moosavi et al. 2019). Facilitating the formation of the SDHA–SDHB subunit and the subsequent assembly of the SDH complex is the other function of SDHAF4. SDH enzymatic activity was significantly decreased in yeast, *Arabidopsis thaliana*, and mammalian cells with deletion of SDHAF4, since this prevented the development of the SDHA–SDHB intermediate (Belt et al. 2018). Several reports have demonstrated the connection between the onset of several diseases and SDHAF4 failure; nevertheless, the exact mechanism by which SDHAF4 contributes to SDH function remains unclear. In conclusion, the SDHA–SDHAF4 dimer's creation signifies SDHA's first maturation.

2.8 SDHB Maturation

The insertion of Fe-S clusters is necessary for SDHB development. The 4Fe–4S centre is expected to be formed by a scaffold protein complex made up of HSC20 (JAC1 in *S. cerevisiae* orthologs), HSPA9 (Ssq1 in *S. cerevisiae* orthologs) and ISU1. The 2Fe–2S centre is thought to be derived from the Grx 5: GSH complex. Clusters are transferred to SDHB via the scaffold protein, which binds to SDHB. Following the coimmunoprecipitation of SDHB, the intermediate was visualised using blue native-polyacrylamide gel electrophoresis (BN–PAGE) (Maio et al. 2014). Furthermore, SDH activity was reduced when HSC20 or HSPA9 expression was downregulated using short interfering RNA (siRNA) (Maio and Rouault 2016). Further research is necessary since the process leading to 3Fe–3S biogenesis is unknown.

SDHB maturation involves both SDHAF1 (LYRM8, MC2DN2) and SDHAF3 (ACN9, DC11, LYRM10) (Ghezzi et al. 2009). Cells with no SDHAF1 or SDHAF3 show reduced SDH activity and improper SDH assembly (Ghezzi et al. 2009). Therefore, the insertion of Fe–S clusters is associated with both the SDHAF1 and SDHB assembly factors (Meyer et al. 2019). It has been suggested that the first stage of Fe–S cluster insertion is as follows: Using its own LYR motif, SDHAF1 drives the HSPA9–HSC20–ISCU complex to approach the LYR-binding sites in SDHB. Lastly, the Fe–S clusters are transferred from the ISU scaffold protein to SDHB. SDHAF1 contacts SDHB through non-LYR-binding sites. In cells lacking SDHD, SDHAF1 draws down the SDHA–SDHB heterodimer, but it is unable to pull down SDHB mutants (Na et al. 2014). These findings verify that SDHB is impacted by SDHAF1 during SDH assembly. The specifics of SDHAF3's engagement are still unknown, though.

Furthermore, when Fe–S clusters are transferred, they are very vulnerable to destruction from oxygen exposure. Thus, shielding the Fe–S clusters from oxidative stress is the other specific role of SDHAF1 and SDHAF3 (Sun et al. 2005; Imlay 2006). Overall, further research is needed to determine how SDHAF1 and SDHAF3 affect SDHB assembly. We also know very little about the involvement of the heme group in the function of SDH and very little about the biogenesis and potential assembly components for the membrane-buried SDHC and SDHD.

2.9 Disorders of the Mitochondria Linked to SDHx Variations

Only 2% of cases of mitochondrial diseases are caused by SDH deficiency, making mitochondrial diseases a widespread group of inherited disorders with diverse aetiologies (Lin et al. 2021). Patients with SDHx mutations typically experienced childhood-onset symptoms; of the patients in this study, only 3.2% (2/67) had symptoms that started before the age of 40, and 9.8% (6/67) had symptoms that started between the ages of 16 and 25, which is consistent with patterns of defects in the mitochondrial respiratory chain (Fernandez-Vizarra and Zeviani 2021). Typically, the earliest

Table 2.2 Known complex II subunits and assembly factors in mitochondrial disease

Case	Subunits	M/F	Age of onset	Clinical diagnosis	References
1	SDHA	M	15Y	Cardiomyopathy, development delay, ataxia	Courage et al. (2017)
2	SDHA	M	8 m	Cardiomyopathy	
3	SDHA	F	9 m	Cardiomyopathy	
4	SDHA	F	9 m	Development delay, Leigh syndrome	Levitas et al. (2010)
5	SDHA	F	2 m	Respiratory distress, congestive heart failure	
6	SDHD	F	4 m	Intellectual disability, developmental regression, visual impairment, nystagmus, optic atrophy, hypertrichosis	Lin et al. (2021)
7	SDHD	M	4 m	Dystonia, nystagmus, cardiac abnormalities	
8	SDHAF1	M	12 m	Microcephaly, truncal hypotonia, limb spasticity, ataxia, mild cognitive impairment	Helman et al. (2016)
9	SDHAF1	M	11 m	Microcephaly, truncal hypotonia, limb spasticity, ataxia, seizure	
10	SDHAF1	M	12 m	Truncal hypotonia, limb spasticity, died of respiratory failure	

signs of SDH deficiency are shown in organs with high mitochondrial energy metabolism, such as the heart, liver, kidneys, nerves and muscles (Fullerton et al. 2020). Thus far, it has been determined that SDH is linked to the following conditions: ischemia-reperfusion injury, insulin resistance, high-altitude sickness, inflammation, neurodegenerative disease and infantile leukoencephalopathy. The majority of people with these illnesses have SDHA gene variations, according to the literature. Apart from the prevalent kind of SDHA mutations, there have been reports of variants linked to SDHB, SDHD and SDHAF1 (Table 2.2).

2.10 SDHA

Mitochondrial illnesses of Mendelian inheritance are thought to be mostly caused by abnormalities in the SDHA gene, which is the most often discovered pathogenic mutation in SDH deficiency (Rutter et al. 2010). According to reports, the majority of variations exhibit an autosomal recessive inheritance pattern (Fullerton et al. 2020). Only one variant has been found to exhibit a dominant inheritance pattern. Patients with this variant were diagnosed with symptoms such as osteoarthritis, dystonia, hypotonia, cardiomyopathy, development regression and delay in development. A few patients also had severe vision issues, including double vision and visual loss (Fullerton et al. 2020). A recent study demonstrated that the heart muscle of model mice with myocardial infarction had lower levels of SDHAF4 expression. It also demonstrated that the loss of SDHAF4 in the cardiac tissue inhibited complex II assembly, leading to complex II deficiency, which in turn caused excessive mitochondrial fission, and mitophagy, which in turn caused progressive DCM and fatal heart failure. This result confirms that SDH plays a critical role in the onset of cardiomyopathy (Wang et al. 2022). In a recent study it has also revealed that nonsense mutation in SDHA leads to a disease known as Leigh syndrome (Parfait et al. 2000). This variant's pathophysiology is caused by a leucine substitution in the translation initiation codon sequence, which results in an inactive protein lacking N-terminal residues. However, Leigh syndrome has several major causes besides SDH deficiency (mostly caused by SDHA mutation) (Parfait et al. 2000). Levitas et al. initially reported the link between SDHA variations and cardiomyopathies in relation to missense mutations. Within the first eight months of life, fifteen patients from two consanguineous Bedouin families developed DCM, and the patients' muscle samples showed a noticeably lower level of SDH activity (Levitas et al. 2010).

2.11 SDHB

Only five harmful SDHB variants have been identified so far, and patients with SDHB mutations are typically diagnosed with carcinogenesis as opposed to those with SDHA mutations (Fullerton et al. 2020). The most commonly reported mutation was linked to dystonia, ataxia, limb spasticity and truncal hypotonia. A biochemical investigation revealed that patient muscle cells and fibroblasts had less SDHB expression (Alston et al. 2012). Interestingly, a little decrease in SDHA expression was also discovered in certain samples, suggesting that the SDHB mutation had an impact on the stability of SDHB and the interaction between SDHA and SDHB (Ardissone et al. 2015). It has been hypothesised that this mutation may impede the Fe-S cluster domain's ability to transfer electrons, which would ultimately result in reduced SDH activity given the structure of SDH (Shokes 2006). Furthermore, a patient exhibiting severe symptoms was determined to have a mutation by Parneet K. et al. They hypothesised that this mutation changed the polar interactions in residues (Kaur et al. 2020). Structural analysis suggested that the alteration was related to the 4Fe–4S cluster, which affects electron transfer, or the insertion of Fe–S clusters, even though these researchers did not perform additional biochemical analyses to measure the protein levels during pathogenesis (Sun et al. 2005). Among all reported cases of SDHB-related SDH deficiency, the youngest patient identified with the most recent pathogenic SDHB variant was a 2-month-old boy who died (Ece Solmaz et al. 2022). He was homozygous for the SDHB mutation, experienced seizures and sepsis, and showed markedly reduced SDH activity based on histochemical analysis of a muscle biopsy (Ece Solmaz et al. 2022). Although he died from sepsis, the connection between complex II deficiency and his clinical symptoms—and whether the SDHB mutation contributed to the sepsis onset—remains unclear. Even though the boy passed away from sepsis, it is unclear how the complex II deficiency relates to the boy's clinical presentation and whether the SDHB mutation was linked to the onset of sepsis.

2.12 SDHD

The literature has reported on two forms of SDHD. Since SDHD is a membrane protein with five helices, as was previously indicated, mutations may impact SDHD's capacity to bind to the SDHA–SDHB heterodimer. According to Jackson et al., the first patient with SDHD deficiency was diagnosed at the age of three months. The patient then experienced progressive neurological deterioration and passed away at the age of 10 (Courage et al. 2017). To put it briefly, mutations in SDHD can lead to mitochondrial respiratory chain illnesses by impairing the expression of the disease or causing the protein to be misfolded. This can further disrupt complex II assembly.

Taken together, current research has identified mutations only in SDHA, SDHB, and SDHD in patients; no mutations have yet been documented in SDHC or in other assembly factors. Due to their distinct roles in SDH complex formation and enzymatic function, these mutant proteins often result in either severe pathology or embryonic lethality. Although complex II deficiency has been linked to these mutations, the wide range of clinical manifestations suggests that SDH subunits and their assembly factors may play diverse roles across different tissues. However, the reliance on basic case report data provides only a limited understanding of the underlying pathogenic mechanisms. Further research is essential to elucidate the functions of SDHx mutations. Owing to their distinct role in the formation of the SDH complex and their enzymatic activity, these mutant proteins cause either severe pathology or embryonic death. Despite the fact that complex II deficiency has been linked to mutations, the erratic clinical manifestations suggest that SDH subunits and assembly factors may play diverse roles in various organs. In any event, a limited understanding of the pathogenic pathways can be achieved using only basic case report data. There must be greater efforts made to learn more about the function of SDHx mutants.

2.13 SDH's Developing Function in Epigenomic Regulation

Gene mutations or deletions encoding SDH subunits are mostly responsible for the clinical disorders linked to SDH dysfunction that have been previously reported. Interestingly, it has been demonstrated that pathological phenotypes are caused by epigenetic control via DNA methylation of these genes' promoter regions. According to a case study, reduced gene transcription as a result of hypermethylation of the CpG islands of the SDHC promoter led to GIST (Urbini et al. 2015). In this instance, decreased SDH activity and decreased protein levels of SDHB and SDHC were the outcomes of hypermethylation of the SDHC promoter. Multiple patients with PGL and GIST have been shown to have SDHC promoter methylation in the absence of gene mutation. Thus, new information suggests a connection between neoplasias and alterations related to the epigenomic control of genes encoding SDHC subunits (Richter et al. 2016). Oncometabolites are metabolites whose aberrant accumulation encourages the development of malignant transformation. Fumarate, succinate and α-ketoglutarate are TCA cycle intermediates that are on the list of possible oncometabolites. According to recent research, succinate alters the epigenomic landscape, which results in pathogenic conditions (Kaochar and Tu 2012). Because of its structural similarities to α-ketoglutarate, succinate is a competitive inhibitor of α-ketoglutarate-dependent dioxygenases. Within the extensive family of Jumanji domain-containing proteins, α-ketoglutarate-dependent dioxygenases (>60 in mammals), succinate inhibits the proteins known as 'ten-eleven translocation' (TET) and Jumanji domain-containing (JMJC). Additionally, it has been demonstrated that succinate inhibits the prolyl 4-hydroxylases-2 and -3, members of another dioxygenase family of proteins that stabilise Hif1α post-translationally (Favier et al. 2015). Further evidence supporting the role of succinate in epigenetics came from genome-wide methylation analyses of PGL and GIST patients. These investigations corroborated the aforementioned in vitro research by demonstrating that SDH-malfunctional GISTs have a worldwide decrease in 5hmC and an increase in DNA methylation. Only 19 hypermethylated sites were found in KIT-mutant tumours, compared to approximately 400 hypermethylated sites in SDH-mutant GIST patients (Letouzé et al. 2013; Killian et al. 2013). A genome-wide investigation of methylation in PGL and PHEO cancers with mutations in the SDHx, VHL, NF1 and RET genes demonstrated that the hypermethylated phenotype was related to mutations in SDHx. This is highlighted by the survival of patients, which is adversely correlated with methylation, with the prognosis linked to SDHx mutation-associated hypermethylation being substantially worse than that for other PGL/PHEO-susceptible genes (Killian et al. 2013). This suggests that DNA methylation changes with clinical manifestation have succinate as an underlying cause. Combining these results shows that chromatin remodelling and DNA methylation changes are linked to pathologies caused by malfunctioning SDH, suggesting that mitochondria-mediated regulation of apoptosis, ROS generation and accumulation of HIF-α are not the only molecular mechanisms underlying the pathogenesis of diseases associated with SDH dysfunction. This theory is supported by the finding that PGL carcinogenesis is likewise associated with mutations in epigenetic regulatory proteins (Toledo et al. 2016).

2.14 Conclusion

In conclusion, by supporting vital cellular functions, SDH plays a critical part in preserving normal health. SDH plays a crucial role in the TCA cycle and ETC, which are both essential to the creation of ATP, which is the energy source. The maintenance of the metabolic requirements of the body's numerous tissues and organs depends on this bioenergetic function.

When considering disorders linked to SDH dysfunction, the importance of SDH for health becomes clear. Deficits or mutations in SDH have been connected to a number of clinical illnesses, such as cancers and neurological diseases. Gaining insight into the molecular pathways behind SDH is crucial to improving our understanding of these disorders and creating focused treatment plans. SDH subunits experience a variety of modifications, such as phosphorylation, acetylation, succinylation and SUMOylation, which impact SDH function in addition to the regulation of assembly factors. Sadly, there is a lack of proof to validate the PTM of assembly factors, which creates a vacuum in our knowledge of PTM involvement in the SDHAF-related SDH assembly process. All things considered, current research has shed light on the intricate function of the SDH complex and its assembly components. However, further information is required to fully comprehend SDH assembly and its significance in determining cell fate.

References

Alabduladhem TO, Bordoni B (2022) Physiology, Krebs cycle. StatPearls [Internet]. StatPearls

Alston CL, Davison JE, Meloni F, Westhuizen FHV, He L, Hornig-Do HT, Peet AC, Gissen P, Goffrini P, Ferrero I, Wassmer E, McFarland R, Taylor RW (2012) Recessive germline SDHA and SDHB mutations causing leukodystrophy and isolated mitochondrial complex II deficiency. J Med Genet 49(9):569–577

Ardissone A, Invernizzi F, Nasca A, Moroni I, Farina L, Ghezzi D (2015) Mitochondrial leukoencephalopathy and complex II deficiency associated with a recessive SDHB mutation with reduced penetrance. Mol Genet Metab Rep 5:51–54

Arnold PK, Finley LWS (2023) Regulation and function of the mammalian tricarboxylic acid cycle. J Biol Chem 299(2):102838

Barletta JA, Hornick JL (2012) Succinate dehydrogenase-deficient tumors: diagnostic advances and clinical implications. Adv Anat Pathol 19(4):193–203

Belt K, Van Aken O, Murcha M, Millar AH, Huang S (2018) An assembly factor promotes assembly of flavinated SDH1 into the succinate dehydrogenase complex. Plant Physiol 177(4):1439–1452

Bezawork-Geleta A, Rohlena J, Dong L, Pacak K, Neuzil J (2017) Mitochondrial Complex II: at the crossroads. Trends Biochem Sci 42(4):312–325

Boikos SA, Pappo AS, Killian JK, LaQuaglia MP, Weldon CB, George S, Trent JC, von Mehren M, Wright JA, Schiffman JD, Raygada M, Pacak K, Meltzer PS, Miettinen MM, Stratakis C, Janeway KA, Helman LJ (2016a) Molecular subtypes of KIT/PDGFRA wild-type gastrointestinal stromal tumors: a report from the National Institutes of Health Gastrointestinal Stromal Tumor Clinic. JAMA Oncol 2(7):922–928

Boikos SA, Xekouki P, Fumagalli E, Faucz FR, Raygada M, Szarek E, Ball E, Kim SY, Miettinen M, Helman LJ, Carney JA, Pacak K, Stratakis CA (2016b) Carney triad can be (rarely) associated with germline succinate dehydrogenase defects. Eur J Hum Genet 24(4):569–573

Bricker DK, Taylor EB, Schell JC, Orsak T, Boutron A, Chen YC, Cox JE, Cardon CM, Van Vranken JG, Dephoure N, Redin C, Boudina S, Gygi SP, Brivet M, Thummel CS, Rutter J (2012) A mitochondrial pyruvate carrier required for pyruvate uptake in yeast, Drosophila, and humans. Science 337(6090):96–100

Casey R, Garrahy A, Tuthill A, O'Halloran D, Joyce C, Casey MB, O'Shea P, Bell M (2014) Universal genetic screening uncovers a novel presentation of an SDHAF2 mutation. J Clin Endocrinol Metab 99(7):E1392–E1396

Cecchini G (2003) Function and structure of complex II of the respiratory chain. Annu Rev Biochem 72:77–109

Courage C, Jackson CB, Hahn D, Euro L, Nuoffer JM, Gallati S, Schaller A (2017) SDHA mutation with dominant transmission results in complex II deficiency with ocular, cardiac, and neurologic involvement. Am J Med Genet A 173(1):225–230

Du Z, Zhou X, Lai Y, Xu J, Zhang Y, Zhou S, Feng Z, Yu L, Tang Y, Wang W, Yu L, Tian C, Ran T, Chen H, Guddat LW, Liu F, Gao Y, Rao Z, Gong H (2023) Structure of the human respiratory complex II. Proc Natl Acad Sci USA 120(18):e2216713120

Ece Solmaz A, Pariltay E, Talim B, Onay H (2022) A novel bi-allelic variant in the SDHB gene causes a severe mitochondrial complex II deficiency: a case report. Clin Neurol Neurosurg 212:107039

Favier J, Amar L, Gimenez-Roqueplo AP (2015) Paraganglioma and phaeochromocytoma: from genetics to personalized medicine. Nat Rev Endocrinol 11(2):101–111
Fernandez-Vizarra E, Zeviani M (2021) Mitochondrial disorders of the OXPHOS system. FEBS Lett 595(8):1062–1106
Finsterer J (2008) Leigh and Leigh-like syndrome in children and adults. Pediatr Neurol 39(4):223–235
Fujita N, Aishima S, Iguchi T, Nishihara Y, Yamamoto H, Taketomi A, Oda Y, Honda H, Tsuneyoshi M (2010) Down-regulation of artery in moderately differentiated hepatocellular carcinoma related to tumor development. Hum Pathol 41(6):838–647
Fullerton M, McFarland R, Taylor RW, Alston CL (2020) The genetic basis of isolated mitochondrial complex II deficiency. Mol Genet Metab 131(1–2):53–65
Ghezzi D, Goffrini P, Uziel G, Horvath R, Klopstock T, Lochmüller H, D'Adamo P, Gasparini P, Strom TM, Prokisch H, Invernizzi F, Ferrero I, Zeviani M (2009) SDHAF1, encoding a LYR complex-II specific assembly factor, is mutated in SDH-defective infantile leukoencephalopathy. Nat Genet 41(6):654–656
Gill AJ (2012) Succinate dehydrogenase (SDH) and mitochondrial driven neoplasia. Pathology 44(4):285–292
Gill AJ, Hes O, Papathomas T, Šedivcová M, Tan PH, Agaimy A, Andresen PA, Kedziora A, Clarkson A, Toon CW, Sioson L, Watson N, Chou A, Paik J, Clifton-Bligh RJ, Robinson BG, Benn DE, Hills K, Maclean F, Niemeijer ND, Vlatkovic L, Hartmann A, Corssmit EP, van Leenders GJ, Przybycin C, McKenney JK, Magi-Galluzzi C, Yilmaz A, Yu D, Nicoll KD, Yong JL, Sibony M, Yakirevich E, Fleming S, Chow CW, Miettinen M, Michal M, Trpkov K (2014) Succinate dehydrogenase (SDH)-deficient renal carcinoma: a morphologically distinct entity: a clinicopathologic series of 36 tumors from 27 patients. Am J Surg Pathol 38(12):1588–1602
Haddad A, Mohiuddin SS (2022) Biochemistry, citric acid cycle. StatPearls [internet]. StatPearls
Hao HX, Khalimonchuk O, Schraders M, Dephoure N, Bayley JP, Kunst H, Devilee P, Cremers CW, Schiffman JD, Bentz BG, Gygi SP, Winge DR, Kremer H, Rutter J (2009) SDH5, a gene required for flavination of succinate dehydrogenase, is mutated in paraganglioma. Science 325(5944):1139–1142
Helman G, Caldovic L, Whitehead MT, Simons C, Brockmann K, Edvardson S, Bai R, Moroni I, Taylor JM, Van Haren K; SDH Study Group; Taft RJ, Vanderver A, van der Knaap MS (2016) Magnetic resonance imaging spectrum of succinate dehydrogenase–related infantile leukoencephalopathy. Ann Neurol 79(3):379–386
Horsefield R, Yankovskaya V, Sexton G, Whittingham W, Shiomi K, Omura S, Byrne B, Cecchini G, Iwata S (2006) Structural and computational analysis of the quinone-binding site of complex II (succinate-ubiquinone oxidoreductase): a mechanism of electron transfer and proton conduction during ubiquinone reduction. J Biol Chem 281(11):7309–7316
Imlay JA (2006) Iron-sulphur clusters and the problem with oxygen. Mol Microbiol 59(4):1073–1082
Janeway KA, Kim SY, Lodish M, Nosé V, Rustin P, Gaal J, Dahia PL, Liegl B, Ball ER, Raygada M, Lai AH, Kelly L, Hornick JL; NIH Pediatric and Wild-Type GIST Clinic; O'Sullivan M, de Krijger RR, Dinjens WN, Demetri GD, Antonescu CR, Fletcher JA, Helman L, Stratakis CA (2011) Defects in succinate dehydrogenase in gastrointestinal stromal tumors lacking KIT and PDGFRA mutations. Proc Natl Acad Sci USA 108(1):314–318
Kantorovich V, King KS, Pacak K (2010) SDH-related pheochromocytoma and paraganglioma. Best Pract Res Clin Endocrinol Metab 24(3):415–424
Kaochar S, Tu BP (2012) Gatekeepers of chromatin: Small metabolites elicit big changes in gene expression. Trends Biochem Sci 37(11):477–483
Kaur P, Sharma S, Kadavigere R, Girisha KM, Shukla A (2020) Novel variant p. (Ala102Thr) in SDHB causes mitochondrial complex II deficiency: case report and review of the literature. Ann Hum Genet 84(4):345–351
Killian JK, Kim SY, Miettinen M, Smith C, Merino M, Tsokos M, Quezado M, Smith WI Jr, Jahromi MS, Xekouki P, Szarek E, Walker RL, Lasota J, Raffeld M, Klotzle B, Wang Z, Jones L, Zhu Y, Wang Y, Waterfall JJ, O'Sullivan MJ, Bibikova M, Pacak K, Stratakis C, Janeway

KA, Schiffman JD, Fan JB, Helman L, Meltzer PS (2013) Succinate dehydrogenase mutation underlies global epigenomic divergence in gastrointestinal stromal tumor. Cancer Discov 3(6):648–657

Letouzé E, Martinelli C, Loriot C, Burnichon N, Abermil N, Ottolenghi C, Janin M, Menara M, Nguyen AT, Benit P, Buffet A, Marcaillou C, Bertherat J, Amar L, Rustin P, De Reyniès A, Gimenez-Roqueplo AP, Favier J (2013) SDH mutations establish a hypermethylator phenotype in paraganglioma. Cancer Cell 23(6):739–752

Levitas A, Muhammad E, Harel G, Saada A, Caspi VC, Manor E, Beck JC, Sheffield V, Parvari R (2010) Familial neonatal isolated cardiomyopathy caused by a mutation in the flavoprotein subunit of succinate dehydrogenase. Eur J Hum Genet 18(10):1160–1165

Lin S, Fasham J, Al-Hijawi F, Qutob N, Gunning A, Leslie JS, McGavin L, Ubeyratna N, Baker W, Zeid R, Turnpenny PD, Crosby AH, Baple EL, Khalaf-Nazzal R (2021) Consolidating biallelic SDHD variants as a cause of mitochondrial complex II deficiency. Eur J Hum Genet 29(10):1570–1576

Maio N, Rouault TA (2016) Mammalian Fe-S proteins: definition of a consensus motif recognized by the co-chaperone HSC20. Metallomics 8(10):1032–1046

Maio N, Singh A, Uhrigshardt H, Saxena N, Tong W-H, Rouault TA (2014) Cochaperone binding to LYR motifs confers specificity of iron sulfur cluster delivery. Cell Metab 19(3):445–457

Martucci VL, Emaminia A, del Rivero J, Lechan RM, Magoon BT, Galia A, Fojo T, Leung S, Lorusso R, Jimenez C, Shulkin BL, Audibert JL, Adams KT, Rosing DR, Vaidya A, Dluhy RG, Horvath KA, Pacak K (2015) Succinate dehydrogenase gene mutations in cardiac paragangliomas. Am J Cardiol 115(12):1753–1759

McNeil MB, Fineran PC (2013) The conserved RGxxE motif of the bacterial FAD assembly factor SdhE is required for succinate dehydrogenase flavinylation and activity. Biochemistry 52(43):7628–7640

McNeil MB, Hampton HG, Hards KJ, Watson BNJ, Cook GM, Fineran PC (2014) The succinate dehydrogenase assembly factor, SdhE, is required for the flavinylation and activation of fumarate reductase in bacteria. FEBS Lett 588(3):414–421

Meyer EH, Welchen E, Carrie C (2019) Assembly of the complexes of the oxidative phosphorylation system in land plant mitochondria. Annu Rev Plant Biol 70:23–50

Moosavi B, Berry EA, Zhu XL, Yang WC, Yang GF (2019) The assembly of succinate dehydrogenase: a key enzyme in bioenergetics. Cell Mol Life Sci 76(20):4023–4042

Mukherjee S, Ghosh A (2020) Molecular mechanism of mitochondrial respiratory chain assembly and its relation to mitochondrial diseases. Mitochondrion 53:1–20

Murphy MP, O'Neill LAJ (2018) Krebs cycle reimagined: the emerging roles of succinate and itaconate as signal transducers. Cell 174(4):780–784

Na U, Yu W, Cox J, Bricker DK, Brockmann K, Rutter J, Thummel CS, Winge DR (2014) The LYR factors SDHAF1 and SDHAF3 mediate maturation of the iron-sulfur subunit of succinate dehydrogenase. Cell Metab 20(2):253–266

Niemeijer ND, Papathomas TG, Korpershoek E, de Krijger RR, Oudijk L, Morreau H, Bayley JP, Hes FJ, Jansen JC, Dinjens WN, Corssmit EP (2015) Succinate dehydrogenase (SDH)-deficient pancreatic neuroendocrine tumor expands the SDH-related tumor spectrum. J Clin Endocrinol Metab 100(10):E1386–E1393

Ozluk Y, Taheri D, Matoso A, Sanli O, Berker NK, Yakirevich E, Balasubramanian S, Ross JS, Ali SM, Netto GJ (2015) Renal carcinoma associated with a novel succinate dehydrogenase A mutation: a case report and review of literature of a rare subtype of renal carcinoma. Hum Pathol 46(12):1951–1955

Paik JY, Toon CW, Benn DE, High H, Hasovitz C, Pavlakis N, Clifton-Bligh RJ, Gill AJ (2014) Renal carcinoma associated with succinate dehydrogenase B mutation: a new and unique subtype of renal carcinoma. J Clin Oncol 32(6):e10–e13

Pantaleo MA, Astolfi A, Di Battista M, Heinrich MC, Paterini P, Scotlandi K, Santini D, Catena F, Manara MC, Nannini M, Maleddu A, Saponara M, Lolli C, Formica S, Biasco G (2009) Insulin-like growth factor 1 receptor expression in wild-type GISTs: a potential novel therapeutic target. Int J Cancer 125(12):2991–2994

Parfait B, Chretien D, Rötig A, Marsac C, Munnich A, Rustin P, Parfait B (2000) Compound heterozygous mutations in the flavoprotein gene of the respiratory chain complex II in a patient with leigh syndrome. Hum Genet 106:236–243

Pasini B, McWhinney SR, Bei T, Matyakhina L, Stergiopoulos S, Muchow M, Boikos SA, Ferrando B, Pacak K, Assie G, Baudin E, Chompret A, Ellison JW, Briere JJ, Rustin P, Gimenez-Roqueplo AP, Eng C, Carney JA, Stratakis CA (2008) Clinical and molecular genetics of patients with the Carney-Stratakis syndrome and germline mutations of the genes coding for the succinate dehydrogenase subunits SDHB, SDHC, and SDHD. Eur J Hum Genet 16(1):79–88

Protasoni M, Zeviani M (2021) Mitochondrial structure and bioenergetics in normal and disease conditions. Int J Mol Sci 22(2)

Rasheed M, Tarjan G (2018) Succinate dehydrogenase complex: an updated review. Arch Pathol Lab Med 142(12):1564–1570

Richter S, Klink B, Nacke B, de Cubas AA, Mangelis A, Rapizzi E, Meinhardt M, Skondra C, Mannelli M, Robledo M, Menschikowski M, Eisenhofer G (2016) Epigenetic mutation of the succinate Dehydrogenase C promoter in a patient with two paragangliomas. J Clin Endocrinol Metab 101(2):359–363

Rogala J, Zhou M (2023) Hereditary succinate dehydrogenase-deficient renal cell carcinoma. Semin Diagn Pathol

Rutter J, Winge DR, Schiffman JD (2010) Succinate dehydrogenase – assembly, regulation and role in human disease. Mitochondrion 10(4):393–401

Schipani A, Nannini M, Astolfi A, Pantaleo MA (2023) SDHA germline mutations in SDH-deficient GISTs: a current update. Genes (Basel) 14(3)

Shokes JE (ed) (2006) X-ray absorption spectroscopic analysis of metal-containing proteins and peptides. (Doctoral dissertation, University of Georgia).

Sun F, Huo X, Zhai Y, Wang A, Xu J, Su D, Bartlam M, Rao Z (2005) Crystal structure of mitochondrial respiratory membrane protein complex II. Cell 121(7):1043–1057

Tivendale ND, Belt K, Berkowitz O, Whelan J, Millar AH, Huang S (2021) Knockdown of succinate dehydrogenase assembly factor 2 Induces reactive oxygen species–mediated auxin hypersensitivity causing pH-dependent root elongation. Plant Cell Physiol 62(7):1185–1198

Toledo RA, Qin Y, Cheng ZM, Gao Q, Iwata S, Silva GM, Prasad ML, Ocal IT, Rao S, Aronin N, Barontini M, Bruder J, Reddick RL, Chen Y, Aguiar RC, Dahia PL (2016) Recurrent mutations of chromatin-remodeling genes and kinase receptors in pheochromocytomas and paragangliomas. Clin Cancer Res 22(9):2301–2310

Udager AM, Mehra R (2016) Morphologic, molecular, and taxonomic evolution of renal cell carcinoma: a conceptual perspective with emphasis on updates to the 2016 World Health Organization classification. Arch Pathol Lab Med 140(10):1026–1037

Urbini M, Astolfi A, Indio V, Heinrich MC, Corless CL, Nannini M, Ravegnini G, Biasco G, Pantaleo MA (2015) SDHC methylation in gastrointestinal stromal tumors (GIST): a case report. BMC Med Genet 16:87

Van Coster R, Seneca S, Smet J, Van Hecke R, Gerlo E, Devreese B, Van Beeumen J, Leroy JG, De Meirleir L, Lissens W (2003) Homozygous Gly555Glu mutation in the nuclear-encoded 70 kDa flavoprotein gene causes instability of the respiratory chain complex II. Am J Med Genet A 120a(1):13–18

Wang X, Zhang X, Cao K, Zeng M, Fu X, Zheng A, Zhang F, Gao F, Zou X, Li H, Li M, Lv W, Xu J, Long J, Zang W, Chen J, Gao F, Ding J, Liu J, Feng Z (2022) Cardiac disruption of SDHAF4-mediated mitochondrial complex II assembly promotes dilated cardiomyopathy. Nat Commun 13(1):3947

Welander J, Söderkvist P, Gimm O (2011) Genetics and clinical characteristics of hereditary pheochromocytomas and paragangliomas. Endocr Relat Cancer 18(6):R253–R276

Xekouki P, Szarek E, Bullova P, Giubellino A, Quezado M, Mastroyannis SA, Mastorakos P, Wassif CA, Raygada M, Rentia N, Dye L, Cougnoux A, Koziol D, Sierra Mde L, Lyssikatos C, Belyavskaya E, Malchoff C, Moline J, Eng C, Maher LJ 3rd, Pacak K, Lodish M, Stratakis

CA (2015) Pituitary adenoma with paraganglioma/pheochromocytoma (3PAs) and succinate dehydrogenase defects in humans and mice. J Clin Endocrinol Metab 100(5):E710–E719
Xiao W, Zhang Q, Zhao S, Chen D, Gao N, Huang M, Ye X (2023) Citric acid secretion from rice roots contributes to reduction and immobilization of Cr (VI) by driving microbial sulfur and iron cycle in paddy soil. Sci Total Environ 854:158832
Zafreen L, Walker-Kopp N, Huang L-S, Berry EA (2016) *In-vitro*, SDH5-dependent flavinylation of immobilized human respiratory complex II flavoprotein. Arch Biochem Biophys 604:47–56
Zhu Y, Liu C, Zhang Y, Shi Q, Kong Y, Wang M, Xia X, Zhang F (2019) Identification and resolution of drug-related problems in a tertiary hospital respiratory unit in China. Int J Clin Pharm 41(6):1570–1577

Role of Succinate Dehydrogenase in Cellular Respiration

3

Viswapriya Viswalingam and Dileep Kumar

Abstract

Succinate dehydrogenase, also known as electron transport chain complex II, has recently experienced a notable and focused resurgence in research. Despite being the least explored among the mitochondrial respiratory complexes, this complex has gained renewed attention owing to its identified role in human diseases. This heightened examination has unveiled succinate dehydrogenase as an intriguing apparatus, with its regulation and assembly involving newly emerging factors. Managing within the citric acid cycle, SDH acquires electrons and passes them through its four subunits, namely SDHA, SDHB, SDHC, and SDHD, enabling electron transfer in the ETC as complex II. The electrons from reduced ubiquinone and $FADH_2$ are subsequently taken into the complex III to produce a sustainable flow of ATPs, thus supplying energy for the cell. The precise regulation of this enzyme through its diverse complexes enables cells to carry out crucial functions such as cellular respiration, hypoxic response, and gene expression. The distinct functionalities of various subunits within the SDH complex can account for these metabolic alterations. Therefore, various human disorders are caused due to mutations found in these factors and the structural components of the complex. However, deviations in SDH activity can lead to diseases and development of carcinogenicity due to disruptions in electron flow, heightened oxygen toxicity, and accumulation of succinate. Through this chapter, we delve deeper into the insights of mechanisms that underlie the pathology resulting from these SDH mutations.

V. Viswalingam
Department of Pharmaceutical Chemistry, Poona College of Pharmacy, Bharati Vidyapeeth (Deemed to be University), Pune, Maharashtra, India

D. Kumar (✉)
Department of Pharmaceutical Chemistry, Manipal College of Pharmaceutical Sciences, Manipal Academy of Higher Education, Manipal, Karnataka, India

V. M. Patil et al. (eds.), *Deciphering The Role of Succinate Dehydrogenase in Drug Discovery*, https://doi.org/10.1007/978-981-96-6831-1_3

Keywords

Cellular respiration · Electron transport · Mitochondrial complex II · Oxidative stress · SDHC · Tumorigenesis

Abbreviations

AMS	Acute mountain sickness
ECAR	Extracellular acidification rate
ETC	Electron transport chain
GIST	Gastrointestinal stromal tumor
IMM	Inner mitochondrial membrane
lncRNA	Long noncoding RNA
NAD	Nicotinamide adenine dinucleotide
PCC	Pheochromocytoma
PGL	Paraganglioma
RAS	Renin-angiotensin system
RCC	Renal cell carcinoma
SAM	S-adenosyl methionine
SDH	Succinate dehydrogenase
T2D	Type 2 diabetes mellitus

3.1 Introduction

Succinate dehydrogenase (SDH) is a critical mitochondrial enzyme involved in metabolic processes through Krebs cycle and the electron transport chain (ETC). Although its foremost function is in catalyzing the transformation of succinate to fumarate through oxidation in the Krebs cycle, it has also been known to have importance in reducing ubiquinone to ubiquinol in the ETC (Rutter et al. 2010). SDH deficiency has been rarely reported, and all known cases have stemmed from mutations in the SDHA subunit-encoding gene (Brière et al. 2005). Mutation in SDHA gene was documented in two siblings exhibiting a combination of Leigh syndrome and SDH deficiency (Bourgeron et al. 1995). Along with which, a study indicated that almost 23% of patients with defects in respiratory-chain enzymes had fractional deficiencies in activity of succinate dehydrogenase when performed on muscle biopsies. Typically, this reduction could be identified through histochemical methods in biopsies (Vladutiu and Heffner 2000).

Germinal mutations in SDH have been linked to an increased incidence of various neuroendocrine tumors, including hereditary paraganglioma (PGL) and pheochromocytoma (PCC) (Chen et al. 2015). SDH-B mutations predominantly lead to type-4 PGLs (PGL4), including familial adrenal PCC. PGLs that are malignant due to SDH-B mutations frequently manifest as extra-adrenal or metastatic tumors, often occurring in regions like the abdomen and thorax (Srirangalingam et al. 2017).

Furthermore, mutations in SDH-B are linked to serous ovarian cancer, gastrointestinal stromal tumor (GIST), and familial renal cell carcinoma (RCC) (Thompson and Denny 2019). The SDH-C gene has noted to have over 30 mutations linked to an increased risk of type-3 PGL (PGL3). Tumors related to SDH-D usually appear as noncancerous type-1 PGLs (PGL1) and initiate in the head and neck regions (HNPGLs). Additionally, individuals with Cowden syndrome, which is defined by hamartomas and an inclination toward certain cancers, have been found to possess mutations in SDH-D. Fascinatingly, the SDH-D gene, situated on chromosome 11q23, commonly undergoes mutations in a variety of cancer types, such as those affecting the ovary, breast, nasopharynx, bladder, and lung. SDH-D mutations have also been found in individuals with GIST (Gammon et al. 2016; Hobert et al. 2012). It is however worth mentioning that the association between SDH-A mutations and PGL/PCC is notably weak, with only one heterozygous germline SDH-A mutation identified in a rare malignancy known as catecholamine-secreting abdominal paraganglioma (Martucci and Pacak 2014). Additionally, hereditary type-2 PGL (PGL2) is linked to germline mutations that deactivate the SDHAF2/SDH5 gene. This gene encodes a mitochondrial protein vital for the flavination of the SDH-A subunit. Yet, it is still unclear if mutations in SDH are directly accountable for the progression of papillary thyroid cancers, adrenal neuroblastoma, or colon cancer.

In specific types of cancer cells, SDH had demonstrated tumor suppression by inhibiting hypoxia inducible factor 1α (HIF-1α) through decreased levels of succinate (Selak et al. 2005). Moreover, the components of SDH can engage with RNA regulatory networks that involve various categories of noncoding RNAs, RNA-editing enzymes, RNA-modifying enzymes, transcription factors, and various small molecules. The consequences of changes in RNA and the unequal presence of noncoding RNAs can act in either a tumor-suppressive or oncogenic manner, ultimately disrupting gene expression, cell signaling pathways, and cellular metabolism (Barbieri and Kouzarides 2020). Notably, complex dysfunction is often caused by targeting SDH by noncoding RNAs (Puisségur et al. 2011; Zhang et al. 2019a). Furthermore, noncoding RNAs, under the control of RNA-editing and RNA-modifying enzymes, along with transcription factors linked to several cancers, can impact SDH (Alarcón et al. 2015; Pinto et al. 2018).

SDH-x mutations lead to compromised or completely lost enzymatic activity, significantly affecting electron flow to ubiquinone, a crucial component to transfer electron across the ETC (Bénit et al. 2022). The main result of these changes is the hindered process of oxygen reduction, leading to heightened production of the superoxide anion—a reactive oxygen species (ROS)—within the mitochondria. Heightened levels of radical dotO_2 signify an escalation in oxidative stress, causing protein carbonylation, DNA damage (converting deoxyguanosine to 8-OH-deoxyguanosine), a decline in membrane potential pf mitochondria, and its structural modifications. Notably, it has been suggested that a solitary Val-69 mutation in SDH-C could trigger a tumor-like characteristic in murine fibroblasts and facilitate tumorigenesis due to oxidative stress stemming from oxygen radicals (Cardaci and Ciriolo 2012; Slane et al. 2006).

Abnormal accumulation of succinate is yet another consequence of mutations in SDH genes. The buildup of succinate converts into an oncometabolite, disrupting α-KG-dependent dioxygenase enzymes as a result of structural similarity to α-KG (Xu et al. 2011). Like D-2HG generated by IDH mutations, succinate predominantly affects the functioning of epigenetic enzymes. PGL/PCC tumors with SDH mutations display reduced 5hmC levels, causing a hypermethylated DNA pattern and increased histone H3K27 methylation. This suggests that succinate hampers the demethylation processes mediated by both TET and KDM. In line with this, chromaffin cells from SDHB knockout mice have been noted with higher 5mC: 5hmC ratio compared to wild cell types, which is in turn pointed toward the exogenous addition of α-KG, indicating that succinate inhibits α-KG-dependent dioxygenases in a competitive manner (De Vivo et al. 2015; Phasaludeen et al. 2022; Yong et al. 2020). Additionally, the diminished levels of 5hmC were linked to the TET1 enzyme being excluded from the nucleus in tumors lacking SDH. But, tumors holding SDH-B mutations exhibit a higher methylation profile across CpG islands compared to other SDH-mutated cancers, suggesting that SDH-B activity prevails over other SDH subunits, nurturing higher succinate accumulation and more pronounced inhibition of α-KG-dependent dioxygenases (Ricketts et al. 2022). Moreover, SDH-B mutations confer a unique, more insistent, and metastatic phenotype. Furthermore, suppressing SDH-B has demonstrated its role in promoting the transition from epithelial to mesenchymal states (EMT) in mouse ovarian cancer cells. This process involves heightened histone methylation, specifically influenced by the increased expression of genes related to S-adenosyl methionine (SAM) metabolism, methyl donor for histone, and DNA methyltransferase enzymes (Aspuria et al. 2014; Morandi et al. 2017).

Much like cancers with mutations in IDH, tumors deficient in SDH present a "pseudohypoxic" response activated by HIF. Succinate can obstruct PHD activity, ultimately stabilizing HIF-1α. The effects of succinate-induced HIF-1α stabilization in cancers with SDH mutations closely mirror those resulting from fumarate buildup in FH-associated tumors (Macklin et al. 2020; Yang et al. 2013). Complex II, which consists of four subunits of protein, is encoded by nuclear genes. Among these, SDHA and SDHB are situated toward the inner matrix of membrane, hosting the succinate binding site, three FeS clusters, and a flavoprotein bound to FAD cofactor (Goetzman et al. 2023; Huang et al. 2019; Karavaeva and Sousa 2023). On the other hand, SDHC and SDHD, the remaining two subunits, are lipophilic and secure the complex toward the inner membrane. Collaboratively, SDHC and SDHD create the CoQ binding site, enabling the transfer of electrons from complex II to CoQ. Notably, mutations in SDHD have been observed in patients with paragangliomas and pheochromocytomas. Studies involving condensed SDHD cDNA in Chinese hamster fibroblasts revealed elevated superoxide generation, suggesting a potential link between SDHD mutations and carcinogenesis. Given SDHD's critical role in anchoring complex II, transferring electrons to CoQ, and its association with pathologies, investigating SDHD dysfunction can shed light on the workings of complex II in metabolism and aid in developing new therapeutic approaches.

Unfortunately, there are currently no appropriate models for studying any subunit of complex II (Dekker et al. 2003; Slane et al. 2006).

SDH deficiencies are relatively uncommon in humans (Rustin and Rötig 2002). However, there are significant variations in clinical manifestations among individuals with mutations in any of the four genes that encode the SDH subunits. Mutations in the flavoprotein subunit, SDHA, were the initial nuclear gene mutations identified in the mid-1990s, leading to Leigh syndrome, a distinct form of mitochondrial encephalopathy (Bourgeron et al. 1995). Lately, mutations in the subunits B, C, and D have been interrelated to the development of paragangliomas—generally benign, vascularized tumors found in the head and neck—or more specifically phaeochromocytomas. The carotid body, frequently affected by these tumors, is a highly vascular organ located at the common carotid artery bifurcation and plays a role in sensing blood oxygen levels. Because somatic mutations in these genes are observed in tumors, they are classified as tumor-suppressor genes (Sen et al. 2020; van Nederveen et al. 2003).

To address the consequences initiated by SDH mutations or metabolic irregularities, researchers are currently evaluating a range of compounds, including inhibitors and activators of SDH, in preclinical models and ongoing clinical trials (Moreno et al. 2020). Within this, we delineate the physiological and pathological mechanisms relevant to SDH, encompassing associated diseases such as cancer. Furthermore, we delve into inventive methods elucidating how RNA networks influence the state of SDH and offer encouraging approaches for directing interventions toward the SDH complex.

3.2 Succinyl Dehydrogenase in Mitochondrial Electron Transport Chain

SDH induces a pivotal reaction within the tricarboxylic acid cycle, which is itself an enzyme with multiple subunits that link the oxidation of succinate to reduction of quinone. Interestingly, this enzyme complex is part of the succinate and quinone oxidoreductase superfamily, showcasing a multitude of shared functional and structural features. This encompasses mitochondrial complex II as well as bacterial quinol and fumarate oxidoreductases (Arias-Cartin et al. 2012), although SDH serves a dual purpose in both the citric acid cycle and the respiratory ETC. However, in the citric acid cycle, it specifically oxidizes succinate into fumarate. Structurally, SDH bears resemblance to an enzyme involved in anaerobic bacterial respiration, specifically fumarate reductase, which catalyzes the reverse reaction (Hägerhäll 1997; Rutter et al. 2010).

The four subunits of SDH enzyme have each one encoded by a specific nuclear gene. These genes are SDHA (with 15 exons), SDHB (with 8 exons), SDHC (with 6 exons), and SDHD (with 4 exons). They are located on chromosomes 5p15, 1p35-p36.1, 1q21, and 11q23, respectively (Rustin et al. 2002). The SDH complex, alternatively termed succinate: ubiquinone oxidoreductase or mitochondrial complex II, holds significant importance as an enzyme linking the citric acid cycle with the

ETC. Within the TCA cycle, SDH facilitates the sixth step by oxidizing succinate to fumarate and concurrently reducing ubiquinone to ubiquinol (Moosavi et al. 2019). Comprising four subunits, the enzyme is heterotetrameric. The active units consist of the flavoprotein SDH-A and the iron sulfur protein SDH-B, serving as catalysts. In contrast, the hydrophobic subunits anchoring to the membrane, SDH-C and SDH-D, participate in binding ubiquinone to facilitate the electron transport chain reactions (Calisto et al. 2021).

The production of ATP in mitochondria relies on the collaborative functioning of the ETC. Facilitating the movement of an electrochemical gradient toward oxidizing NADH and $FADH_2$ ultimately leads to the reduction of oxygen to produce water (Ahmad et al. 2023). The ETC comprises four complex structures and two portable carriers for electrons (coenzyme Q and cytochrome c), all located within or associated with the inner mitochondrial membrane (IMM) (Zhao et al. 2019). Electrons, generated from the oxidation of NADH through complex I from the TCA cycle via complex II (succinate dehydrogenase [SDH]), move along the ETC (Janeway et al. 2011). This transfer is accompanied by the proton pumping, leading to the establishment of a proton-based gradient across the IMM. Ultimately, complex V (ATP synthase) utilizes the regulated proton movement across the electrochemical gradient to fuel the ATP synthesis process.

The organization of the ETC complexes poses a challenge for cells in coordinating the synthesis and interactions of individual subunits. These subunits are transcribed and translated from separate genomes in distinct sections (Tang et al. 2020). The goal is to assemble them into complex membrane-bound structures. The complexity is heightened by the properties of the ETC complexes themselves (Ghifari et al. 2023). Some subunits are hydrophobic and likely to aggregate due to their embedding in the inner mitochondrial membrane (IMM) during assembly. Additionally, these complexes have redox-active cofactors that can cause undesired reactions if not properly contained within the native complex. To address this, specific factors play a crucial role in assisting the assembly of these complexes. They facilitate cofactor insertion, prevent unproductive interactions, and stabilize their assembly intermediates (Fernández-Vizarra et al. 2009). In contrast to the more detailed yet functional structures of mammalian ETC, complexes I, III, and IV, which consist of 44, 11, and 14 subunits, respectively, involve proteins from both mitochondrial and nuclear genomes. SDH is less intricate, being formed from just four genes that are encoded in the nucleus. Despite this seemingly straightforward quaternary structure, recent insights emphasize that SDH, like all other ETC complexes, relies on assembly factors for its formation (Fernández-Vizarra et al. 2009).

3.3 Assembly Factors of SDH

The process of assembling the ETC complexes within a cell presents a formidable challenge, particularly in eukaryotes. The complexity arises from the fact that complexes I, III, IV, and V of the ETC encompass subunits coded by both the nuclear and mitochondrial genomes (Clemente-Suárez et al. 2023; Li et al. 2006; Watson and McStay 2020). This necessitates a meticulous coordination of their synthesis and assembly processes to form functional complexes crucial for energy production. Consequently, the cell allocates a significant number of proteins with specific roles dedicated to the assembly of these intricate complexes (Cappadocia and Lima 2018; Gu and Jan Fada 2020; Hove-Jensen et al. 2016).

Among the complexes, complex I has gained attention, showcasing an expanding list of assembly factors critical for its proper formation. Complex IV, also known as cytochrome oxidase, is equally demanding, relying on over 20 factors to orchestrate its assembly and functionality. In contrast, the assembly of complex II, or SDH, has remained less understood for a significant period. Only a handful of factors were initially recognized, leaving a gap in comprehending this complex's assembly process (Li et al. 2006; Mansilla et al. 2018; Zhou et al. 2020).

Two essential factors, Tcm62 and Flx1, have been identified, providing crucial insights into SDH assembly. Tcm62 was initially discovered in a mutant with diminished SDH activity. Further studies demonstrated its indispensable role in the SDH assembly, shedding light on its crucial involvement in maintaining mitochondrial respiratory function (Dibrov et al. 1998; Klanner et al. 2000). While its sequence similarity to Hsp60 chaperones suggests a potential role in protein folding, the inability to bind ATP sets Tcm62 apart, hinting at unique mechanistic functions. Additionally, connections were drawn between Tcm62 and protein stability under stress, as well as its potential involvement in apoptosis regulation (Caruso Bavisotto et al. 2020; Saibil 2013).

On the other hand, Flx1, initially associated with maintaining the FAD/FMN ratio in mitochondria, later emerged as a player in FAD transport. Intriguingly, its role seems to extend to the post-transcriptional regulation of SDH1, a critical subunit of SDH. However, the precise mechanism of Flx1's involvement in SDH assembly remains an ongoing area of research (Kim and Winge 2013).

Two significant breakthroughs in 2009 marked a pivotal leap in understanding SDH assembly, the discovery of SDHAF1 and SDH5. SDHAF1 emerged as a crucial factor vital for stable assembly and adequate functionality of the SDH complex (Ohlenbusch et al. 2012). Mutations in SDHAF1 were notably associated with infantile leukoencephalopathy, underscoring the critical role of SDH assembly in human health. Equally significant was the discovery of SDH5, a factor playing a pivotal role in the covalent insertion of FAD into SDH1, a key subunit of SDH. A mutation in SDH5 was linked to hereditary paraganglioma, further emphasizing the relevance of SDH assembly in the context of human health (Dwight et al. 2017; Maio et al. 2016).

These discoveries have propelled the understanding of SDH assembly, yet many questions remain unanswered. The intricate interplay of multiple factors and their precise biochemical functions in SDH assembly warrant continued investigation (Xie et al. 2020). As research in this field progresses, the identification of additional assembly factors is highly anticipated, promising to unravel the molecular underpinnings of various enigmatic human diseases linked to SDH dysfunction.

3.4 Regulation and Control of SDH

The functioning of SDH's catalytic activity is influenced through post-translational modifications such as phosphorylation and acetylation at specific sites, as well as inhibition at the active site. In mice SDH1, reversible acetylation at multiple Lys residues has been demonstrated to decrease the catalytic activity of SDH1 (Cimen et al. 2010). The deacetylase SIRT3 plays a central role in regulating the level of acetylation of SDH1. Furthermore, in mammalian cells, the Sdh1 subunit of SDH undergoes phosphorylation, which, similar to acetylation, seems to reduce its activity (Tomitsuka et al. 2009). In vitro experiments have signified the phosphorylation of SDH1 at two tyrosine residues with Fgr tyrosine kinase. However, any physiological importance of this modification is yet not uncovered (Salvi et al. 2007).

Another means of modulating SDH catalytic activity involves Krebs cycle intermediates, including oxaloacetate, which serves as a strong inhibitor. Succinate, on the other hand, triggers the detachment of oxaloacetate from SDH, thereby activating the enzyme. This inhibitory effect of oxaloacetate likely contributes to the well-established regulation of SDH activity based on the metabolism of mitochondria.

3.4.1 Allosteric Regulation and Substrate Availability

Recent reports have described mutations in various subunits, demonstrating the effects of reducing or eliminating SDH function in plant cells. These reports suggest that the main factors contributing to the various dysfunction phenotypes related to SDH involve differences in levels of organic acids, alterations in respiration rates within specific tissues, and fluctuations in mitochondrial ROS production. SDH1, a flavoprotein, is where succinate binds and serves as the point for allosteric activation of SDH by ATP (Fig. 3.1) (Belt et al. 2018; Jacoby et al. 2012). Additionally, it hosts the first cofactor of SDH, a flavin linked to a histidine residue. Arabidopsis has two genes for SDH1: SDH1-1 (At5g66760) and SDH1-2 (At2g18450). Deleting SDH1-2 does not affect Arabidopsis growth, likely due to its minimal transcript levels and unclear protein identification. Conversely, the knockout of SDH1-1 results in failed development of gametophyte, pollen abortion, and decreased seed production. When SDH1-1 is reduced through RNA interference, there is a notable impact on pollen and seed development (León et al. 2007; Li et al. 2021; Li et al. 2022).

Plants containing heterozygous SDH1-1/sdh1-1 genotype exhibit lower SDH activity but demonstrate improved photosynthesis and nitrogen-limited growth due to modified conductance of stomata and nitrogen assimilation (Clavijo-Herrera et al. 2018). These plants also have higher stomatal aperture and density in leaves. Metabolite analysis shows a twofold increase in succinate and decreased levels of certain amino acids like proline and glutamine in these mutants. It is believed that metabolic changes caused by reduced SDH1 could be the main reason behind the major alterations in photosynthesis, stomatal functions, and assimilation of nitrogen (Fuentes et al. 2011).

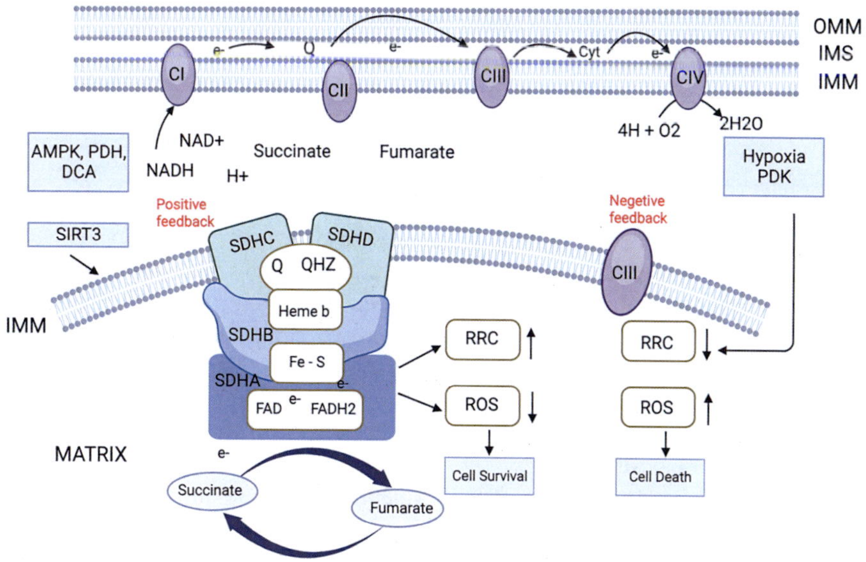

Fig. 3.1 A diagrammatic representation of allosteric regulation of SDH. Mitochondrial respiratory complex II controls the reserve respiratory capacity and cell viability. While mitochondrial electron flow through the ETC shows the integral role of complex II, it is yet an essential component of tricarboxylic acid cycle too. It connects substrate utilization to mitochondrial ETC and respiration. Hypoxia and pyruvate dehydrogenase kinase negatively regulate complex II activity, leading to reduced reserve respiratory capacity and reactive oxygen species leading to cell demise

3.5 Role of Complex II Subunit SDHD for Cell Growth

The major source of ATP is the mitochondria, primarily through the electron transport chain (ETC). Succinate dehydrogenase is a unique component which plays an imperative role in both the Krebs cycle and the ETC (Cecchini 2003; Sun et al. 2005). Within the Krebs cycle, complex II facilitates the conversion of succinate to fumarate through oxidation (Cecchini 2003; Sun et al. 2005). ETC serves as one of the important access points for electrons, taking electrons from succinate and passing them into ubiquinone (CoQ) units. Consequently, any disruptions in the function of complex II can significantly impact maintaining the balance of energy (Sun et al. 2005). Alterations in the components of complex II have been detected in individuals experiencing mitochondrial respiratory impairments and various cancer types (Astuti et al. 2001; Baysal et al. 2000; Bourgeron et al. 1995). Hence, a thorough comprehension of complex II in energy balance and potential thereof as a treatment target is crucial.

Complex II comprises four protein subunits, all of which are determined by genes located in the cell nucleus. SDHA and SDHB, two of which, are located on the inner membrane's matrix side and possess the succinate binding site, three FeS clusters, and a flavoprotein bound to an FAD cofactor (Bezawork-Geleta et al. 2017;

Owens et al. 2012). The remaining components, SDHC and SDHD, are hydrophobic in nature and play a role in anchoring the complex toward the inner membrane. SDHC and SDHD form the CoQ binding site of complex II and facilitate the transfer of electrons from complex II to CoQ. Notably, mutations in SDHD have been observed in patients with paragangliomas and pheochromocytomas. Expressing condensed SDHD cDNA in Chinese hamster fibroblasts led to increased levels of superoxide production, indicating a potential link between SDHD mutations and the onset of cancer (Owens et al. 2012). Considering the critical role of SDHD involves the attachment of complex II to the inner membrane, thereby facilitating electron transfer to CoQ, eventually contributing to associated pathologies. Also developing models for investigating SDHD dysfunction could prove highly beneficial for gaining insights into role of complex II in metabolism and progressing toward innovative therapeutic strategies. However, as of now, appropriate models for molecular investigations of any complex II subunit are lacking.

In this study, a CRISPR/Cas9 approach was effectively utilized to induce mutations in SDHD, a subunit of complex II situated in the inner-membrane region (Cecchini 2003), within HEK293 cells. This paved the way for establishing the initial model, focusing on the role of this pivotal subunit within complex II. The alterations at the 118-base pair site resulted in insertions or deletions (INDELs) that disrupted or modified the entire downstream amino acid sequence following the PAM site (Bandara et al. 2021; van Senten et al. 2022). Remarkably, the procedure of mutagenesis did not yield any detectable off-target genetic changes. The DNA sequence of SDHD was altered, resulting in a complete disruption of SDHD protein expression at the amino acid level in the mutant. Additionally, due to the SDHD mutation, the expression of another subunit of complex II (SDHB) significantly decreased, indicating an interdependence in the expression of various subunits within this complex (Bakare et al. 2021; Moreno et al. 2020). Interestingly, the mutation in SDHD also led to a notable reduction in the expression of a subunit of complex I (NDUFB8), a subunit of complex IV (MT-CO1), and a subunit of complex V (ATP5A) in the mutant, suggesting a correlation in protein expression between complex II and other complexes (Bandara et al. 2021; Piekutowska-Abramczuk et al. 2018). However, the mutation in SDHD did not affect the expression of the subunit UQCRC2 in complex III, indicating its independence from complex II with respect to protein expression.

Considering the role of complex II in both the ETC and the Krebs cycle (Cecchini 2003; Miyadera et al. 2003), as well as the association between SDHD mutations and disease pathologies (Baysal et al. 2000), a comprehensive understanding of SDHD's contribution to cellular metabolism is imperative. The SDHD mutation disrupted respiration throughout the ETC, suppressed glycolysis and overall ATP synthesis, and impeded cellular growth. Notably, immediate treatment with the synthetic ubiquinone analog idebenone significantly enhanced complex II-mediated mitochondrial respiration, oxidative phosphorylation (OXPHOS) capacity, and proliferation of cells.

As anticipated, the mutation in SDHD hindered complex II-mediated respiration. Moreover, it substantially reduced complex I-mediated respiration, raising the hypothesis that, since complex II propels the Krebs cycle in a clockwise direction by converting succinate to fumarate, the SDHD mutation might have slowed down and/or disrupted the Krebs cycle's production of NADH, the substrate for complex I (Lendvai et al. 2014). The SDHD mutation also compromised capacity of OXPHOS, maximal respiration, and synthesis of ATP. In summary, impairing the SDHD subunit of complex II was adequate to hinder net ETC energy production.

Glycolysis, when assessed by extracellular acidification rate (ECAR), has seen to have significantly decreased in HEK293ΔSDHD compared to the parental cells. This reduction in glycolysis could be attributed to a decreased demand for pyruvate due to a disrupted Krebs cycle resulting from the SDHD mutation, representing a potential negative feedback mechanism on glycolysis (Arnold and Finley 2023; Pal et al. 2022). Furthermore, the inhibition of mitochondrial respiration with oligomycin did not cause a shift in glycolysis within the mutant cells, reinforcing the notion of compromised mitochondrial respiration. Alongside the metabolic suppression, significantly slower growth was observed in the mutant cells in culture media compared to the parental cells, likely due to metabolic impairments in the mutant cells. The SDHD mutation has shown an increased apoptosis and necrosis, which might have resulted due to compromised glycolysis, and mitochondrial respiration could have weakened the cells and their membranes, making the mutant cells more susceptible to apoptosis and necrosis. Remarkably, the long-term generation of ROS decreased as an outcome of the SDHD mutation (Fogg et al. 2011; Zorov et al. 2014). Normally, electrons that deviate from the typical ETC pathway and leak out are ultimately shifted directly to O_2 for generating ROS. In the case of SDHD mutation, it is quite feasible that the hindrance of the typical forward flow in the Krebs's cycle may have diminished the generation of NADH and $FADH_2$. These compounds serve as vital electron donors for the ETC (Alabduladhem and Bordoni 2023; Bouillaud 2023). Hence, the restricted provision of electrons could have resulted in diminished electron transfer along the complex and thereby a reduction in electron leakage from the ETC, ultimately having led to a decrease in the production of ROS.

Finally, investigations were carried out to assess the potential for alleviating some of the observed traits caused by the SDHD mutation. Idebenone, a short chained benzoquinone, which is more hydrophilic than ubiquinone, holds promising ability to act as a curative agent for various conditions related to oxidative stress and mitochondrial dysfunction (Erb et al. 2012; Fiebiger et al. 2013; Heitz et al. 2012; Palecek et al. 2015). Administering different doses of idebenone enhanced the oxygen consumption facilitated by complex II and improved the capacity of OXPHOS, which suggests that idebenone, a ubiquinone analog, has the potential to act as a substitute for complex II in donating electrons, thereby enhancing the function of the electron transport chain (ETC). Interestingly, a long-term treatment with 1 μM of idebenone increased cell proliferation in HEK293ΔSDHD cells, while 5 μM did not, thus known for exerting cell type specific effects on cell proliferation (Seo et al. 2015; Zhang et al. 2018b). The elevated cell growth observed in the mutant HEK293ΔSDHD cells following idebenone treatment is likely a result of more efficient use of substrates. However, it is important to consider the cell type when evaluating the impact of mitochondrial-targeted therapies on different biological processes (Sousa et al. 2023).

3.6 Diseases Caused by SDH Malfunction

3.6.1 Cancer

In initial research, a connection was established between dysfunction in the SDH complex and cancer. This association was substantiated through specific studies demonstrating that mutations in SDHB (Guzy et al. 2008), SDHC (Ishii et al. 2005), and SDHD (Guzy et al. 2008) amplified the release of superoxide anions, causing oxidative damage. Consequently, affected cells underwent either apoptosis or transformation. Lately, SDH has been categorized as a tumor suppressor primarily due to two notable anomalies it experiences, ultimately fostering oncogenesis. First, the inactivation of SDH resulted in the accumulation of succinate, a competitive inhibitor of HIF-α prolyl hydroxylase domain (PHD), leading to HIF stabilization (Zimna and Kurpisz 2015). Stabilized HIF, in turn, spurred malignant cell proliferation by stimulating angiogenesis and ROS production (Selak et al. 2005; Zimna and Kurpisz 2015). Elevated succinate levels also induced ROS production by hindering α-ketoglutarate-dependent enzymes and amplifying the reverse electron flow from complex II to complex I (Zhang et al. 2019b). Additionally, succinate accumulation promoted increased methylation of histone by directly binding to and inhibiting histone demethylase JumonjiD3, thus intensifying epigenetic alterations and the potential for cancerous transformation.

Mutations identified in SDHB, SDHC, and SDHD have been linked to the expansion of two uncommon tumors within the autonomic nervous system, recognized as paraganglioma (PGL) and pheochromocytoma (PCC) (Bardella et al. 2011; Baysal et al. 2000; Niemann and Müller 2000). These precise genetic mutations are acknowledged to heighten the production of ROS (Guzy et al. 2008), resulting in DNA damage and the initiation of tumorigenesis (Wojtovich and Foster 2014). Notably, mutations in SDHB (Guzy et al. 2008) and SDHD increase ROS production, triggering the onset of PGL and PCC by stabilizing HIF activity. Studies have established that mutations in all four subunits of genes encoding SDH lead to PCC and PGL by inhibiting the pathway of histone demethylation (Letouzé et al. 2013). Intriguingly, the same investigation revealed that SDHB mutations exerted a more pronounced inhibitory effect on histone methylases, enabling heightened hypermethylation. This might elucidate the greater malignancy of PGL and PCC caused by SDHB mutations compared to those induced by mutations in other SDH-encoding subunits (Barbara and Stratakis 2009). More recently, a cohort of researchers observed a connection between mutations in SDHAF2 alone whereas PGL and PCC, contrary to previous suggestions that SDHAF2 does not induce any through the inhibition of the histone demethylation pathway (Bayley et al. 2010). However, the specific mechanism remains unknown. PGL and PCC are the types of tumors which are most often linked to inherited or germline mutations, particularly in the mentioned SDH subunits. This highlights the significant impact that both inherited and somatic mutations of the SDH complex have on the development of cancer and underscores the importance of providing genetic counseling to individuals with inherited mutations.

PGL and PCC are primarily linked to SDH mutations, but other types of cancers can also emerge from either SDH mutations or diminished SDH functions. For instance, suppressing SDHB led to heightened levels of HIF-1α and adenosine monophosphate-activated protein kinase, promoting metastasis in ovarian cancer (Chen et al. 2014). Similarly, Warburg effect and increased expression of epithelial–mesenchymal transition based markers has been observed due to decreased SDHB in hepatocellular carcinoma escalated its malignancy through the (Tseng et al. 2018). In colorectal cancer, normal SDHB was found to boost the expression of tumor suppressors like phosphatase and tensin homolog (PTEN), caveolin-1, and cullin-5 to halt the cell cycle, whereas resultant mutation shows increased cell division (D. Zhang et al. 2013). Carney–Stratakis syndrome, arising from mutations in germline of SDH-A-B-C-D, has been associated with GIST and PGL (Janeway et al. 2011). Particularly, SDH deficiency has spurred DNA hypermethylation near fibroblast growth factor 4 and tyrosine-protein kinase kit oncogenes, thus activating them and triggering GIST (Flavahan et al. 2019). Although there are limited mechanisms available, research has connected the dots between position of SDHB and SDHD mutations in renal cell carcinoma and thyroid tumors, SDHB defects in pituitary adenomas, and significant SDHB loss in hemangioblastoma. Notably, the precise mechanisms for many cancers remain unclear, particularly in the analysis that links hemangioblastoma with SDHB mutation, because there are limited subjects of hemangioblastoma, with only 19 out of 35 patients testing positive for SDHB mutation (Roh et al. 2019). The study refrains from conducting additional experiments such as epigenetic analysis. Overall, these challenges hinder a comprehensive understanding of the association between SDHB inactivation and hemangioblastoma, and similar complexities emerge in investigating other SDH mutations and various cancers. In summary, the role of SDH in PGL, PCC, and other cancers is diverse, but advanced exploration has the potential to unveil therapeutic targets of SDH against cancer (Fig. 3.2).

3.6.2 Altitude Sickness

Acute mountain sickness (AMS) occurs due to reduced oxygen pressure at elevated altitudes, resulting in tissue hypoxia (Luks et al. 2017). Studies have investigated the impact of AMS-induced hypoxia on mitochondrial function, revealing a consensus that oxidative damage escalates at high altitudes (Bakonyi and Radak 2004). This is attributed to increased production of ROS in complex I and III of the ETC due to diminished electron flow (Bakonyi and Radak 2004; Guzy and Schumacker 2006). Notably, Lu et al. conducted a comparison between AMS-resistant and individuals with AMS-susceptibility found that those resistant to AMS exhibited a lowered level of plasma succinate levels by regulating SDHA, SDHB, and succinate-CoA ligase (Lu et al. 2018). Succinate-CoA ligase, a vital enzyme, is pivotal for the conversion of succinyl-CoA to succinate through its subunits: succinate-CoA ligase GDP and ADP-forming subunit α (SUCLG1), succinate-CoA ligase GDP-forming subunit β (SUCLG2), and succinate-CoA ligase ADP-forming subunit β (SUCLA2)

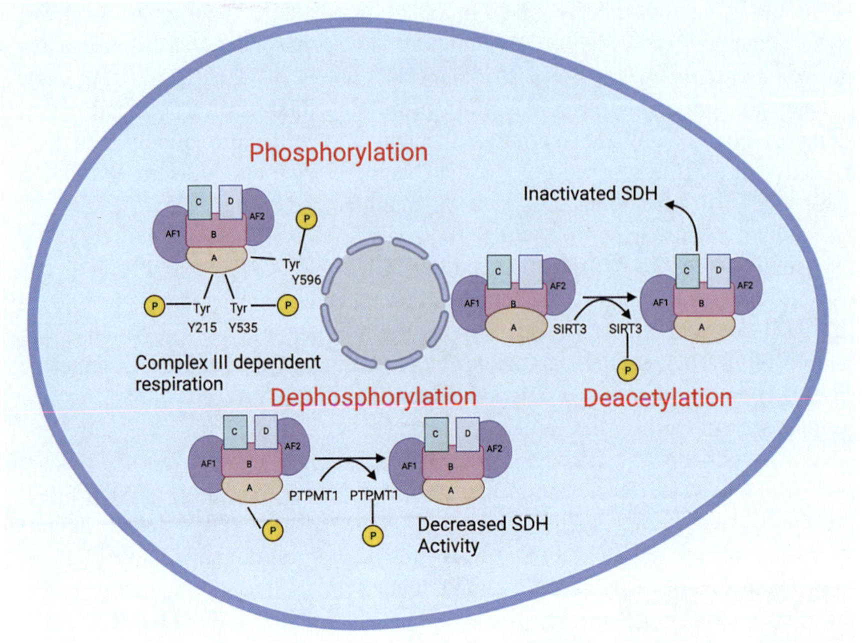

Fig. 3.2 Diseases caused by malfunctioning of SDH, concurrently by the mechanism of phosphorylation, deacetylation, and dephosphorylation

(Chinopoulos et al. 2019; Landsverk et al. 2014). The study suggested that exposure to high altitude silenced the mentioned genes, although the exact mechanisms remain to be elucidated (Lu et al. 2018). Similarly, in the cases of PGL, a parallel interplay between the environment and genetics akin to AMS is observed. Prolonged exposure to high altitudes leads to an SDHB mutation causing PGL in patients (Cerecer-Gil et al. 2010). This underlines a clear connection among high-altitude conditions, SDH, AMS, and cancer, highlighting the potential for further research to deepen our understanding of this intricate association.

3.6.3 Neurodegenerative Disease

SDH plays essential role in regulating the flow of electrons (Zhang et al. 2018a). Specifically, SDH activity hinders the oxidation of electron carriers like nicotinamide adenine dinucleotide (NAD), resulting in reduced electron flow to complex III and ubiquinone, subsequently leading to ROS production (Rustin et al. 2002). Mutations in SDHA have been identified as factors in disrupting this mechanism, contributing to the onset of Leigh syndrome (Parfait et al. 2000; Zhu et al. 1998). Other mutations in SDH subunits have been allied with neurodegenerative disorders. For example, SDHA mutations have been linked to ataxia, and mutations in SDHA, SDHB, SDHD, and SDHAF1 have been implicated in leukodystrophy

(Alston et al. 2012; Birch-Machin et al. 2000; Ghezzi et al. 2009; Jackson et al. 2014). A recent review highlighted succinate's association with the mammalian target of rapamycin (mTOR), a kinase implicated in various neurodegenerative diseases (Jahrling and Laberge 2015). This connection piques interest in further exploring succinate's roles in neurodegenerative conditions. Notably, SDHAF4 has shown considerable accumulation of succinate, enhancement of mitochondrial SDH activity, reduced ROS production, and protection against neurodegeneration in *Drosophila* (Van Vranken et al. 2014).

3.6.4 Diabetes

The mechanism involving succinate and insulin release proposes that elevated succinate levels within mitochondria result in the production of mevalonic acid, ultimately triggering insulin release in pancreatic islet (Fahien and MacDonald 2002). Analyses of proteins in individuals with type 2 diabetes mellitus (T2D) experiencing prolonged high blood sugar levels revealed that heightened glucose levels inhibit the electron transport chain (ETC). This inhibition is caused by reduced expression of SDH, fumarate hydratase, and citrate synthase within the mitochondria, subsequently leading to reduced insulin release from β-cells of pancreatic (Haythorne et al. 2019). While succinate levels decline within the mitochondria, it has been observed that circulating succinate levels increase in T2D patients (Sadagopan et al. 2007). This circulating succinate can accumulate in the kidneys of diabetic individuals, resulting in the stimulation of GPR91 by succinate, a receptor which stimulates the renin-angiotensin system (RAS) (Toma et al. 2008). The activation of the RAS in diabetic rats has been associated with nephropathy and hypertension (Gurley and Coffman 2007).

3.6.5 Ischemia–Reperfusion Injury

Accumulation of succinate is a recognized phenomenon during cardiac ischemia, and it is subsequently metabolized during reperfusion, contributing to oxidative damage due to heightened production of ROS (Chouchani et al. 2014). A prominent model explaining this process involves the heightened activity of glycolysis and the citric acid cycle, a well-known occurrence during ischemia. The conversion of glutamate to 2-oxoglutarate is facilitated by this process, which is then transformed into succinate through succinate-CoA ligase. In the case of Barth syndrome, a genetic disorder linked to cardiomyopathy, there is a loss of SDH in cardiac tissue, resulting in an increased production of ROS. Although succinate is not the direct cause of cardiac ischemia or Barth syndrome, their impact on succinate has adverse implications. Further exploration into the ramifications of these conditions on succinate is essential.

3.7 Future Prospects

While succinate was traditionally seen as an intermediary in the citric acid cycle, it is now clear that its significance extends beyond energy metabolism. Succinate is now recognized to have a vital role in gene expression and communication between cells, in addition to its energy-related functions. Recently, there has been substantial evidence highlighting the role of succinate buildup in the progression of carcinogenesis. This understanding has directed to the categorization of succinate and SDH as an oncometabolite and a tumor suppressor respectively.

First, examining the interactions between SDH pseudogenes like SDHAP1/2/3 and SDHA/B/C/D subunits along with additional molecules is of significant interest. The application of state-of-the-art cryo-electron spectroscopy is anticipated to enhance our comprehension of these interactions, offering valuable insights into cancer and disease development. Unexpectedly, in ovarian cancer, SDHAP1 has been categorized as a long noncoding RNA (lncRNA), suggesting that SDHAP1/2/3 may have an independent role amidst the metabolic reactions associated with SDH enzymes.

Exploring the interplay between the SDH complex and noncoding RNAs is still in its infancy. Only a limited number of miRNAs have been thoroughly investigated to date. A comprehensive screening of noncoding RNAs utilizing RNA-sequencing should be implemented in this investigation. Moreover, research on RNA editors of SDH, such as editing of C to U using APOBEC1 and APOBEC3A, has gained increasing attention recently. Beyond the direct APOBEC1/3A – SDH interaction, RNA editors can also modulate noncoding RNA levels in both cancers and noncancer diseases (Nishikura 2016). Therefore, understanding how RNA editors indirectly regulate SDH via the editing of noncoding RNAs warrants future investigation. Additionally, it is intriguing to study whether succinate accumulation resulting from SDH inactivation can influence the activity of RNA editors. Notably, the SDH – RNA network interaction is dynamic, considering that some RNAs may act as messengers, being transmitted from one type of cell to another. As an example, miRNAs found within exosomes circulating freely in cells can communicate signals as they travel from their source to a designated destination (Thomou et al. 2017). Succinate is known to shuttle from hypoxic retina to oxygen-enriched tissue for efficient electron transfer quite similar to circulation of miRNAs (Bisbach et al. 2020). Therefore, this new module suggests the possibility of applying the SDH/RNA interaction in different microenvironments where RNA and succinate are remotely diffused.

Numerous fatalities among patients have been noticed due to fungal infections (Brown et al. 2012), because most fungicides are designed to battle against the human fungal pathogens by inhibiting cellular mitochondrial respiration, including activities of SDH in fungi (Duvenage et al. 2019). However, SDH inhibitors targeting fungal infection getting integrated into human applications have been challenging due to the resemblance in SDH structures between human and fungal cells. Hence, while addressing SDH inhibitors in antifungal treatment for humans, two major concerns need to be tackled. Initially, the SDH inhibitors used could potentially transfer to human cells, elevating the risk of tumor formation by disabling the tumor suppressor SDH within human cells. Additionally, certain pathogens, such as lipopolysaccharide-induced succinate accumulation from Gram-negative bacteria, activate HIF-1α and augment the discharge of interleukin-1β in human macrophages. Moreover, pathogen-based succinate is known to induce tumor initiation and progression in human beings (Tretter et al. 2016; Zhao et al. 2017).

RNA-based therapeutics have recently garnered increasing attention due to their precision and specificity compared to traditional small molecules (Wang et al. 2020). However, addressing the efficiency of RNA delivery and off-target effects of RNA therapeutics is crucial before advancing to clinical trials. It is imperative to target cells using cutting-edge nanoparticles and ligand-conjugated carriers in order to enhance site-specific RNA delivery (Nogrady 2019). Concurrently, scrutinizing the circulation and functions of RNA-based therapeutics while still treating nontargeted cells (e.g., nontumor cells in cancer) can functionalize the application therapeutics of RNA in clinical trials (Guo et al. 2012). In future, assessing the feasibility of employing RNA-based therapeutics, for instance targeting RNA editors, noncoding segments of RNA, and RNA modifiers, to combat cancer and diseases induced by SDH mutations or dysfunction is paramount.

3.8 Conclusion

The SDH complex has garnered a renewed surge of interest in recent years, primarily sparked by its recognition in human disease and the identification of fresh assembly components. The precise control of this enzyme through its complexes SDHA, SDHB, SDHC, and SDHD enables cells to execute vital processes like cellular respiration, response to low oxygen levels (hypoxic response), and regulation of gene expression. The distinctive roles of different subunits within the SDH complex contribute to these metabolic adjustments. Mutations occurring within these subunits and the structural constituents of the complex can result in a wide spectrum of human disorders. As with many significant findings, these recent revelations generate more inquiries than solutions. It is foreseeable that in the upcoming years, we will gain fresh perspectives on SDH assembly, including the discovery of novel factors and a deeper understanding of the underlying mechanisms. Moreover, we anticipate a more comprehensive comprehension of how SDH and mutations related to it contribute to human ailments. This progress paves the way for potential targeted therapeutic approaches. Undoubtedly, this marks an exhilarating period for scholars delving into the intricacies of the SDH complex.

References

Ahmad M, Wolberg A, Kahwaji CI (2023) Biochemistry, electron transport chain. In: StatPearls. StatPearls. http://www.ncbi.nlm.nih.gov/books/NBK526105/

Alabduladhem TO, Bordoni B (2023) Physiology, Krebs cycle. In: StatPearls. StatPearls. http://www.ncbi.nlm.nih.gov/books/NBK556032/

Alarcón CR, Lee H, Goodarzi H, Halberg N, Tavazoie SF (2015) N6-methyl-adenosine (m6A) marks primary microRNAs for processing. Nature 519(7544):482–485. https://doi.org/10.1038/nature14281

Alston CL, Davison JE, Meloni F, van der Westhuizen FH, He L, Hornig-Do H-T, Peet AC, Gissen P, Goffrini P, Ferrero I, Wassmer E, McFarland R, Taylor RW (2012) Recessive germline SDHA and SDHB mutations causing leukodystrophy and isolated mitochondrial complex II deficiency. J Med Genet 49(9):569–577. https://doi.org/10.1136/jmedgenet-2012-101146

Arias-Cartin R, Grimaldi S, Arnoux P, Guigliarelli B, Magalon A (2012) Cardiolipin binding in bacterial respiratory complexes: structural and functional implications. Biochim Biophys Acta 1817(10):1937–1949. https://doi.org/10.1016/j.bbabio.2012.04.005

Arnold PK, Finley LWS (2023) Regulation and function of the mammalian tricarboxylic acid cycle. J Biol Chem 299(2):102838. https://doi.org/10.1016/j.jbc.2022.102838

Aspuria P-JP, Lunt SY, Väremo L, Vergnes L, Gozo M, Beach JA, Salumbides B, Reue K, Wiedemeyer WR, Nielsen J, Karlan BY, Orsulic S (2014) Succinate dehydrogenase inhibition leads to epithelial-mesenchymal transition and reprogrammed carbon metabolism. Cancer Metab 2:21. https://doi.org/10.1186/2049-3002-2-21

Astuti D, Latif F, Dallol A, Dahia PLM, Douglas F, George E, Sköldberg F, Husebye ES, Eng C, Maher ER (2001) Gene mutations in the succinate dehydrogenase subunit SDHB cause susceptibility to familial Pheochromocytoma and to familial Paraganglioma. Am J Hum Genet 69(1):49–54

Bakare AB, Lesnefsky EJ, Iyer S (2021) Leigh syndrome: a tale of two genomes. Front Physiol 12. https://www.frontiersin.org/articles/10.3389/fphys.2021.693734

Bakonyi T, Radak Z (2004) High altitude and free radicals. J Sports Sci Med 3(2):64–69

Bandara AB, Drake JC, Brown DA (2021) Complex II subunit SDHD is critical for cell growth and metabolism, which can be partially restored with a synthetic ubiquinone analog. BMC Mol Cell Biol 22:35. https://doi.org/10.1186/s12860-021-00370-w

Barbara P, Stratakis CA (2009) SDH mutations in tumourigenesis and inherited endocrine tumours. J Intern Med 266(1):19–42. https://doi.org/10.1111/j.1365-2796.2009.02111.x

Barbieri I, Kouzarides T (2020) Role of RNA modifications in cancer. Nat Rev Cancer 20(6):303–322. https://doi.org/10.1038/s41568-020-0253-2

Bardella C, Pollard PJ, Tomlinson I (2011) SDH mutations in cancer. Biochim Biophys Acta 1807(11):1432–1443. https://doi.org/10.1016/j.bbabio.2011.07.003

Bayley J-P, Kunst HPM, Cascon A, Sampietro ML, Gaal J, Korpershoek E, Hinojar-Gutierrez A, Timmers HJLM, Hoefsloot LH, Hermsen MA, Suárez C, Hussain AK, Vriends AHJT, Hes FJ, Jansen JC, Tops CM, Corssmit EP, de Knijff P, Lenders JWM et al (2010) SDHAF2 mutations in familial and sporadic paraganglioma and phaeochromocytoma. Lancet Oncol 11(4):366–372. https://doi.org/10.1016/S1470-2045(10)70007-3

Baysal BE, Ferrell RE, Willett-Brozick JE, Lawrence EC, Myssiorek D, Bosch A, van der Mey A, Taschner PE, Rubinstein WS, Myers EN, Richard CW, Cornelisse CJ, Devilee P, Devlin B (2000) Mutations in SDHD, a mitochondrial complex II gene, in hereditary paraganglioma. Science 287(5454):848–851. https://doi.org/10.1126/science.287.5454.848

Belt K, Van Aken O, Murcha M, Millar AH, Huang S (2018) An assembly factor promotes assembly of flavinated SDH1 into the succinate dehydrogenase complex 1 [OPEN]. Plant Physiol 177(4):1439–1452. https://doi.org/10.1104/pp.18.00320

Bénit P, Goncalves J, El Khoury R, Rak M, Favier J, Gimenez-Roqueplo A-P, Rustin P (2022) Succinate dehydrogenase, succinate, and superoxides: a genetic, epigenetic, metabolic, environmental explosive crossroad. Biomedicines 10(8):1788. https://doi.org/10.3390/biomedicines10081788

Bezawork-Geleta A, Rohlena J, Dong L, Pacak K, Neuzil J (2017) Mitochondrial complex II: at the crossroads. Trends Biochem Sci 42(4):312–325. https://doi.org/10.1016/j.tibs.2017.01.003

Birch-Machin MA, Taylor RW, Cochran B, Ackrell BA, Turnbull DM (2000) Late-onset optic atrophy, ataxia, and myopathy associated with a mutation of a complex II gene. Ann Neurol 48(3):330–335

Bisbach CM, Hass DT, Robbings BM, Rountree AM, Sadilek M, Sweet IR, Hurley JB (2020) Succinate can shuttle reducing power from the hypoxic retina to the O2-rich pigment epithelium. Cell Rep 31(5):107606. https://doi.org/10.1016/j.celrep.2020.107606
Bouillaud F (2023) Inhibition of succinate dehydrogenase by pesticides (SDHIs) and energy metabolism. Int J Mol Sci 24(4):4. https://doi.org/10.3390/ijms24044045
Bourgeron T, Rustin P, Chretien D, Birch-Machin M, Bourgeois M, Viegas-Péquignot E, Munnich A, Rötig A (1995) Mutation of a nuclear succinate dehydrogenase gene results in mitochondrial respiratory chain deficiency. Nat Genet 11(2):144–149. https://doi.org/10.1038/ng1095-144
Brière J-J, Favier J, El Ghouzzi V, Djouadi F, Bénit P, Gimenez A-P, Rustin P (2005) Succinate dehydrogenase deficiency in human. Cell Mol Life Sci 62(19–20):2317–2324. https://doi.org/10.1007/s00018-005-5237-6
Brown GD, Denning DW, Gow NAR, Levitz SM, Netea MG, White TC (2012) Hidden killers: human fungal infections. Sci Transl Med 4(165):165rv13. https://doi.org/10.1126/scitranslmed.3004404
Calisto F, Sousa FM, Sena FV, Refojo PN, Pereira MM (2021) Mechanisms of energy transduction by charge translocating membrane proteins. Chem Rev 121(3):1804–1844. https://doi.org/10.1021/acs.chemrev.0c00830
Cappadocia L, Lima CD (2018) Ubiquitin-like protein conjugation: structures, chemistry, and mechanism. Chem Rev 118(3):889–918. https://doi.org/10.1021/acs.chemrev.6b00737
Cardaci S, Ciriolo MR (2012) TCA cycle defects and cancer: when metabolism tunes redox state. Int J Cell Biol 2012:161837. https://doi.org/10.1155/2012/161837
Caruso Bavisotto C, Alberti G, Vitale AM, Paladino L, Campanella C, Rappa F, Gorska M, Conway de Macario E, Cappello F, Macario AJL, Marino Gammazza A (2020) Hsp60 post-translational modifications: functional and pathological consequences. Front Mol Biosci 7:95. https://doi.org/10.3389/fmolb.2020.00095
Cecchini G (2003) Function and structure of complex II of the respiratory chain. Annu Rev Biochem 72:77–109. https://doi.org/10.1146/annurev.biochem.72.121801.161700
Cerecer-Gil NY, Figuera LE, Llamas FJ, Lara M, Escamilla JG, Ramos R, Estrada G, Hussain AK, Gaal J, Korpershoek E, de Krijger RR, Dinjens WNM, Devilee P, Bayley JP (2010) Mutation of SDHB is a cause of hypoxia-related high-altitude paraganglioma. Clin Cancer Res 16(16):4148–4154. https://doi.org/10.1158/1078-0432.CCR-10-0637
Chen L, Liu T, Zhang S, Zhou J, Wang Y, Di W (2014) Succinate dehydrogenase subunit B inhibits the AMPK-HIF-1α pathway in human ovarian cancer in vitro. J Ovarian Res 7:115. https://doi.org/10.1186/s13048-014-0115-1
Chen Y, Cui Y, Shen B, Niu Y, Zhao X, Wang L, Wang J, Li W, Zhou Q, Ji W, Sha J, Huang X (2015) Germline acquisition of Cas9/RNA-mediated gene modifications in monkeys. Cell Res 25(2):262–265. https://doi.org/10.1038/cr.2014.167
Chinopoulos C, Batzios S, van den Heuvel LP, Rodenburg R, Smeets R, Waterham HR, Turkenburg M, Ruiter JP, Wanders RJA, Doczi J, Horvath G, Dobolyi A, Vargiami E, Wevers RA, Zafeiriou D (2019) Mutated SUCLG1 causes mislocalization of SUCLG2 protein, morphological alterations of mitochondria and an early-onset severe neurometabolic disorder. Mol Genet Metab 126(1):43–52. https://doi.org/10.1016/j.ymgme.2018.11.009
Chouchani ET, Pell VR, Gaude E, Aksentijević D, Sundier SY, Robb EL, Logan A, Nadtochiy SM, Ord ENJ, Smith AC, Eyassu F, Shirley R, Hu C-H, Dare AJ, James AM, Rogatti S, Hartley RC, Eaton S, Costa ASH et al (2014) Ischaemic accumulation of succinate controls reperfusion injury through mitochondrial ROS. Nature 515(7527):431–435. https://doi.org/10.1038/nature13909
Cimen H, Han M-J, Yang Y, Tong Q, Koc H, Koc EC (2010) Regulation of succinate dehydrogenase activity by SIRT3 in mammalian mitochondria. Biochemistry 49(2):304–311. https://doi.org/10.1021/bi901627u
Clavijo-Herrera J, Van Santen E, Gómez C (2018) Growth, water-use efficiency, stomatal conductance, and nitrogen uptake of two lettuce cultivars grown under different percentages of blue and red light. Horticulturae 4(3):3. https://doi.org/10.3390/horticulturae4030016

Clemente-Suárez VJ, Redondo-Flórez L, Beltrán-Velasco AI, Ramos-Campo DJ, Belinchón-deMiguel P, Martinez-Guardado I, Dalamitros AA, Yáñez-Sepúlveda R, Martín-Rodríguez A, Tornero-Aguilera JF (2023) Mitochondria and brain disease: a comprehensive review of pathological mechanisms and therapeutic opportunities. Biomedicines 11(9):9. https://doi.org/10.3390/biomedicines11092488

De Vivo DC, Paradas C, DiMauro S (2015) Chapter 41 – mitochondrial Encephalomyopathies. In: Darras BT, Jones HR, Ryan MM, De Vivo DC (eds) Neuromuscular disorders of infancy, childhood, and adolescence, 2nd edn. Academic Press, pp 796–833. https://doi.org/10.1016/B978-0-12-417044-5.00041-X

Dekker P, Hogendoorn P, Kuipers-Dijkshoorn N, Prins F, Duinen S, Taschner P, van der Mey A, Cornelisse C (2003) SDHD mutations in head and neck paragangliomas result in destabilization of complex II in the mitochondrial respiratory chain with loss of enzymatic activity and abnormal mitochondrial morphology. J Pathol 201:480–486. https://doi.org/10.1002/path.1461

Dibrov E, Fu S, Lemire BD (1998) The Saccharomyces cerevisiae TCM62 gene encodes a chaperone necessary for the assembly of the mitochondrial succinate dehydrogenase (complex II). J Biol Chem 273(48):32042–32048. https://doi.org/10.1074/jbc.273.48.32042

Duvenage L, Munro CA, Gourlay CW (2019) The potential of respiration inhibition as a new approach to combat human fungal pathogens. Curr Genet 65(6):1347–1353. https://doi.org/10.1007/s00294-019-01001-w

Dwight T, Na U, Kim E, Zhu Y, Richardson AL, Robinson BG, Tucker KM, Gill AJ, Benn DE, Clifton-Bligh RJ, Winge DR (2017) Analysis of SDHAF3 in familial and sporadic pheochromocytoma and paraganglioma. BMC Cancer 17(1):497. https://doi.org/10.1186/s12885-017-3486-z

Erb M, Hoffmann-Enger B, Deppe H, Soeberdt M, Haefeli RH, Rummey C, Feurer A, Gueven N (2012) Features of Idebenone and related short-chain Quinones that rescue ATP levels under conditions of impaired mitochondrial complex I. PLoS One 7(4):e36153. https://doi.org/10.1371/journal.pone.0036153

Fahien LA, MacDonald MJ (2002) The succinate mechanism of insulin release. Diabetes 51(9):2669–2676. https://doi.org/10.2337/diabetes.51.9.2669

Fernández-Vizarra E, Tiranti V, Zeviani M (2009) Assembly of the oxidative phosphorylation system in humans: what we have learned by studying its defects. Biochim Biophys Acta 1793(1):200–211. https://doi.org/10.1016/j.bbamcr.2008.05.028

Fiebiger SM, Bros H, Grobosch T, Janssen A, Chanvillard C, Paul F, Dörr J, Millward JM, Infante-Duarte C (2013) The antioxidant idebenone fails to prevent or attenuate chronic experimental autoimmune encephalomyelitis in the mouse. J Neuroimmunol 262(1–2):66–71. https://doi.org/10.1016/j.jneuroim.2013.07.002

Flavahan WA, Drier Y, Johnstone SE, Hemming ML, Tarjan DR, Hegazi E, Shareef SJ, Javed NM, Raut CP, Eschle BK, Gokhale PC, Hornick JL, Sicinska ET, Demetri GD, Bernstein BE (2019) Altered chromosomal topology drives oncogenic programs in SDH-deficient GISTs. Nature 575(7781):229–233. https://doi.org/10.1038/s41586-019-1668-3

Fogg VC, Lanning NJ, MacKeigan JP (2011) Mitochondria in cancer: at the crossroads of life and death. Chin J Cancer 30(8):526–539. https://doi.org/10.5732/cjc.011.10018

Fuentes D, Meneses M, Nunes-Nesi A, Araújo WL, Tapia R, Gómez I, Holuigue L, Gutiérrez RA, Fernie AR, Jordana X (2011) A deficiency in the Flavoprotein of Arabidopsis mitochondrial complex II results in elevated photosynthesis and better growth in nitrogen-limiting conditions1[W][OA]. Plant Physiol 157(3):1114–1127. https://doi.org/10.1104/pp.111.183939

Gammon A, Jasperson K, Champine M (2016) Genetic basis of Cowden syndrome and its implications for clinical practice and risk management. Appl Clin Genet 9:83–92. https://doi.org/10.2147/TACG.S41947

Ghezzi D, Goffrini P, Uziel G, Horvath R, Klopstock T, Lochmüller H, D'Adamo P, Gasparini P, Strom TM, Prokisch H, Invernizzi F, Ferrero I, Zeviani M (2009) SDHAF1, encoding a LYR complex-II specific assembly factor, is mutated in SDH-defective infantile leukoencephalopathy. Nat Genet 41(6):654–656. https://doi.org/10.1038/ng.378

Ghifari AS, Saha S, Murcha MW (2023) The biogenesis and regulation of the plant oxidative phosphorylation system. Plant Physiol 192(2):728–747. https://doi.org/10.1093/plphys/kiad108
Goetzman E, Gong Z, Zhang B, Muzumdar R (2023) Complex II biology in aging, health, and disease. Antioxidants 12(7):7. https://doi.org/10.3390/antiox12071477
Gu H, Jan Fada B (2020) Specificity in ubiquitination triggered by virus infection. Int J Mol Sci 21(11):4088. https://doi.org/10.3390/ijms21114088
Guo P, Haque F, Hallahan B, Reif R, Li H (2012) Uniqueness, advantages, challenges, solutions, and perspectives in therapeutics applying RNA nanotechnology. Nucleic Acid Ther 22(4):226–245. https://doi.org/10.1089/nat.2012.0350
Gurley SB, Coffman TM (2007) The renin-angiotensin system and diabetic nephropathy. Semin Nephrol 27(2):144–152. https://doi.org/10.1016/j.semnephrol.2007.01.009
Guzy RD, Schumacker PT (2006) Oxygen sensing by mitochondria at complex III: the paradox of increased reactive oxygen species during hypoxia. Exp Physiol 91(5):807–819. https://doi.org/10.1113/expphysiol.2006.033506
Guzy RD, Sharma B, Bell E, Chandel NS, Schumacker PT (2008) Loss of the SdhB, but not the SdhA, subunit of complex II triggers reactive oxygen species-dependent hypoxia-inducible factor activation and tumorigenesis. Mol Cell Biol 28(2):718–731. https://doi.org/10.1128/MCB.01338-07
Hägerhäll C (1997) Succinate: Quinone oxidoreductases. Variations on a conserved theme. Biochim Biophys Acta 1320(2):107–141. https://doi.org/10.1016/s0005-2728(97)00019-4
Haythorne E, Rohm M, van de Bunt M, Brereton MF, Tarasov AI, Blacker TS, Sachse G, Silva dos Santos M, Terron Exposito R, Davis S, Baba O, Fischer R, Duchen MR, Rorsman P, MacRae JI, Ashcroft FM (2019) Diabetes causes marked inhibition of mitochondrial metabolism in pancreatic β-cells. Nat Commun 10:2474. https://doi.org/10.1038/s41467-019-10189-x
Heitz FD, Erb M, Anklin C, Robay D, Pernet V, Gueven N (2012) Idebenone protects against retinal damage and loss of vision in a mouse model of Leber's hereditary optic neuropathy. PLoS One 7(9):e45182. https://doi.org/10.1371/journal.pone.0045182
Hobert JA, Mester JL, Moline J, Eng C (2012) Elevated plasma succinate in PTEN, SDHB, and SDHD mutation–positive individuals. Genet Med 14(6):6. https://doi.org/10.1038/gim.2011.63
Hove-Jensen B, Andersen KR, Kilstrup M, Martinussen J, Switzer RL, Willemoës M (2016) Phosphoribosyl diphosphate (PRPP): biosynthesis, enzymology, utilization, and metabolic significance. Microbiol Mol Biol Rev 81(1):e00040–e00016. https://doi.org/10.1128/MMBR.00040-16
Huang S, Braun H-P, Gawryluk RMR, Millar AH (2019) Mitochondrial complex II of plants: subunit composition, assembly, and function in respiration and signaling. Plant J 98(3):405–417. https://doi.org/10.1111/tpj.14227
Ishii T, Yasuda K, Akatsuka A, Hino O, Hartman PS, Ishii N (2005) A mutation in the SDHC gene of complex II increases oxidative stress, resulting in apoptosis and tumorigenesis. Cancer Res 65(1):203–209
Jackson CB, Nuoffer J-M, Hahn D, Prokisch H, Haberberger B, Gautschi M, Häberli A, Gallati S, Schaller A (2014) Mutations in SDHD lead to autosomal recessive encephalomyopathy and isolated mitochondrial complex II deficiency. J Med Genet 51(3):170–175. https://doi.org/10.1136/jmedgenet-2013-101932
Jacoby RP, Li L, Huang S, Pong Lee C, Millar AH, Taylor NL (2012) Mitochondrial composition, function and stress response in PlantsF. J Integr Plant Biol 54(11):887–906. https://doi.org/10.1111/j.1744-7909.2012.01177.x
Jahrling JB, Laberge R-M (2015) Age-related neurodegeneration prevention through mTOR inhibition: potential mechanisms and remaining questions. Curr Top Med Chem 15(21):2139–2151
Janeway KA, Kim SY, Lodish M, Nosé V, Rustin P, Gaal J, Dahia PLM, Liegl B, Ball ER, Raygada M, Lai AH, Kelly L, Hornick JL, O'Sullivan M, de Krijger RR, Dinjens WNM, Demetri GD, Antonescu CR, Fletcher JA et al (2011) Defects in succinate dehydrogenase in gastrointestinal stromal tumors lacking KIT and PDGFRA mutations. Proc Natl Acad Sci USA 108(1):314–318. https://doi.org/10.1073/pnas.1009199108

Karavaeva V, Sousa FL (2023) Modular structure of complex II: an evolutionary perspective. Biochim Biophys Acta 1864(1):148916. https://doi.org/10.1016/j.bbabio.2022.148916
Kim HJ, Winge DR (2013) Emerging concepts in the Flavinylation of succinate dehydrogenase. Biochim Biophys Acta 1827(5):627–636. https://doi.org/10.1016/j.bbabio.2013.01.012
Klanner C, Neupert W, Langer T (2000) The chaperonin-related protein Tcm62p ensures mitochondrial gene expression under heat stress. FEBS Lett 470(3):365–369. https://doi.org/10.1016/s0014-5793(00)01322-3
Landsverk ML, Zhang VW, Wong L-JC, Andersson HC (2014) A SUCLG1 mutation in a patient with mitochondrial DNA depletion and congenital anomalies. Mol Genet Metab Rep 1:451–454. https://doi.org/10.1016/j.ymgmr.2014.09.007
Lendvai N, Pawlosky R, Bullova P, Eisenhofer G, Patocs A, Veech RL, Pacak K (2014) Succinate-to-fumarate ratio as a new metabolic marker to detect the presence of SDHB/D-related Paraganglioma: initial experimental and ex vivo findings. Endocrinology 155(1):27–32. https://doi.org/10.1210/en.2013-1549
León G, Holuigue L, Jordana X (2007) Mitochondrial complex II is essential for gametophyte development in arabidopsis. Plant Physiol 143(4):1534–1546. https://doi.org/10.1104/pp.106.095158
Letouzé E, Martinelli C, Loriot C, Burnichon N, Abermil N, Ottolenghi C, Janin M, Menara M, Nguyen AT, Benit P, Buffet A, Marcaillou C, Bertherat J, Amar L, Rustin P, De Reyniès A, Gimenez-Roqueplo A-P, Favier J (2013) SDH mutations establish a hypermethylator phenotype in paraganglioma. Cancer Cell 23(6):739–752. https://doi.org/10.1016/j.ccr.2013.04.018
Li Y, Park J-S, Deng J-H, Bai Y (2006) Cytochrome c oxidase subunit IV is essential for assembly and respiratory function of the enzyme complex. J Bioenerg Biomembr 38(5–6):283–291. https://doi.org/10.1007/s10863-006-9052-z
Li C, Liu C-Q, Zhang H-S, Chen C-P, Yang X-R, Chen L-F, Liu Q-S, Guo J, Sun C-H, Wang P-R, Deng X-J (2021) LPS1, encoding iron-Sulfur subunit SDH2-1 of succinate dehydrogenase, affects leaf senescence and grain yield in Rice. Int J Mol Sci 22(1):Article 1. https://doi.org/10.3390/ijms22010157
Li Y, Belt K, Alqahtani SF, Saha S, Fenske R, Van Aken O, Whelan J, Millar AH, Murcha MW, Huang S (2022) The mitochondrial LYR protein SDHAF1 is required for succinate dehydrogenase activity in arabidopsis. Plant J 110(2):499–512. https://doi.org/10.1111/tpj.15684
Lu H, Wang R, Li W, Xie H, Wang C, Hao Y, Sun Y, Jia Z (2018) Plasma proteomic study of acute mountain sickness susceptible and resistant individuals. Sci Rep 8:1265. https://doi.org/10.1038/s41598-018-19818-9
Luks AM, Swenson ER, Bärtsch P (2017) Sports-related lung disease: acute high-altitude sickness. Eur Respir Rev 26(143). https://doi.org/10.1183/16000617.0096-2016
Macklin PS, Yamamoto A, Browning L, Hofer M, Adam J, Pugh CW (2020) Recent advances in the biology of tumour hypoxia with relevance to diagnostic practice and tissue-based research. J Pathol 250(5):593–611. https://doi.org/10.1002/path.5402
Maio N, Ghezzi D, Verrigni D, Rizza T, Bertini E, Martinelli D, Zeviani M, Singh A, Carrozzo R, Rouault TA (2016) Disease-causing SDHAF1 mutations impair transfer of Fe-S clusters to SDHB. Cell Metab 23(2):292–302. https://doi.org/10.1016/j.cmet.2015.12.005
Mansilla N, Racca S, Gras DE, Gonzalez DH, Welchen E (2018) The complexity of mitochondrial complex IV: an update of cytochrome c oxidase biogenesis in plants. Int J Mol Sci 19(3):662. https://doi.org/10.3390/ijms19030662
Martucci VL, Pacak K (2014) Pheochromocytoma and Paraganglioma: diagnosis, genetics, management, and treatment. Curr Probl Cancer 38(1):7–41. https://doi.org/10.1016/j.currproblcancer.2014.01.001
Miyadera H, Shiomi K, Ui H, Yamaguchi Y, Masuma R, Tomoda H, Miyoshi H, Osanai A, Kita K, Ōmura S (2003) Atpenins, potent and specific inhibitors of mitochondrial complex II (succinate-ubiquinone oxidoreductase). Proc Natl Acad Sci USA 100(2):473–477. https://doi.org/10.1073/pnas.0237315100

Moosavi B, Berry EA, Zhu X-L, Yang W-C, Yang G-F (2019) The assembly of succinate dehydrogenase: a key enzyme in bioenergetics. Cell Mol Life Sci 76(20):4023–4042. https://doi.org/10.1007/s00018-019-03200-7

Morandi A, Taddei ML, Chiarugi P, Giannoni E (2017) Targeting the metabolic reprogramming that controls epithelial-to-mesenchymal transition in aggressive tumors. Front Oncol 7. https://www.frontiersin.org/articles/10.3389/fonc.2017.00040

Moreno C, Santos RM, Burns R, Zhang WC (2020) Succinate dehydrogenase and ribonucleic acid networks in cancer and other diseases. Cancers 12(11):3237. https://doi.org/10.3390/cancers12113237

Niemann S, Müller U (2000) Mutations in SDHC cause autosomal dominant paraganglioma, type 3. Nat Genet 26(3):268–270. https://doi.org/10.1038/81551

Nishikura K (2016) A-to-I editing of coding and non-coding RNAs by ADARs. Nat Rev Mol Cell Biol 17(2):83–96. https://doi.org/10.1038/nrm.2015.4

Nogrady B (2019) The challenge of delivering RNA-interference therapeutics to their target cells. Nature 574(7778):S8–S9. https://doi.org/10.1038/d41586-019-03071-9

Ohlenbusch A, Edvardson S, Skorpen J, Bjornstad A, Saada A, Elpeleg O, Gärtner J, Brockmann K (2012) Leukoencephalopathy with accumulated succinate is indicative of SDHAF1 related complex II deficiency. Orphanet J Rare Dis 7(1):69. https://doi.org/10.1186/1750-1172-7-69

Owens KM, Aykin-Burns N, Dayal D, Coleman MC, Domann FE, Spitz DR (2012) Genomic instability induced by mutant succinate dehydrogenase subunit D (SDHD) is mediated by O2−• and H2O2. Free Radic Biol Med 52(1):160–166. https://doi.org/10.1016/j.freeradbiomed.2011.10.435

Pal S, Sharma A, Mathew SP, Jaganathan BG (2022) Targeting cancer-specific metabolic pathways for developing novel cancer therapeutics. Front Immunol 13. https://www.frontiersin.org/articles/10.3389/fimmu.2022.955476

Palecek T, Fikrle M, Nemecek E, Bauerova L, Kuchynka P, Louch WE, Spicka I, Rysava R (2015) Contemporary treatment of amyloid heart disease. Curr Pharm Des 21(4):491–506. https://doi.org/10.2174/138161282104141204143657

Parfait B, Chretien D, Rötig A, Marsac C, Munnich A, Rustin P (2000) Compound heterozygous mutations in the flavoprotein gene of the respiratory chain complex II in a patient with Leigh syndrome. Hum Genet 106(2):236–243. https://doi.org/10.1007/s004390051033

Phasaludeen B, Emerald BS, Ansari SA (2022) The epigenetic–metabolic interplay in gliomagenesis. Open Biol 12(4):210350. https://doi.org/10.1098/rsob.210350

Piekutowska-Abramczuk D, Assouline Z, Mataković L, Feichtinger RG, Koňaříková E, Jurkiewicz E, Stawiński P, Gusic M, Koller A, Pollak A, Gasperowicz P, Trubicka J, Ciara E, Iwanicka-Pronicka K, Rokicki D, Hanein S, Wortmann SB, Sperl W, Rötig A et al (2018) NDUFB8 mutations cause mitochondrial complex I deficiency in individuals with Leigh-like Encephalomyopathy. Am J Hum Genet 102(3):460–467. https://doi.org/10.1016/j.ajhg.2018.01.008

Pinto Y, Buchumenski I, Levanon EY, Eisenberg E (2018) Human cancer tissues exhibit reduced A-to-I editing of miRNAs coupled with elevated editing of their targets. Nucleic Acids Res 46(1):71–82. https://doi.org/10.1093/nar/gkx1176

Puisségur M-P, Mazure NM, Bertero T, Pradelli L, Grosso S, Robbe-Sermesant K, Maurin T, Lebrigand K, Cardinaud B, Hofman V, Fourre S, Magnone V, Ricci JE, Pouysségur J, Gounon P, Hofman P, Barbry P, Mari B (2011) miR-210 is overexpressed in late stages of lung cancer and mediates mitochondrial alterations associated with modulation of HIF-1 activity. Cell Death Differ 18(3):465–478. https://doi.org/10.1038/cdd.2010.119

Ricketts CJ, Killian JK, Vocke CD, Wang Y, Merino MJ, Meltzer PS, Linehan WM (2022) Kidney tumors associated with germline mutations of FH and SDHB show a CpG Island methylator phenotype (CIMP). PLoS One 17(12):e0278108. https://doi.org/10.1371/journal.pone.0278108

Roh TH, Yim H, Roh J, Lee KB, Park SH, Jeong S-Y, Kim S-H, Kim J-H (2019) The loss of succinate dehydrogenase B expression is frequently identified in hemangioblastoma of the central nervous system. Sci Rep 9:5873. https://doi.org/10.1038/s41598-019-42338-z

Rustin P, Rötig A (2002) Inborn errors of complex II–unusual human mitochondrial diseases. Biochim Biophys Acta 1553(1–2):117–122. https://doi.org/10.1016/s0005-2728(01)00228-6

Rustin P, Munnich A, Rötig A (2002) Succinate dehydrogenase and human diseases: new insights into a well-known enzyme. Eur J Hum Genet 10(5):289–291. https://doi.org/10.1038/sj.ejhg.5200793

Rutter J, Winge DR, Schiffman JD (2010) Succinate dehydrogenase – assembly, regulation and role in human disease. Mitochondrion 10(4):393–401. https://doi.org/10.1016/j.mito.2010.03.001

Sadagopan N, Li W, Roberds SL, Major T, Preston GM, Yu Y, Tones MA (2007) Circulating succinate is elevated in rodent models of hypertension and metabolic disease. Am J Hypertens 20(11):1209–1215. https://doi.org/10.1016/j.amjhyper.2007.05.010

Saibil H (2013) Chaperone machines for protein folding, unfolding and disaggregation. Nat Rev Mol Cell Biol 14(10):630–642. https://doi.org/10.1038/nrm3658

Salvi M, Morrice NA, Brunati AM, Toninello A (2007) Identification of the flavoprotein of succinate dehydrogenase and aconitase as in vitro mitochondrial substrates of Fgr tyrosine kinase. FEBS Lett 581(29):5579–5585. https://doi.org/10.1016/j.febslet.2007.11.005

Selak MA, Armour SM, MacKenzie ED, Boulahbel H, Watson DG, Mansfield KD, Pan Y, Simon MC, Thompson CB, Gottlieb E (2005) Succinate links TCA cycle dysfunction to oncogenesis by inhibiting HIF-alpha prolyl hydroxylase. Cancer Cell 7(1):77–85. https://doi.org/10.1016/j.ccr.2004.11.022

Sen I, Young WF, Kasperbauer JL, Polonis K, Harmsen WS, Colglazier JJ, DeMartino RR, Oderich GS, Kalra M, Bower TC (2020) Tumor-specific prognosis of mutation-positive patients with head and neck paragangliomas. J Vasc Surg 71(5):1602–1612.e2. https://doi.org/10.1016/j.jvs.2019.08.232

Seo Y, Park J, Kim M, Lee HK, Kim J-H, Jeong J-H, Namkung W (2015) Inhibition of ANO1/TMEM16A chloride channel by Idebenone and its cytotoxicity to cancer cell lines. PLoS One 10(7):e0133656. https://doi.org/10.1371/journal.pone.0133656

Slane BG, Aykin-Burns N, Smith BJ, Kalen AL, Goswami PC, Domann FE, Spitz DR (2006) Mutation of succinate dehydrogenase subunit C results in increased O2.-, oxidative stress, and genomic instability. Cancer Res 66(15):7615–7620. https://doi.org/10.1158/0008-5472.CAN-06-0833

Sousa T, Moreira PI, Cardoso S (2023) Current advances in mitochondrial targeted interventions in Alzheimer's disease. Biomedicines 11(9):9. https://doi.org/10.3390/biomedicines11092331

Srirangalingam U, LeCain M, Tufton N, Akker SA, Drake WM, Metcalfe K (2017) Four generations of SDHB-related disease: complexities in management. Fam Cancer 16(2):279–282. https://doi.org/10.1007/s10689-016-9946-9

Sun F, Huo X, Zhai Y, Wang A, Xu J, Su D, Bartlam M, Rao Z (2005) Crystal structure of mitochondrial respiratory membrane protein complex II. Cell 121(7):1043–1057. https://doi.org/10.1016/j.cell.2005.05.025

Tang JX, Thompson K, Taylor RW, Oláhová M (2020) Mitochondrial OXPHOS biogenesis: co-regulation of protein synthesis, import, and assembly pathways. Int J Mol Sci 21(11):3820. https://doi.org/10.3390/ijms21113820

Thomou T, Mori MA, Dreyfuss JM, Konishi M, Sakaguchi M, Wolfrum C, Rao TN, Winnay JN, Garcia-Martin R, Grinspoon SK, Gorden P, Kahn CR (2017) Adipose-derived circulating miRNAs regulate gene expression in other tissues. Nature 542(7642):450–455. https://doi.org/10.1038/nature21365

Thompson AM, Denny WA (2019) Chapter four – inhibitors of enzymes in the electron transport chain of mycobacterium tuberculosis. In: Chibale K (ed) Annual reports in medicinal chemistry, vol 52. Academic Press, pp 97–130. https://doi.org/10.1016/bs.armc.2019.05.001

Toma I, Kang JJ, Sipos A, Vargas S, Bansal E, Hanner F, Meer E, Peti-Peterdi J (2008) Succinate receptor GPR91 provides a direct link between high glucose levels and renin release in murine and rabbit kidney. J Clin Invest 118(7):2526–2534. https://doi.org/10.1172/JCI33293

Tomitsuka E, Kita K, Esumi H (2009) Regulation of succinate-ubiquinone reductase and fumarate reductase activities in human complex II by phosphorylation of its flavoprotein subunit. Proc Jpn Acad Seri B Phys Biol Sci 85(7):258–265. https://doi.org/10.2183/pjab.85.258

Tretter L, Patocs A, Chinopoulos C (2016) Succinate, an intermediate in metabolism, signal transduction, ROS, hypoxia, and tumorigenesis. Biochim Biophys Acta 1857(8):1086–1101. https://doi.org/10.1016/j.bbabio.2016.03.012

Tseng P-L, Wu W-H, Hu T-H, Chen C-W, Cheng H-C, Li C-F, Tsai W-H, Tsai H-J, Hsieh M-C, Chuang J-H, Chang W-T (2018) Decreased succinate dehydrogenase B in human hepatocellular carcinoma accelerates tumor malignancy by inducing the Warburg effect. Sci Rep 8:3081. https://doi.org/10.1038/s41598-018-21361-6

van Nederveen FH, Dannenberg H, Sleddens HFBM, de Krijger RR, Dinjens WNM (2003) P53 alterations and their relationship to SDHD mutations in parasympathetic Paragangliomas. Mod Pathol 16(9):9. https://doi.org/10.1097/01.MP.0000084111.03922.4D

van Senten JR, Møller TC, Moo EV, Seiersen SD, Bräuner-Osborne H (2022) Use of CRISPR/Cas9-edited HEK293 cells reveals that both conventional and novel protein kinase C isozymes are involved in mGlu5a receptor internalization. J Biol Chem 298(10). https://doi.org/10.1016/j.jbc.2022.102466

Van Vranken JG, Bricker DK, Dephoure N, Gygi SP, Cox JE, Thummel CS, Rutter J (2014) SDHAF4 promotes mitochondrial succinate dehydrogenase activity and prevents neurodegeneration. Cell Metab 20(2):241–252. https://doi.org/10.1016/j.cmet.2014.05.012

Vladutiu GD, Heffner RR (2000) Succinate dehydrogenase deficiency: a qualitative and quantitative assessment in muscle. Arch Pathol Lab Med 124(12):1755–1758. https://doi.org/10.5858/2000-124-1755-SDD

Wang F, Zuroske T, Watts JK (2020) RNA therapeutics on the rise. Nat Rev Drug Discov 19(7):441–442. https://doi.org/10.1038/d41573-020-00078-0

Watson SA, McStay GP (2020) Functions of cytochrome c oxidase assembly factors. Int J Mol Sci 21(19):7254. https://doi.org/10.3390/ijms21197254

Wojtovich AP, Foster TH (2014) Optogenetic control of ROS production. Redox Biol 2:368–376. https://doi.org/10.1016/j.redox.2014.01.019

Xie N, Zhang L, Gao W, Huang C, Huber PE, Zhou X, Li C, Shen G, Zou B (2020) NAD+ metabolism: pathophysiologic mechanisms and therapeutic potential. Signal Transduct Target Ther 5(1):Article 1. https://doi.org/10.1038/s41392-020-00311-7

Xu W, Yang H, Liu Y, Yang Y, Wang P, Kim S-H, Ito S, Yang C, Wang P, Xiao M-T, Liu L, Jiang W, Liu J, Zhang J, Wang B, Frye S, Zhang Y, Xu Y, Lei Q et al (2011) Oncometabolite 2-Hydroxyglutarate is a competitive inhibitor of α-Ketoglutarate-dependent dioxygenases. Cancer Cell 19(1):17–30. https://doi.org/10.1016/j.ccr.2010.12.014

Yang M, Soga T, Pollard PJ (2013) Oncometabolites: linking altered metabolism with cancer. J Clin Invest 123(9):3652–3658. https://doi.org/10.1172/JCI67228

Yong C, Stewart GD, Frezza C (2020) Oncometabolites in renal cancer: Warburg's hypothesis re-examined. Nat Rev Nephrol 16(3):156–172. https://doi.org/10.1038/s41581-019-0210-z

Zhang D, Wang W, Xiang B, Li N, Huang S, Zhou W, Sun Y, Wang X, Ma J, Li G, Li X, Shen S (2013) Reduced succinate dehydrogenase B expression is associated with growth and de-differentiation of colorectal cancer cells. Tumour Biol 34(4):2337–2347. https://doi.org/10.1007/s13277-013-0781-4

Zhang Y, Liu J, Chen X-Q, Oliver Chen C-Y (2018a) Ubiquinol is superior to ubiquinone to enhance coenzyme Q10 status in older men. Food Funct 9(11):5653–5659. https://doi.org/10.1039/c8fo00971f

Zhang J, Zhang J, Li T, Zhang N, Tang S, Tao Z, Zhou X, Sun X, Chen H (2018b) Effect of idebenone on bone marrow mesenchymal stem cells in vitro. Mol Med Rep 17(4):5376–5383. https://doi.org/10.3892/mmr.2018.8506

Zhang WC, Wells JM, Chow K-H, Huang H, Yuan M, Saxena T, Melnick MA, Politi K, Asara JM, Costa DB, Bult CJ, Slack FJ (2019a) miR-147b-mediated TCA cycle dysfunction and pseudohypoxia initiate drug tolerance to EGFR inhibitors in lung adenocarcinoma. Nat Metab 1(4):460–474. https://doi.org/10.1038/s42255-019-0052-9

Zhang Y, Zhang M, Zhu W, Yu J, Wang Q, Zhang J, Cui Y, Pan X, Gao X, Sun H (2019b) Succinate accumulation induces mitochondrial reactive oxygen species generation and promotes status epilepticus in the Kainic acid rat model. Redox Biol 28:101365. https://doi.org/10.1016/j.redox.2019.101365

Zhao T, Mu X, You Q (2017) Succinate: an initiator in tumorigenesis and progression. Oncotarget 8(32):53819–53828. https://doi.org/10.18632/oncotarget.17734

Zhao R-Z, Jiang S, Zhang L, Yu Z-B (2019) Mitochondrial electron transport chain, ROS generation and uncoupling (review). Int J Mol Med 44(1):3–15. https://doi.org/10.3892/ijmm.2019.4188

Zhou Y, Mintz KJ, Cheng L, Chen J, Ferreira BCLB, Hettiarachchi SD, Liyanage PY, Seven ES, Miloserdov N, Pandey RR, Quiroga B, Blackwelder PL, Chusuei CC, Li S, Peng Z, Leblanc RM (2020) Direct conjugation of distinct carbon dots as Lego-like building blocks for the assembly of versatile drug nanocarriers. J Colloid Interface Sci 576:412–425. https://doi.org/10.1016/j.jcis.2020.05.005

Zhu Z, Yao J, Johns T, Fu K, De Bie I, Macmillan C, Cuthbert AP, Newbold RF, Wang J, Chevrette M, Brown GK, Brown RM, Shoubridge EA (1998) SURF1, encoding a factor involved in the biogenesis of cytochrome c oxidase, is mutated in Leigh syndrome. Nat Genet 20(4):337–343. https://doi.org/10.1038/3804

Zimna A, Kurpisz M (2015) Hypoxia-inducible Factor-1 in physiological and pathophysiological angiogenesis: applications and therapies. Biomed Res Int 2015:549412. https://doi.org/10.1155/2015/549412

Zorov DB, Juhaszova M, Sollott SJ (2014) Mitochondrial reactive oxygen species (ROS) and ROS-induced ROS release. Physiol Rev 94(3):909–950. https://doi.org/10.1152/physrev.00026.2013

4 Succinate Dehydrogenase: An Important Target in Drug Discovery for Tropical and Neglected Diseases

Natália Ferreira de Sousa, Pablo Rayff da Silva,
Marcus Tullius Scotti, and Luciana Scotti

Abstract

Enzymes related to energy metabolism are important targets for the design of new drugs in a wide range of diseases, as they are related to energy production and the redox metabolism of organisms. Among these enzymes, the complex corresponding to succinate dehydrogenase stands out, which is present in the citric acid cycle and the respiratory electron transfer chain. In the study of diseases that affect humans, it is suggested that the enzyme also differs from other mitochondrial dehydrogenases, thanks to its exclusive redox properties. Furthermore, neglected diseases are a group of illnesses that predominantly affect populations in low-income countries and their main transmitters are parasites, susceptible to energy imbalance. From this perspective, the present work aims to evaluate the complex II enzyme succinate dehydrogenase as an important therapeutic target for neglected diseases. A literature review was carried out through a search in the "Web of Sciences" database, in which the results demonstrate that trypanosomiases were the diseases with the highest occurrence in articles (33), being justified by the high prevalence of this disease in the world, the second most common disease was schistosomiasis, followed by Leishmaniasis. Therefore, it can be concluded that succinate dehydrogenase is an important target for the design of new drugs for neglected diseases.

Keywords

Mitochondria · Energy metabolism · drug discovery · Neglected diseases

N. F. de Sousa · P. R. da Silva · M. T. Scotti · L. Scotti (✉)
Postgraduate Program of Natural and Synthetic Bioactive Products (PgPNSB), Health Sciences Center, Federal University of Paraíba, João Pessoa, PB, Brazil

V. M. Patil et al. (eds.), *Deciphering The Role of Succinate Dehydrogenase in Drug Discovery*, https://doi.org/10.1007/978-981-96-6831-1_4

Abbreviations

AChE	Acetylcholinesterase
AdoMet	S-adenosylmethionine
ADP	Adenosine diphosphate
AMP	Adenosine monophosphate
AP	Acid phosphatase
ASATs	Aspartate aminotransferases
AST	Aspartate aminotransferase
AT	Tegument antigens
ATP	Adenosine triphosphate
CoA	Coenzyme A
DHOD	Dihydroorotate dehydrogenase
DLDH	Dihydrolipoamide dehydrogenase
DNA	Deoxyribonucleic acid
ECM	Extracellular matrix
FEPG	Fecal eggs per gram
FRD	Soluble fumarate reductase
FRDg	Soluble NADH-dependent fumarate reductase
GAPDH	Glyceraldehyde-3-phosphate dehydrogenase
LDH	Lactate dehydrogenase
LEPG	Liver egg parasite load per gram
MRC	Mitochondrial respiratory chain
MV-FRD	Methylviologen-fumarate reductase
NOS	Nitric oxide synthase
PDB	Protein Data Bank
PEPCK	Phosphoenolpyruvate carboxykinase
RNA	Ribonucleic acid
ROS	Reactive oxygen species
SDH	Succinate dehydrogenase
SDS	Sodium dodecyl sulfate
SOD	Superoxide dismutase
TCA	Tricarboxylic acid cycle
WHO	World Health Organization

4.1 Introduction

Neglected diseases are a group of illnesses that predominantly affect populations in low-income countries, often receiving little attention compared to investment in research, prevention, and treatment (Yamey et al. 2018; Boutayeb 2007). These diseases such as toxoplasmosis, Chagas disease, malaria, Leishmaniasis, schistosome, dengue, zika, and chikungunya are examples listed by the World Health Organization (WHO), often associated with conditions of poverty, lack of basic sanitation, and limited access to services of health (Karbalaei and Keikha 2022; Relman and Choffnes 2011).

As a result, these groups of illnesses contribute significantly to damage to the socioeconomic cycle, especially for individuals in conditions of greater vulnerability (Lindoso and Lindoso 2009). Therefore, control and prevention strategies can be considered based on knowledge of the physiology of microorganisms that cause parasitic diseases and the involvement of enzymes vital to their metabolism, which constitute important therapeutic targets for the use of drugs aimed at treating these diseases (Zare et al. 2023).

Among important pharmacological sites, there is complex II, also called the enzyme succinate dehydrogenase (SDH), which plays a crucial role in the tricarboxylic acid cycle (TCA) and in the electron transport chain during cellular respiration. This enzyme is located in the inner mitochondrial membrane, responsible for catalyzing the oxidation of succinate to fumarate by transferring electrons through ubiquinone. Furthermore, SDH is also co-responsible for the formation of a proton gradient, essential for ADP synthesis (Pecsi et al. 2014).

Structurally, SDH is composed of four subunits (SDHA-D or SDH 1-4) that together form a complex that can be divided into two main domains, one being enzymatic, located in the matrix, and one membrane domain. In mammalian cells, SDHA and SDHB have matrix-facing domains that demarcate the succinate binding site and the FAD cofactor and the three 2Fe-se clusters (SDHB). The remaining SDHC and SDHD domains form the membrane anchoring domain and contain the heme prosthetic group (Sun et al. 2005; Silva et al. 2023). It is worth highlighting that enzymatic regulation is sensitive to the cellular energy needs of cells from different organisms, including parasites and other pathogenic microorganisms (Wang et al. 2023).

Furthermore, it is noteworthy that SDH is a vital enzyme in the life cycle of parasites, such as protozoa and viruses. As stated, this enzyme complex directly interferes with the mitochondrial transport chain (Aspuria et al. 2014). As a consequence, the interruption of ATP formation implies limitations in the processes involving energy supply for differentiation, intracellular replication, and the transition between morphological forms and immunological regulation (Mounkoro et al. 2021; Xiao et al. 2022). Therefore, this research aims to explore how selective inhibition of this enzyme can become a specific therapeutic strategy against emerging disease parasites, aiming to interrupt their metabolism and life cycle.

4.2 Succinate Dehydrogenase and Neglected Diseases

Studies related to the importance of succinate dehydrogenase for the discovery of promising compounds in neglected diseases were obtained through literature research through the Web of Science database (https://clarivate.com/). The search was carried out between the months of October and December of the year 2023 and was carried out using the keywords: Schistosoma, Trypanosoma, Leishmania, and succinate dehydrogenase. All articles found were included in the search. The total number of articles found corresponded to 54 articles covering the years 1988–2023. Three diseases were most prevalent, corresponding to Chagas disease (33 articles), followed by schistosomiasis (14 articles) and Leishmaniasis (7 articles). The articles found are described in the following topic.

4.2.1 Trypanosomiasis

The work developed by Bombaça and collaborators talks about the development of synthetic derivatives N, N-di-alkylnaphthoimidazolium derived from β-lapachone (Fig. 4.1) in inhibiting the mitochondrial electron transport of *Trypanosoma cruzi* (Bombaça et al. 2021). In this research, the biological evaluation of the compounds was carried out through the analysis of the direct effect on trypomastigotes, as well as the evaluation of the inhibition of the amastigote form and toxicity analyses. Energy metabolism was assessed using high-resolution oxygraphy and ATP production. The activity of the succinate dehydrogenase enzyme was evaluated by the biochemical test with the epimastigote form, based on the reduction of ferrocytochrome c at 550 nm ($\varepsilon = 19\ mM^{-1}\ cm^{-1}$) and measured in a reaction mixture consisting of 100 mM buffer potassium phosphate (pH 7.4), 150 μM bovine heart cytochrome c, 3 mM succinic acid, and 1 mM KCN.

Molecular docking analyses were carried out with the aim of confirming the results obtained in the evaluation of the inhibitory potential of the compound N4 on mitochondrial physiology, and evaluating a possible affinity with the target succinate dehydrogenase. The software used for the analyses under study corresponded to Gold 5.7 (Cambridge, United Kingdom), with the design and minimization of the structure carried out using DFT (density functional theory) methods with the Becke-3-Lee Yang Parr method (B3LYP) and standard base set 6-31G*, available in Spartan'18 software (Wavefunction, Inc., Irvine, USA). Succinate dehydrogenase (complex II) was obtained by homology on the Swiss Model platform, with the docking results generated in PyMol (Delano Scientific LLC, Schrödinger, New York, USA). The elaboration of a homology model was adopted due to the absence of crystallography. The active site region comprised the binding sites for FAD and ubiquinone. For the ubiquinone site, it was seen that compound N4 can interact mainly at the ubiquinone binding site in mitochondrial complex II through van der Waals forces with the amino acid residues Ile-30, Met-39, Ile-43, Tyr-91, Arg-93,

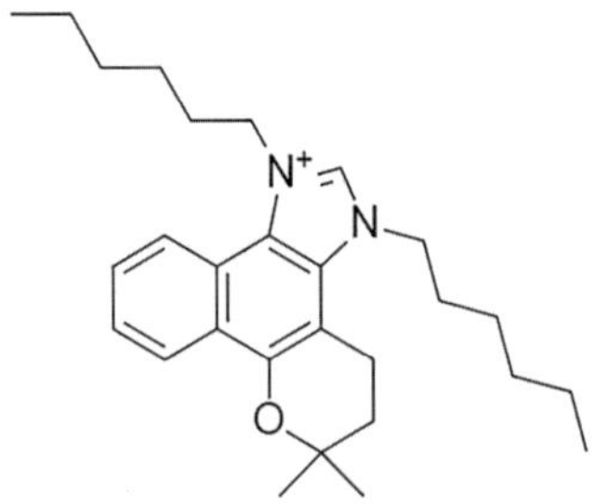

Fig. 4.1 Chemical structure of the compound N4 derived from naphtoimidazole. (Source: Bombaça et al. 2021)

Pro-169, Trp-173, and Ile-218 with a distance of 3.50, 2.30, 3.50, 1.70, 3.00, 3.40, 2.90, and 2.40 Å, respectively. Assessment of affinity for the FAD site was evaluated through molecular docking simulation with the enzyme dihydroorotate dehydrogenasc (DHOD).

The results demonstrate that compound N4 has a strong trypanocidal effect in in vitro tests and propose that this effect is dependent on the mitochondrial production of reactive oxygen species (ROS), involving damage to complexes II-III in the first hours and extensive mitochondrial damage (morphological and functional) after 24 h, differently from what was observed for other naphthoimidazoles. According to molecular docking, as van der Waals forces are the main interactions between N4 and enzymes, the insertion of aliphatic chains into the structure of the naphtoimidazole derivative was shown to have an essential role in improving trypanocidal activity.

The action of sesquiterpene lactones eupatoriopicrin and estafietin (Fig. 4.2) isolated from the *Stevia alpina* Griseb species (Asteraceae) and *Stevia maimarensis* (Hieron.) Cabrera (Asteraceae), in the energy metabolism of *Trypanosoma cruzi* was evaluated by Elso and collaborators (2022).

In this study, the authors aimed to evaluate the possible mechanisms of action against *Trypanosoma cruzi*. Since these compounds have already shown inhibitory activity in previous tests, however, in this work only in vitro tests were carried out and these aimed to evaluate the interaction with hemin, which corresponds to a metoloporphyrin complex that contains iron bound to chloride ions (Luo et al. 2014). This interaction was evaluated through the quantitative evaluation of changes in the Soret absorption band of hemin, another important point evaluated corresponded to the interaction of thiol groups, which was determined by spectroscopic methods through its reaction with 5,5′-dithiobis-2-nitrobenzoate. Spectroscopic analyses were also applied to evaluate the effect on cruzipain activity. Other parameters evaluated consisted of the quantitative and qualitative assessment of sterol

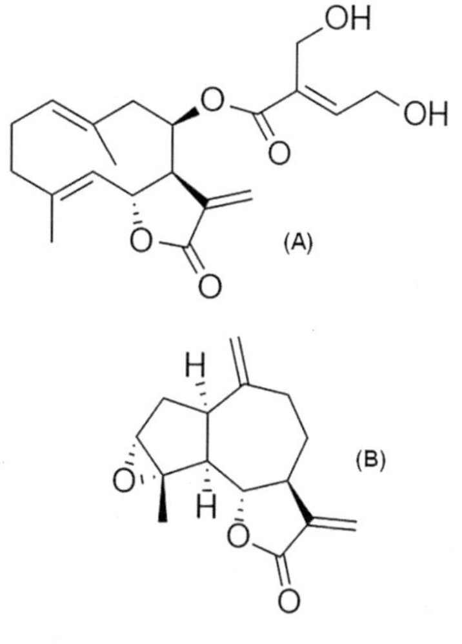

Fig. 4.2 Chemical structure of the sesquiterpene lactones eupatoriopicrin (**a**) and estafietin (**b**). (Source: Elso et al. 2022)

synthesis. Due to the importance of the energetic function for the survival of the parasite mitochondrial functionality was assessed by measuring mitochondrial membrane potential and the activity of the enzymes NADH-cytochrome-c-reductase and succinate-cytochrome c reductase. Furthermore, the state of the antioxidant system was assessed by quantifying the level of free thiols by spectrophotometry, together with the intracellular oxidative state by flow cytometry. Ultrastructural changes were analyzed by transmission electron microscopy.

The results demonstrated that the two compounds belonging to the class of lactonized sesquiterpenes impair the functionality and redox status of the parasite. The compound stafietin caused a greater decrease in the activity of succinate dehydrogenase than eupatoriopicrin, thus affecting the functioning of the respiratory chain and the Krebs cycle. Eupatoriopicrin was responsible for increasing the formation of triglycerides leading to the presence of cytoplasmic lipid droplets. In electron microscopy analyses, it was possible to observe the appearance of large translucent vacuoles in the cytoplasm in the kinetoplast for both compounds.

From this perspective, it can be concluded that both sesquiterpenlactones demonstrated an additive effect on *Trypanosoma cruzi*, supporting the hypothesis that each compound would be acting on different primary targets. Treatment combining eupatoriopicrin and stafietin can be considered a promising pharmacological alternative for the treatment of Chagas disease.

In another work, developed by Negreiros and collaborators the authors hypothesized whether mitochondrial pyruvate transport subunits are essential for the respiration, infectivity, and intracellular replication of Trypanosoma cruzi driven by pyruvate. The studies developed specifically addressed the markings of the heterocomplex formed by mitochondrial pyruvate transporters 1 and 2 (MPC1/MPC2). (Negreiros et al. 2021).

The results demonstrated that endogenous labeling of *T. cruzi* MPC1 (TcMPC1) and TcMPC2 with 3× c-Myc demonstrated that both encoded proteins colocalize with MitoTracker in the mitochondria of epimastigotes. Individual killing of the TcMPC1 and TcMPC2 genes using CRISPR/Cas9 was confirmed by PCR and Southern blot analyses. The assays referring to the epimastigote form TcMPC1-KO and TcMPC2-KO permeabilized by digitonin showed reduced rates of O_2 consumption when pyruvate was used as mitochondrial substrate, while α-ketoglutarate increased their O_2 consumption rates due to an increase in α-ketoglutarate dehydrogenase activity. Defective mitochondrial pyruvate imports resulted in decreased Ca^{2+} uptake. The inhibitors UK5099 and malonate impaired pyruvate-driven oxygen consumption in permeabilized control cells.

Regarding specific results on the involvement of the succinate dehydrogenase enzyme by malonate, it was indicated that pyruvate needs to be converted into succinate to increase respiration. In the TcMPC1-KO and TcMPC2-KO epimastigote form, a small difference in growth was observed in standard or low glucose culture medium. However, the ability of trypomastigotes to infect tissue culture cells and replicate as intracellular amastigotes was diminished in TcMPC-KOs. From this perspective, the authors were able to conclude that MPC1 and MPC2 of *T. cruzi* are essential for cellular respiration in the presence of pyruvate, host cell invasion, and amastigote replication.

This methodology was also visualized in the research developed by Dick and collaborators (2023); however, in this work the authors aimed to evaluate whether iron absorption controls the metabolic change and cell proliferation of *T. cruzi*. To this end, investigations were carried out on mechanisms related to growth and metacyclogenesis, intracellular Fe variations, transferrin endocytosis assays, succinate-cytochrome C oxidoreductase activity, mitochondrial membrane potential, quantification of intracellular ATP, succinate-cytochrome C activity oxidoreductase, glyceraldehyde-3-phosphate dehydrogenase (GAPDH) activity, glucokinase activity, and superoxide dismutase (SOD) activity (Dick et al. 2023).

The tests were carried out using the amastigote, epimastigote, and trypomastigote forms, and at the end of the study it was possible to identify that Fe depletion increased oxidative stress, inhibited mitochondrial function and ATP formation, increased the accumulation of lipids in reservosomes, and inhibited differentiation toward trypomastigotes, with the simultaneous metabolic shift from respiration to glycolysis. Fe depletion contributes to a decrease in the activity of succinate-cytochrome c oxidoreductase, as well as a drop in O_2 consumption in the presence of normal partial pressure of O_2 and lower intracellular ATP, indicating mitochondrial damage, since the impairment of function mitochondrial function appears to be linked to the depletion of Fe protein, especially hemin. With adequate Fe supplementation, the enzyme succinate dehydrogenase is upregulated, as this enzyme is upregulated in the presence of Heme. From this perspective, it can be concluded that the processes modulated by ionic Fe provide energy for the life cycle of *T. cruzi* and the spread of Chagas disease.

Strategies related to genomics and proteomics were adopted by Mattos and collaborators. In this way, the authors aimed to reprogram the metabolism of *T. cruzi* triggered by the interaction of the parasite with the extracellular matrix of the host cell (Mattos et al. 2019). The authors carried out the characterization of the proteome and phosphoproteome of *T. cruzi* after interaction with ECM (MTy), as well as an analysis of the expression of proteins involved in metabolic processes, such as the glycolytic pathway, kinases, proteins related to flagella and microtubules, and proteins associated with the transport and binding of ribonucleic acid (RNA)/deoxyribonucleic acid (DNA), are represented in the set of proteins modified by phosphorylation. The extracellular matrix (ECM) was an important factor in the study, as it was from them that modifications to proteins were carried out through phosphorylation, as well as metabolic switches triggered by this interaction with the ECM were indicated by decreases in the phosphorylation of hexokinase, phosphofructokinase, fructose-2,6-bisphosphatase, phosphoglucomutase, and phosphoglycerate kinase in MTy. Metabolomics analyses were performed by GC-MS derivatization and were also used to evaluate a decrease in pyruvate and lactate and an increase in glucose and succinate contents.

In silico analyses were carried out using phosphopeptides, phosphorylation sites, and number of phosphorus residues using filter tools in Microsoft® Office Excel software. The 303 identified phosphopeptides were used to predict kinase sites using the GPS 2.1 software: improved prediction of kinase-specific phosphorylation sites, following the default and high threshold parameters. After identification of the kinase by the software, only sequences with scores above 20 were selected. For analysis of conservation between phospho-sites of the kinase family, all sequences obtained in the GPS run were selected to construct a sequence logo for each kinase family, using the Weblogo software (//weblogo.berkeley.edu/logo.cgi), following standard parameters.

Among the results observed, incubation of the *T. cruzi* Trypomastigote form with the extracellular matrix (ECM) results in a reduction in the level of protein phosphorylation; this reduction was observed more visibly in enzymes belonging to the kinase classes, while that in the metabolic analysis an increase in the expression of the enzyme succinate lyase was observed. The authors conclude that in addition to changes in the cytoskeleton and activation of the protease, a reprogramming of trypomastigote metabolism is triggered by the interaction of the parasite with the ECM before cell invasion and differentiation into amastigotes, the multiplicative intracellular stage of *T. cruzi* in the host vertebrate.

Metabolomics analyses to propose the origin of certain metabolites such as succinate were carried out by Johnston and collaborators (2019). The main objective of the study consisted in mapping five amino acids derived from glutamine, proline, methionine, cysteine, and arginine using heavy-labeled precursors and LC-MS in the bloodstream form of Trypanosoma brucei through the use of substrates labeled with $U^{-\,13\,C\ e\ LC\text{-}MS}$ (Johnston et al. 2019).

The results demonstrated that the aforementioned amino acids play important roles in the parasite's energy metabolism, as mentioned below: glutamine is used by the blood form of *Trypanosoma brucei* in large quantities, and it is proposed that it serves as a donor of amino groups, 2-oxoglutarate and 2-oxoglutaramate, as well as some glutamate, and that their expression is dependent on the amount of blood forms; proline in parasites is an important source of energy, as it is a pro-cyclic carbon source, and guarantees the viability of these organisms; methionine has a wide range of functions, being an important source of methyl groups when converted into S-adenosylmethionine (AdoMet) and also of aminopropyl groups in the growth of polyamine chains after production of the decarboxylated form; cysteine is responsible for a variety of functions, but its participation in the synthesis of important mediators stands out, such as: it is essential for producing glutathione, trypanothione, and coenzyme A (CoA); arginine stands out as a precursor for the synthesis of polyamines, producing ornithine *via* arginase. Additionally, it generates citrulline and nitric oxide through nitric oxide synthase (NOS).

In general, it was observed that all unexpected metabolites were absent in the fresh medium samples, as well as that only a very low amount of proline is used by the parasite, indicating that the enzymes are present, but only minimally active, perhaps allowing the cells are more resistant to changes in their environment.

Koo and Garg (2019) carried out a literature review on the topic of metabolic programming of macrophage functions and pathogen control. The review shows that macrophages are important markers of the immune system. Succinate, lactate, citrate, arginosuccinate, malate, and aspartate are metabolic markers of classic M1 macrophages (similar to Th1) and can be obtained via the Warburg Glycolysis pathway, as well as via the pentose phosphate pathway, in addition to the synthesis of fatty acids and the aspartate-arginosuccinate derivation. As it is also mentioned in the review, the inhibition of the succinate dehydrogenase enzyme is important in the activation of pro-inflammatory mediators due to the accumulation of succinate and its consequent oxidation generating the accumulation of ROS.

In Table 4.1, other articles that address succinate dehydrogenase as an important target for diseases caused by *Trypanosomas* are cited.

4.2.2 Schistosomiasis

Hamed and collaborators studied possibilities for combating resistance to *Schistosoma* by proposing a vaccine, as the presence of common antigenic fractions between *Schistosoma mansoni* and its intermediate host provides a source for the preparation of a suitable vaccine. Therefore, the objective of this work was to evaluate the nucleoprotein extracted from susceptible or resistant snails for protection against *schistosomiasis* (Hamed et al. 2010).

The methodology used by the authors included the creation of a vaccination schedule that included the administration of a subcutaneous injection of 50 μg of protein from each antigen followed by a new inoculation 15 days later. The analyses consisted of enzymatic tests of markers for different cellular organelles such as succinate dehydrogenase, lactate dehydrogenase (LDH), glucose-6-phosphatase, acid phosphatase, and 5′-nucleotidase. Subsequently, energy parameters were obtained, such as adenosine triphosphate (ATP), adenosine diphosphate (ADP), adenosine monophosphate (AMP), phosphate potentials, inorganic phosphate, amino acids, and isoenzymes (LDH). The work was expanded to record worm and egg counts, determination of oograms in the liver and intestine, and the histopathological pattern of the liver.

The results obtained demonstrate that the nucleoprotein of susceptible snails showed a reduction in worm and egg counts by 70.96% and 51.31%, respectively, while the nucleoprotein of resistant snails showed reductions of 9.67% and 16.77%, respectively. From this perspective, it is possible to identify that the nucleoprotein from susceptible snails was more effective in protecting against schistosomiasis.

Hamed (2011) continued its work on vaccine discovery for *S. mansoni*, aiming to evaluate the role of detergents in enhancing the effect of *S. mansoni* tegument antigens (TA) against challenge infection. The work methodology consisted of using two detergents Triton X-100 (TX-100) and sodium dodecyl sulfate (SDS), which were responsible for the extraction of tegument proteins. Vaccination of mice occurred at a frequency of up to two doses of each preparation (100 μg of protein) at time intervals 15 days before infection. The evaluation was carried out based on mechanisms related to mitochondrial energy metabolism, in a similar way to what was carried out in the first experiment, but this time it aimed to estimate specific

Table 4.1 Other articles that address succinate dehydrogenase as an important target for diseases caused by *Trypanosoma*

Compound	Target	Methodology	Study	Result	References
In the work, the objective did not include the study of specific compounds, but rather the target.	Dihydroorotate dehydrogenase and succinate dehydrogenase.	Analysis of the molecular properties of dihydroorotate dehydrogenase (DHOD), the fourth enzyme of the pyrimidine biosynthetic pathway, as a soluble fumarate reductase (FRD) was carried out.	In vitro study	Dihydroorotate dehydrogenase from *T. cruzi* has methylviologen-fumarate reductase (MV-FRD) activity. Furthermore, *T. cruzi* DHOD was found to catalyze the transfer of electrons from dihydroorotate to fumarate by a Bi-Bi ping-pong mechanism.	Takashima et al. (2002)
Bibliographic review article that the objective did not include the study of specific compounds, but rather the target.	NADH-fumarate reductase and succinate dehydrogenase.	A literature review was carried out which states that the activity of NADH-fumarate reductase in trypanosomatids can be catalyzed by succinate dehydrogenase.	Literature review	The results reported in this study demonstrate that the two activities can be easily separated without any loss in either activity, suggesting that fumarate reductase and succinate dehydrogenase are separate enzymes.	Christmas and Turrens (2000)
Article that the objective did not include the study of specific compounds, but rather the target.	NADH dehydrogenase and NADH oxidase.	Bibliographical review that addresses the intermediary metabolism in *T. cruzi* and the protein catabolism caused by the epimasgota forms of *T. cruzi*.	Literature review	The article addresses the catabolism carried out by *T. cruzi* in secondary metabolism and attributes this function to the target phosphoenolpyruvate carboxykinase (PEPCK).	Cazzulo (1994)
Article that the objective did not include the study of specific compounds, but rather the target.	Aspartate aminotransfers from *T. cruzi*.	Study that evaluates the influence of catalytic parameters on the encoding of the aspartate aminotransferase enzymes (ASATs) of *T. cruzi* and their influence on secondary metabolism.	In vitro study	cASAT is specifically induced in mammalian stages, and the succinate excreted by amastigotes more likely represents a secondary product of an at least partially operative Krebs cycle than an end product of the glycosomal Krebs cycle.	Marciano et al. (2008)

Article that the objective did not include the study of specific compounds, but rather the target.	Dihydroorotate hydrogenase. Succinate dehydrogenase and its products were mentioned in the article.	Evaluation of the kinetic mechanism and catalysis of the enzyme dihydroorotate dehydrogenase by isothermal titration calorimetry.	In vitro study	The results demonstrated that the entropy value was unfavorable and that the enzyme presented saturation. The study demonstrated that thermodynamic parameters improve our understanding of enzyme mechanisms.	Cheleski et al. (2010)
Article that the objective did not include the study of specific compounds, but rather the target.	Substrates for oxygen (O_2) consumption rates. Succinate dehydrogenase and its products were mentioned in the article.	The contribution of different substrates to O_2 consumption rates along the growth curve of *T. cruzi epimastigotes* (Tulahuen 2 and Y strains) was evaluated.	In vitro study	O_2 consumption rates were higher in the late stationary phase, not due to an increase in succinate dehydrogenase activity. Antimycin A and cyanide did not completely inhibit the mitochondrial respiratory chain (MRC). Malonate at 10 or 25 mM was not a potent inhibitor of complex II. Comparing complexes II and III, the first appears to be the main site of H_2O_2 release.	Silva et al. (2011)
Organometallic complexes of osmium (III).	Succinate dehydrogenase, malate dehydrogenase, and pyruvate kinase.	The article mainly addresses the inhibition of forms of *Leishmania donovani* and *T. cruzi*, but succinate dehydrogenase and its products were mentioned in the article.	In vitro study	The metal–drug complexes in the study clearly inhibited the synthesis of DNA, RNA, and proteins, as well as the enzymatic activities of succinate dehydrogenase, malate dehydrogenase, and pyruvate kinase.	Castilla et al. (1996)
Article that the objective did not include the study of specific compounds, but rather the target.	Theoretical study on the enzyme dihydroorotate dehydrogenase. Succinate dehydrogenase and its products were mentioned in the article.	Molecular dynamics study in order to evaluate the physicochemical processes of the interconversion of dihydroorotate into orotate during the catalysis of the enzyme dihydroorotate dehydrogenase.	Theoretical study	The results demonstrate that there is variation in the mechanism between the different types of class 1A dihydroorotate dehydrogenase. These differences arise from the phylogenetic classification of the classes of enzymes that secrete different biochemical properties, which results in multiple mechanisms.	de Farias Silva et al. (2013)

(continued)

Table 4.1 (continued)

Compound	Target	Methodology	Study	Result	References
Heterocycles containing N-oxide.	Inhibition of the epimastigote form of *T. cruzi*. Succinate dehydrogenase and its products were mentioned in the article.	Inhibition assays for the epimastigote form of *T. cruzi* and evaluation of the metabolic profile of heterocycles containing N-oxide.	In vitro study	Compounds belonging to the quinoxaline class and nitrofuran Nf1 showed significant inhibition of mitochondrial dehydrogenase, but with different dose–response profiles.	Benitez et al. (2014)
Article that the objective did not include the study of specific compounds, but rather the target.	Targets involved in glucose metabolism to maintain the life of *T. cruzi*. Succinate dehydrogenase and its products were mentioned in the article.	Bibliographic review on the targets involved in pathways related to glucose metabolism.	Literature review	The article demonstrates that *T. cruzi* metabolizes glucose through two main pathways that correspond to glycolysis and the pentose phosphate pathway. Glucose uptake is carried out through a facilitated transporter and its catabolism is carried out via the glycolytic pathway, resulting in the excretion of reduced products such as succinate and L-alanine, even in the presence of oxygen. All the enzymes of the pentose phosphate pathways and gluconeogenic enzymes are present in the parasite's life cycle.	Maugeri et al. (2011)

Article that the objective did not include the study of specific compounds, but rather the target.	Citrate synthase, NADP-isocitrate dehydrogenase, and succinate dehydrogenase.	The article aimed to highlight the energetic activity in mitochondria of the respective enzymes: NADP-isocitrate dehydrogenase and succinate dehydrogenase as well as demonstrate their kinetic compounding in the two forms of development of *T. cruzi* in the form of insect vector, epimastigotes, and infectious metacyclic trypomastigotes.	In vitro study	The results observed in this work showed a higher mitochondrial metabolism in metacyclic forms, due to the high activity rates of metacyclic citrate synthase, isocitrate dehydrogenase, and succinate dehydrogenase, with the enzyme succinate dehydrogenase presenting specific activities of 8.2 and 39.1 μm/mg for epimastigotes and metacyclic trypomastigotes. In this way, it can be understood that the differentiation between epimastigotes and current trypomastigotes is influenced by changes in the metabolic characteristics of both differentiated forms of *T. cruzi*.	Adroher et al. (1988)
Article that the objective did not include the study of specific compounds, but rather the target.	NADH: cytochrome c reductase, being succinate dehydrogenase and its products were mentioned in the article.	The article talks about NADH oxidation through a pathway sensitive to rotenone and antimycin in the mitochondria of pro-cyclic *T. brucei*.	In vitro study	NADH is preferentially oxidized by a rotenone-sensitive NADH: ubiquione oxidoreductase; this oxidation process can be carried out by the enzyme fumarate reductase present in the mitochondria of *T. brucei*.	Beattie et al. (1994)
Article that the objective did not include the study of specific compounds, but rather the target.	Dihydroorotate dehydrogenase, being succinate dehydrogenase and its products were mentioned in the article.	The article aims to study the structures of the enzyme dihydroorotate dehydrogenase from *T. cruzi* complexed to certain substrates and products, as well as demonstrate characteristics of the atomic resolution of the mechanisms of dihydroorotate oxidation and fumarate reduction.	In vitro study	The results demonstrated that the Cys 130 residue is considered the basis of the active site for enzymes in this category, being well located to abstract a proton from the C5 dihydroorotate and transferring it to external water molecules. Fumarate has a twisted conformation, leading to partial charge separation, and because of this it accepts a proton from Cys130 and C3 (delta+) and a reduced hydride ion.	Inaoka et al. (2008)

(continued)

Table 4.1 (continued)

Compound	Target	Methodology	Study	Result	References
Article that the objective did not include the study of specific compounds, but rather the target.	NADH: cytochrome c oxidase, being succinate dehydrogenase and its products were mentioned in the article.	The article addresses the assessment of mitochondrial bioenergetics and the redox state after alteration of subunit I of the mitochondrial complex corresponding to the genes: ND4, ND5, and ND7 when compromised.	In vitro study	The results demonstrated that no correlation could be established between mitochondrial membrane potentials and ND deletions and that significant differences not associated with ND deletions were observed between parasite isolates, demonstrating that these mutations are not important for controlling production of oxidants. Therefore, it can be seen that complex I has a limited function in *T. cruzi*.	César Carranza et al. (2009)

Article that the objective did not include the study of specific compounds, but rather the target.	Complex II: succinate, ubiquinone reductase.	The article aims to study the mitochondrial complex II egg isolated from *T. cruzi.*	In vitro study	The authors isolated and characterized the mitochondrial Complex II of *T.* cruzi and found that it has a novel organization composed of 12 subunits, six hydrophilic and six hydrophobic, including a heterodimeric iron-sulfur subunit (SDH2N/SDH2C), different from the canonical form found in mammals and bacteria. The enzyme demonstrated low affinity for substrates and inhibitors, reduced catalytic activity and binding to protoheme IX, although lacking the typical histidine ligand in one of the subunits. This unusual structural organization is conserved in other trypanosomatids such as *T. brucei* and Leishmania major, suggesting an adaptive role to the parasitic environment. Due to these unique characteristics, Complex II of *T. cruzi* stands out as a promising target for the development of new chemotherapeutics against diseases such as Chagas disease and Leishmaniasis.	Morales et al. (2009)

(continued)

Table 4.1 (continued)

Compound	Target	Methodology	Study	Result	References
Article that the objective did not include the study of specific compounds, but rather the target.	Soluble NADH-dependent fumarate reductase (FRDg), being succinate dehydrogenase and its products were mentioned in the article.	The article talks about the mechanism of NADH-dependent mitochondrial fumarate reductase involved in the production of succinate excreted by the pro-cyclic *T. brucei*.	In vitro study	Analysis of the resulting mutant cell lines showed that FRDm1 is responsible for 30% of cellular NADH-FRD activity, and that it contributes to the production of between 14% and 44% of the succinate excreted by wild-type cells. Furthermore, FRDg and FRDm1 together account for the total activity of NADH-FRD in procyclics. Thus, the authors propose that FRDm1 is involved in maintaining redox balance in mitochondria, as proposed for the ancestral soluble FRD presumably present in primitive anaerobic cells.	Coustou et al. (2005)
Article that the objective did not include the study of specific compounds, but rather the target.	Soluble NADH-dependent fumarate reductase (FRDg), being succinate dehydrogenase and its products were mentioned in the article.	The article addresses the extramitochondrial localization of the enzyme soluble NADH-dependent fumarate reductase (FRDg) in trypanosomatides	In vitro study	Fumarate reductase was released at digitonin concentrations lower than those required to release isocitrate dehydrogenase or citrate synthase, and these results suggest that in the three trypanosomatid species, the fumarate reductase enzyme is, at least in part, located outside the mitochondrial matrix.	Denicola et al. (2002)

Article that the objective did not include the study of specific compounds, but rather the target.	Dihydrolipoamide dehydrogenase (DLDH), being succinate dehydrogenase and its products were mentioned in the article.	The article addresses the process of reversible inactivation of dihydrolipoamide dehydrogenase by mitochondrial hydrogen peroxide.	In vitro study	An important finding refers that DLDH could only be inactivated by ROS derived from complex III, but not from complex I; and the associated loss of activity due to inactivation could be restored by cysteine and glutathione, indicating that the oxidative inactivation of DLDH by complex III-derived ROS was a reversible process.	Yan et al. (2013)
Article that the objective did not include the study of specific compounds, but rather the target.	Mechanisms of aerobic glucose fermentation in trypanosomatides, succinate dehydrogenase, and its products were mentioned in the article.	The study addresses the metabolic pathways responsible for the mechanisms of aerobic glucose fermentation.	Literature review	The results show that glucose consumption is characterized by the excretion of reduced products such as succinate, pyruvate, ethanol, L-alanine, or lactate (depending on the species) not only in anaerobic conditions, but also in aerobic conditions, but it is also important to mention that the production of fermentative products appears to be due to a relative inefficiency of the respiratory chain, which lacks NADH dehydrogenase and the first phosphorylation site and preferentially uses succinate as a substrate.	Cazzulo (1992)
Peroxynitrite anion ($ONOO^-$)	NADH: cytochrome c oxidase and other enzymes involved in energy metabolism and cellular respiration, with succinate dehydrogenase and its products being mentioned in the article.	The authors studied mechanisms of inactivation of enzymes involved in the metabolic reactions of *T. cruzi* through the peroxynitrite anion.	In vitro study	The results showed that peroxynitrite-mediated inactivation of succinate dehydrogenase and fumarate reductase may be a key mechanism of macrophage-mediated cytotoxicity for *T. cruzi*, through the inhibition of the parasite's energy metabolism.	Rubbo et al. (1994)

metabolic pathways, including the Krebs cycle (via succinate dehydrogenase); glycolysis (lactate dehydrogenase); gluconeogenesis (glucose-6-phosphatase); hydrolytic enzymes (acid phosphatase); and nucleic acid catabolic enzymes (5′-nucleotidase). Serum protein profiles, worm burden, egg count, and spleen and liver relative weight were also investigated. Another point that followed the same methodology adopted in his first work was that the work extended to the study of the histopathological picture of the liver, including granuloma count and total area of infection. The methodology used for this purpose included vaccinated post-challenged mice, and these showed an improvement in selected biochemical parameters and a reduction in worm and egg counts ($P \leq 0.0001$), as well as an improvement in the histological characteristics of the liver.

The results demonstrated that vaccination with tegumentary schistosome antigens protected against challenge infections to varying degrees. This protection was achieved through reductions in worm counts, egg burden, granuloma size and frequency, as well as in certain metabolic parameters that revealed the tendency of cells to restore their physiological function and assist in egg-mediated immunological killing, improving the hepatic histopathological response. Detergents played a significant role in enhancing the protective effect of selected antigens, and TX-100 was more effective than the others tested. Further work is still needed in the molecular identification of proteins elicited by TX-100 for possible protection against schistosomiasis.

The repositioning of drugs to discover new therapeutic alternatives for schistosomiasis was addressed by Calixto and collaborators (Calixto et al. 2018). The methodology was developed by searching for targets in genetic libraries such as the TDR Targets Database version 5.0 (//tdrtargets.org/); subsequently a search for approved drugs for the targets under study was carried out using the DrugBank platform version 4.0, then a search was carried out for the interaction between medicines identified with other protein targets and the possible side effects of this relationship. This activity was carried out on the Web tool promiscuous version 3.0 (//bioinformatics.charite.de/promiscuous/), which contains information about 25,000 medications, 1100 different side effects, and 12,000 protein targets. The last step involved using the PubMed database (http://www.ncbi.nlm.nih.gov/pubmed) to assess whether the identified drugs had already been tested against *Schistosoma* spp.

The results obtained demonstrated that 59 target proteins were identified, of which 18 had one or more approved drugs. The results identified 20 potential drugs for treating schistosomiasis, all approved for use in humans. Of the 20 drugs classified as having potential for treatment, 8 of them were classified as having high schistosomicidal potential, which correspond to: thiabendazole, albendazole, fomepizole, desflurane, isoflurane, alendronate, and diazoxide. However, it is important to highlight that there are no guarantees that the identified compounds will act on energy metabolism targets of *S. mansoni*. Therefore, the authors of this study presented the prospect of carrying out in vitro tests to confirm the effect of thiabendazole on the target of energy metabolism. The results demonstrate that this drug acts against SDH, an essential enzyme for the Krebs cycle. Thus, the importance of the succinate dehydrogenase target is mentioned as an important target for schistosomiasis therapy.

Wilson et al. (2019) carried out a bibliographic review on the metabolic programming of macrophages and its implications in granulomatous disease, where the role of PPAR-γ signaling and lipid metabolism was discussed, which are currently the best represented metabolic pathways in this context, as well as the dysregulated lipid metabolism in the progression of granulomatous disease. This work aimed to highlight the metabolic mechanisms of granuloma immune cell fate and generate research questions for the identification of potential therapeutic targets in the future.

The authors demonstrated the involvement of granuloma in schistosomiasis, with this organism being associated with specific metabolic changes in hepatic macrophages, where granuloma formation around schistosome eggs was associated with differential metabolism of lipids and cholesterol. Overall, *S. mansoni* is characterized by modifications in the PPAR-γ network of lipid metabolism, during which uptake of oxLDL cholesterol via ApoC1 and ApoC3 is directly suppressed, while uptake via CD14 and CD36 can be increased due to signaling PPAR-γ.

Zhu et al. (2023) conducted a literature review on trained immunity as an important route to obtaining vaccines, as recently emerging trained immunity (also called innate immunological memory) has been increasingly recognized as providing a new perspective for the development of vaccines targeting immunity innate (Zhu et al. 2023). In this work, the authors demonstrated that trained immunity carries out metabolic reprogramming and can influence succinate levels due to lipid metabolism.

In Table 4.2, other articles that address succinate dehydrogenase as an important target for diseases caused by schistosomes are cited.

4.2.3 Leishmaniasis

Mesquita and collaborators (2015) evaluated the leishmanicidal and cytotoxic effects of inhibitors of the metabolic pathway of microorganisms. In these studies, the leishmanicidal activity of inhibitors of the metabolic pathway enzymes (IDs) of dihydroorotate dehydrogenase (DHODH), glyceraldehyde 3-phosphate dehydrogenase and cruzine-cysteine protease from *T. cruzi* and scitalone dehydratase, adenosine deaminase, succinate dehydrogenase complex II, and hydroxynaphthalene reductase from *S. cerevisiae* was carried out in extracellular promastigotes and amastigotes of *Leishmania amazonensis* within macrophages (de Mesquita Barbosa et al. 2015).

The compounds are 17 synthetic derivatives and the results demonstrated that compounds ID71, ID130, and ID195 (Fig. 4.3), which are inhibitors of *T. cruzi* DHODH, were toxic against *L. amazonensis* promastigotes. However, ID71 and ID130 were not selective for the parasite over the host as they affect macrophage survival. The most promising compound was ID195, and it was toxic against promastigotes and intracellular amastigotes; however, when compared to amphotericin B, the compound ID195 was less toxic against the parasite. Compound ID195 decreased the infection rate of macrophages infected by *L. amazonensis* (IC_{50},

Table 4.2 Other articles that address succinate dehydrogenase as an important target for diseases caused by schistosomes

Compound	Target	Methodology	Study	Result	References
In the work, the objective did not include the study of specific compounds, but rather the target.	Succinate dehydrogenase	The study aimed to evaluate whether *S. mansoni* sporocysts contain rhodoquinone and produce succinate by reducing fumarate.	In vitro study	The results demonstrated that complex II of the electron transport chain in schistosomes has the kinetic properties of a dedicated fumarate reductase rather than those of a succinate dehydrogenase. The presence of such an enzyme, together with the substantial amounts of rhodoquinone, shows that in *S. mansoni* sporocysts succinate is produced through the reduction of fumarate. Therefore, the energy metabolism of schistosomes does not differ in principle from most other parasitic helminths, which are known to rely heavily on fumarate reduction.	Van Hellemond et al. (1997)
In the work, the objective did not include the study of specific compounds, but rather the target.	Succinate dehydrogenase	The authors aimed to study the cloning, expression, and evaluation of the protective immunity of the complete cDA that encodes the enzyme succinate dehydrogenase from the organism *Schistosoma japonicum*.	In vitro study	The protein obtained by cloning showed 32 kD and high levels of specific antibodies and a significant reduction in the parasite LEPG, FEPG, and intrauterine eggs. Therefore, it can be concluded that the model developed can be a promising candidate for the production of an antischistosomal vaccine.	Yu et al. (2007)
Citrus reticulata compounds.	SDH; LDH and its isoenzymes; G-6-Pase; AP and 5′-nucleotidase. Liver function enzymes; AST	The work aimed to evaluate the antischistosomal potential of compounds from the *Citrus reticulata* plant in the treatment of Miracidium.	In vitro study	The results showed a marked reduction in the activities of the enzymes SDH, LDH, AST, and ALT and a significant increase in G-6-Pase, AP, 5′-nucleotidase, and ALP after *S. mansoni* infection. A notable change in LDH subunits was also observed. Treatment with *C. reticulata* or Mirazid improved all of the above enzyme activities with a notable reduction in egg count and worm load.	Hamed and Hetta (2005)

Organic acids (pyruvate, lactate, succinate, fumarate, malate, acetate, propionate, acetoacetate, and ß-hydroxybutyrate).	Succinate dehydrogenase was one of the enzymes mentioned in the article.	The article aimed to evaluate the profile of organic acid concentrations in the digestive gland and hemolymph of *Biomphalaria glabrata* under estivation.	In vitro study	After a 7-day period of estivation, there was a significant increase in tissue levels of lactate, succinate, malate, and acetate compared to non-estivating snails. After 14 days of estivation, lactate and acetate levels were also significantly elevated. Pyruvate and acetate concentrations in the hemolymph increased significantly after 7 days and acetate concentrations continued to increase significantly up to 14 days of estivation. The other organic acids studied, such as acetoacetate and ß-hydroxybutyrate from ketone bodies or the volatile acid propionate, did not accumulate. Their tissue concentrations, however, increased on the seventh day of estivation and reached normal levels 2 weeks after estivation for some of them. This shows that consideration must be given to how reduced metabolism can be treated under aerobic conditions and what role anaerobic pathways can play in both energy formation and redox balance processes.	Bezerra et al. (1999)
Perirredoxin and albumin.	Enzymes of redox metabolism, with succinate dehydrogenase being one of the enzymes mentioned in the article.	The study aimed to evaluate redox balance mechanisms in *S. mansoni* and evaluate whether they depend on peroxiredoxins and albumin as well as whether they imply peroxiredoxins as new drug targets.	In vitro study	The results demonstrated an increase in lipid peroxides and the generation of several oxidized proteins, and it was also observed that schistosome albumin was induced by oxidative stress. Therefore, peroxiredoxins are essential proteins of the parasite and potential targets for the treatment of schistosomiasis.	Sayed et al. (2006)

(continued)

Table 4.2 (continued)

Compound	Target	Methodology	Study	Result	References
Various compounds present in plant extracts.	The mechanism studied was based on the inhibition of the parasite and succinate dehydrogenase was one of the targets under study.	The study aimed to evaluate the potential of natural compounds as potential therapeutic agents for *Biomphalaria alexandrina*.	In vitro study	The results demonstrated that the highest molluscicidal potency was recorded for *Euphorbia splendens*, *Atriplex stylosa*, and *Guayacum officinalis*. It was also observed that the activities of glycogen phosphorylase, SDH, glucose-6-phosphatase, and AchE in snail tissue homogenate were reduced ($P < 0.001$) in response to treatment with methanolic plant extract, while G-6-PD activity increased ($P < 0.001$). It was concluded that the application of LC_{25} of methanolic extracts of *E. splendens*, *A. stylosa*, and *G. officinalis* can be useful in the control of snails, as it interferes with the biology and physiology of snails.	

LDH lactate dehydrogenase; SDH: succinate dehydrogenase, *AChE* acetylcholinesterase, *AST* aspartate aminotransferase, *G-6-PD* glucose-6-phosphate dehydrogenase, *G-6-Pase* glucose-6-phosphatase, *AP* acid phosphatase, *LEPG* load of hepatic eggs per gram, *FEPG* fecal eggs per gram

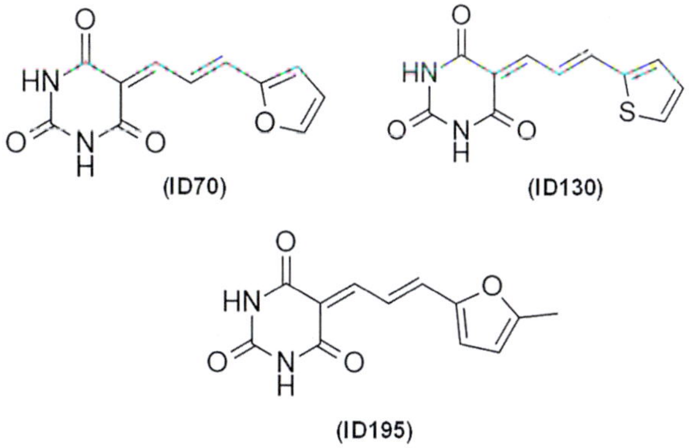

Fig. 4.3 Chemical structure of compounds ID70, ID130, and ID195. (Source: Mesquita and Collaborators 2015)

48.01 ± 3.20 μM) at concentrations that did not significantly affect macrophage survival (IC_{50}, 119 ± 4.12 μM). Furthermore, it was evidenced that a significant decrease in the macrophage infection rate is obtained with concentrations of the compound ID195 (IC_{50}, 48.01 ± 3.20 μM) that are lower than those necessary to inhibit promastigote viability (IC_{50}, 75.13 ± 2.70 μM), suggesting that the ID195 compound is transported across the plasma membrane of macrophages and acts on amastigotes.

8-aminoquinoline analogues of Sitamaquina were designed by Carvalho and collaborators, to evaluate their influence on the oxidative stress caused in the promastigote forms of *Leishmania donovani* through the enzyme succinate dehydrogenase. This was an in vitro study that evaluated enzyme activity and oxygen consumption. The structure of the compound is not included in the article (Carvalho et al. 2011). The results showed that the designed compound causes a dose-dependent inhibition of complex II (succinate dehydrogenase) of the respiratory chain in digitonin-permeabilized promastigotes, along with a drop in intracellular ATP levels and a decrease in mitochondrial electrochemical potential.

In Table 4.3, other articles that address succinate dehydrogenase as an important target for diseases caused by *schistosomes* are cited.

4.3 Discussion

The enzyme succinate dehydrogenase is also known as complex II of the electron transport chain and is part of the mitochondrial respiratory complexes, until recent periods being one of the least previously studied (Goetzman et al. 2023). Recently, there has been a growing number of studies that address this target, mainly focused on understanding the pathogenesis of processes caused by mutations (Rutter et al. 2010; Al Rasheed and Tarjan 2018). This enzyme is reported to be involved in two important pathways of mitochondrial energy metabolism, corresponding to the citric acid cycle and the respiratory electron transfer chain (Martínez-Reyes and

Table 4.3 Other articles that address succinate dehydrogenase as an important target for diseases caused by *schistosomes*

Compound	Target	Methodology	Study	Result	References
In the work, the objective did not include the study of specific compounds, but rather the target	Dihydroorotate dehydrogenase was the main target, with the enzyme succinate dehydrogenase also being addressed.	Theoretical and experimental study that aimed to study the mechanism of dihydroorotate dehydrogenase from *Leishmania major* based on steady-state and pre-stationary kinetic analysis.	In vitro and computational study	The results demonstrate that a hyperbolic concentration of dihydroorotate as well as the release of orotate are established as important steps in the rate of general catalysis and link these events to mechanistic differences between the individual class 1A and class 2 half-reactions to be explored for the development of selective inhibitors.	Reis et al. (2016)
Kielmeyera coriacea extract	Energy metabolism enzymes, with succinate dehydrogenase being an enzyme under study.	The study aimed to evaluate the inhibitory action of *Kielmeyera coriacea* extract on energy metabolism enzymes.	In vitro and in vivo study	The results show that *Kielmeyera coriacea* extract impairs hepatic energy metabolism by acting as a mitochondrial uncoupler and inhibitor of enzymatic activities linked to the respiratory chain.	Zagoto et al. (2006)
Paromomycin	Enzymes of mitochondrial metabolism, including succinate dehydrogenase.	The study evaluated the interference of paromomycin on mitochondrial activity.	In vitro study	The results showed that when added to promastigote cultures for 24–72 h at 150–200 μM (=LC_{50}), paromomycin doubled the generation time, inhibited respiration, and reduced its associated electrical potential difference across mitochondrial membranes. Therefore, it is understood that paromomycin acts on the respiratory chain and has an effect on the cell's energy metabolism.	Maarouf et al. (1997)

Chandel 2020). In the citric acid cycle, this target is responsible for oxidizing succinate into fumarate (Hägerhäll 1997; Bouillaud 2023).

The enzyme succinate dehydrogenase is presented in eukaryotic form and is homologous in structure to an enzyme that catalyzes the reverse reaction during anaerobic respiration in bacteria, which is called fumarate dehydrogenase (Maklashina et al. 2006; Sharma et al. 2019). In its eukaryotic form, this enzyme has four subunits encoded by the nuclear genome (Sun et al. 2005; Yankovskaya et al. 2003; Huang et al. 2019). It is worth highlighting that the succinate dehydrogenase complex is one of the only complexes that does not present subunits encoded by the mitochondrial genome, as well as being the only respiratory complex that does not have the ability to pump protons through the IM during the occurrence of its catalytic cycle (Bénit et al. 2022).

Regarding the physicochemical characteristics, succinate dehydrogenase has two main regions that consist of: a hydrophilic chain which is projected into the matrix compartment, which is formed by two subunits that form the catalytic core (Sdh1, Sdh2 in yeast and SdhA, SdhB in mammals) and a hydrophobic chain that is embedded in the Im with a short segment that projects into the soluble intermembrane space (IMS), and the catalytic activity of the succinate dehydrogenase complex is modulated by phosphorylation and post-translational acetylation, as well as by inhibition of the active site (Zhou et al. 2021; Kregiel 2012; Oyedotun and Lemire 2004).

In the study of diseases that affect humans that are related to the succinate dehydrogenase complex, it suggests that the enzyme not only plays a central role in the Krebs cycle and the respiratory chain but also differs from other mitochondrial dehydrogenases, thanks to its exclusive redox properties (Dröse 2013). Along with ubiquinone, the enzyme succinate dehydrogenase represents a crucial antioxidant enzyme in mitochondria, controlling the superoxide scavenging activity of the cytoplasmic reticulum (Bénit et al. 2022; Fujii et al. 2022). This characteristic directly culminates in the occurrence of mutations in this multifunctional enzyme (Bassal 2023). This proposed mode of action related to oxidative stress justifies succinate dehydrogenase as an important target in the pathogenesis of neglected diseases, as well as its action on worms such as *C. elegans*, and the consequent mutations (Rustin et al. 2002; Goncalves et al. 2020).

A total 53 articles were found that dealt with the enzyme succinate dehydrogenase and neglected diseases; of this total, 33 focused on trypanosomiasis, 14 on schistosomiasis, and 6 on Leishmaniasis. The articles included in vitro studies on energy metabolism and review studies. In silico studies were observed, however at a low frequency, and these, when present, brought molecular docking simulations, as was the case with the work developed by Bombaça and collaborators (Bombaça et al. 2021). When evaluating the antichagasic activity of naphtoimidazole derivatives, it is worth noting that despite the reluctance of the succinate dehydrogenase enzyme in mitochondrial redox metabolism, there is no 3D crystallographic structure available for this target in the Protein Data Bank (PDB) library (https://www.rcsb.org/) for the organism *Trypanosoma cruzi*; in this way the authors obtained the model by homology.

In the PDB library it is possible to find the crystallographic structure for the organism *T. cruzi* referring to dihydroorotate dehydrogenase in complex with succinate (PDB ID: 2DJL) (Pinheiro et al. 2008). For Leishmania species, crystallographic structures are also found referring only to dihydroorotate dehydrogenase and pteridine reductase. The same situation repeats itself when approaching the organism *S. mansoni*. The other computational methodology addressed referred to the genomics and proteomics topics covered in the article by Johnston et al. (2019), where gene bases and metabolome analysis were used.

Another important topic observed was that around ten studies addressed the screening of new inhibitors, with the most relevant and most current ones corresponding to: evaluation of the antichagasic activity of naphtoimidazole derivatives, in the study developed by Bombaça and collaborators (Bombaça et al. 2021), as well as in the evaluation of the antichagasic activity of sesquiterpene lactones in the study developed by Elso and collaborators (Elso et al. 2022) and in the evaluation of the antileishmanicidal potential of hydroxynaphthalene derivatives in the study developed by Mesquita and collaborators (2015). Therefore, it is worth noting that the chemical groups mentioned are important groups in maintaining cytotoxic activity and redox activity (Gohari et al. 2015; Manzoor and Hoda 2020; Olaoye et al. 2022; Santos et al. 2023).

Other important points that were seen in the review refer to the importance of iron metalbolism in regulating the enzyme succinate dehydrogenase (Dick et al. 2023), as well as its participation in the activity of macrophages and the immune system and in the inflammatory process, specifically speaking of the formation of granulomas (Wilson et al. 2019). Another important topic was the evaluation of the activity of succinate dehydrogenase in the development of vaccines and the evaluation of parasite resistance (Hamed et al. 2010; Hamed 2011). This evidence contributes to establishing the enzyme succinate dehydrogenase as an important target in the study and development of new therapeutic alternatives for neglected diseases.

4.4 Conclusion

The enzyme succinate dehydrogenase, known as complex II of the electron transport chain, is part of the mitochondrial respiratory complexes, being present in the acid cycle and the electron transport chain and, therefore, is considered an important target in the discovery of new compounds, as it is responsible for redox metabolism and energy supply to the parasite, thus being crucial for its survival. Within the scope of neglected tropical diseases, its occurrence has been reported as a notable target in trypanosomiasis, schistosomiasis, and Leishmaniasis. Another important factor refers to the fact that this enzyme has its action increased by iron metabolism and by the immune system through the regulation of macrophages. This could also be an important target in the development of vaccines and immunobiologicals.

References

Adroher F-J, Osuna A, Lupiañez JA (1988) Differential energetic metabolism during Trypanosoma cruzi differentiation: I. Citrate synthase, NADP-isocitrate dehydrogenase, and succinate dehydrogenase. Arch Biochem Biophys 267(1):252–261

Aspuria P-JP, Lunt SY, Väremo L, Vergnes L, Gozo M, Beach JA, Salumbides B, Reue K, Wiedemeyer WR, Nielsen J, Karlan BY, Orsulic S (2014) Succinate dehydrogenase inhibition leads to epithelial-mesenchymal transition and reprogrammed carbon metabolism. Cancer Metab 2(1):1–15

Bassal MA (2023) The interplay between dysregulated metabolism and epigenetics in cancer. Biomol Ther 13(6):944

Beattie DS, Obungu VH, Kiaira JK (1994) Oxidation of NADH by a rotenone and antimycin-sensitive pathway in the mitochondrion of procyclic Trypanosoma brucei brucei. Mol Biochem Parasitol 64(1):87–94

Bénit P, Goncalves J, El Khoury R, Rak M, Favier J, Gimenez-Roqueplo AP, Rustin P (2022) Succinate dehydrogenase, succinate, and superoxides: a genetic, epigenetic, metabolic, environmental explosive crossroad. Biomedicines 10(8):1788

Benitez D, Casanova G, Cabrera G, Galanti N, Cerecetto H, Gonzalez M (2014) Initial studies on mechanism of action and cell death of active N-oxide-containing heterocycles in Trypanosoma cruzi epimastigotes *in vitro*. Parasitology 141(5):682–696

Bezerra JCB, Kemper A, Becker W (1999) Profile of organic acid concentrations in the digestive gland and hemolymph of Biomphalaria glabrata under estivation. Mem Inst Oswaldo Cruz 94:779–784

Bombaça ACS, Silva LA, Chaves OA, da Silva LS, Barbosa JMC, da Silva AM, Ferreira ABB, Menna-Barreto RFS (2021) Novel N, N-di-alkylnaphthoimidazolium derivative of β-lapachone impaired Trypanosoma cruzi mitochondrial electron transport system. Biomed Pharmacother 135:111186

Bouillaud F (2023) Inhibition of succinate dehydrogenase by pesticides (SDHIs) and energy metabolism. Int J Mol Sci 24(4):4045

Boutayeb A (2007) Developing countries and neglected diseases: challenges and perspectives. Int J Equity Health 6(1):20

Calixto NM, Dos Santos DB, Bezerra JCB, Silva L d A (2018) In silico repositioning of approved drugs against Schistosoma mansoni energy metabolism targets. PLoS One 13(12):e0203340

Carvalho L, Luque-Ortega JR, López-Martín C, Castanys S, Rivas L, Gamarro F (2011) The 8-aminoquinoline analogue sitamaquine causes oxidative stress in Leishmania donovani promastigotes by targeting succinate dehydrogenase. Antimicrob Agents Chemother 55(9):4204–4210

Castilla JJ, Mesa-Valle CM, Sanchez-Moreno M, Arnedo T, Rosales MJ, Mascaro C, Craciunescu D, Osuna A (1996) *In vitro* activity and biochemical effectiveness of new organometallic complexes of osmium (III) against *Leishmania donovani* and *Trypanosoma cruzi*. Arzneimittelforschung 46(10):990–996

Cazzulo JJ (1992) Aerobic fermentation of glucose by trypanosomatids. FASEB J 6(13):3153–3161

Cazzulo JJ (1994) Intermediate metabolism in *Trypanosoma cruzi*. J Bioenerg Biomembr 26:157–165

César Carranza J, Kowaltowski AJ, Mendonça MAG, de Oliveira TC, Gadelha FR, Zingales B (2009) Mitochondrial bioenergetics and redox state are unaltered in Trypanosoma cruzi isolates with compromised mitochondrial complex I subunit genes. J Bioenerg Biomembr 41:299–308

Cheleski J, Wiggers HJ, Citadini AP, da Costa Filho AJ, Nonato MC, Montanari CA (2010) Kinetic mechanism and catalysis of Trypanosoma cruzi dihydroorotate dehydrogenase enzyme evaluated by isothermal titration calorimetry. Anal Biochem 399(1):13–22

Christmas PB, Turrens JF (2000) Separation of NADH-fumarate reductase and succinate dehydrogenase activities in *Trypanosoma cruzi*. FEMS Microbiol Lett 183(2):225–228

Coustou V, Besteiro S, Riviere L, Biran M, Biteau N, Franconi J-M, Boshart M, Baltz T, Bringaud F (2005) A mitochondrial NADH-dependent fumarate reductase involved in the production of succinate excreted by procyclic *Trypanosoma brucei*. J Biol Chem 280(17):16559–16570

de Farias Silva N, Lameira J, Alves CN, Martí S (2013) Computational study of the mechanism of half-reactions in class 1A dihydroorotate dehydrogenase from Trypanosoma cruzi. Phys Chem Chem Phys 15(43):18863–18871

de Mesquita Barbosa A, dos Santos Costa S, da Rocha JR, Montanari CA, Giorgio S (2015) Evaluation of the leishmanicidal and cytotoxic effects of inhibitors for microorganism metabolic pathway enzymes. Biomed Pharmacother 74:95–100

Denicola A, Rubbo H, Haden L, Turrens JF (2002) Extramitochondrial localization of NADH-fumarate reductase in trypanosomatids. Comp Biochem Physiol B: Biochem Mol Biol 133(1):23–27

Dick CF, Alcantara CL, Carvalho-Kelly LF, Lacerda-Abreu MA, Cunha-e-Silva NL, Meyer-Fernandes JR, Vieyra A (2023) Iron uptake controls Trypanosoma cruzi metabolic shift and cell proliferation. Antioxidants 12(5):984

Dröse S (2013) Differential effects of complex II on mitochondrial ROS production and their relation to cardioprotective pre-and postconditioning. Biochim Biophys Acta 1827(5):578–587

Elso OG, Puente V, Barrera P, Sosa-Escudero MA, Sülsen VP, Lombardo ME (2022) Mode of action of the sesquiterpene lactones eupatoriopicrin and estafietin on Trypanosoma cruzi. Phytomedicine 96:153900

Fujii J, Homma T, Osaki T (2022) Superoxide radicals in the execution of cell death. Antioxidants 11(3):501

Goetzman E, Gong Z, Zhang B, Muzumdar R (2023) Complex II biology in aging, health, and disease. Antioxidants 12(7):1477

Gohari AR, Mosaddegh M, Naghibi F, Eslami-Tehrani B, Pirani A, Hamzeloo-Moghadam M, Read RW (2015) Cytotoxic sesquiterpene lactones from the aerial parts of Inula aucheriana. An Acad Bras Cienc 87:777–785

Goncalves J, Wan Y, Guo X, Rha K, LeBoeuf B, Zhang L, Estler K, Garcia LR (2020) Succinate dehydrogenase-regulated phosphoenolpyruvate carboxykinase sustains copulation fitness in aging C. elegans males. iScience 23(4):100990

Hägerhäll C (1997) Succinate: quinone oxidoreductases: variations on a conserved theme. Biochim Biophys Acta 1320(2):107–141

Hamed MAA (2011) Potency of detergents in enhancing Schistosoma mansoni tegumental antigens. J Infect Dev Ctries 5(03):209–215

Hamed MA, Hetta MH (2005) Efficacy of Citrus reticulata and Mirazid in treatment of Schistosoma mansoni. Mem Inst Oswaldo Cruz 100:771–778

Hamed MA, Ali SA, Aly HF, El-Rigal NS, Rizk MZ (2010) Biomphalaria alexandrina snails as immunogens against Schistosoma mansoni infection in mice. Mem Inst Oswaldo Cruz 105:879–888

Huang S, Braun H, Gawryluk RMR, Millar AH (2019) Mitochondrial complex II of plants: subunit composition, assembly, and function in respiration and signaling. Plant J 98(3):405–417

Inaoka DK, Sakamoto K, Shimizu H, Shiba T, Kurisu G, Nara T, Aoki T, Kita K, Harada S (2008) Structures of Trypanosoma cruzi dihydroorotate dehydrogenase complexed with substrates and products: atomic resolution insights into mechanisms of dihydroorotate oxidation and fumarate reduction. Biochemistry 47(41):10881–10891

Johnston K, Kim D-H, Kerkhoven EJ, Burchmore R, Barrett MP, Achcar F (2019) Mapping the metabolism of five amino acids in bloodstream form Trypanosoma brucei using U-13C-labelled substrates and LC–MS. Biosci Rep 39(5):BSR20181601

Karbalaei M, Keikha M (2022) Chikungunya, zika, and dengue: three neglected re-emerging Aedes-borne diseases. Ann Med Surg:81

Koo S, Garg NJ (2019) Metabolic programming of macrophage functions and pathogens control. Redox Biol 24:101198

Kregiel D (2012) Succinate dehydrogenase of Saccharomyces cerevisiae–the unique enzyme of TCA cycle–current knowledge and new perspectives. In: Canuto RA (ed) Dehydrogenases, pp 211–234

Lindoso JAL, Lindoso AABP (2009) Neglected tropical diseases in Brazil. Rev Inst Med Trop Sao Paulo 51:247–253

Luo Y, Jiang L, Kang K, Fei D, Meng X, Nan CC, Pan SH, Zhao MR, Zhao MY (2014) Hemin inhibits NLRP3 inflammasome activation in sepsis-induced acute lung injury, involving heme oxygenase-1. Int Immunopharmacol 20(1):24–32

Maarouf M, de Kouchkovsky Y, Brown S, Petit PX, Robert-Gero M (1997) In vivo interference of paromomycin with mitochondrial activity of Leishmania. Exp Cell Res 232(2):339–348

Maklashina E, Iverson TM, Sher Y, Kotlyar V, Andréll J, Mirza O, Hudson JM, Armstrong FA, Rothery RA, Weiner JH, Cecchini G (2006) Fumarate reductase and succinate oxidase activity of Escherichia coli complex II homologs are perturbed differently by mutation of the flavin binding domain. J Biol Chem 281(16):11357–11365

Manzoor S, Hoda N (2020) A comprehensive review of monoamine oxidase inhibitors as Anti-Alzheimer's disease agents: a review. Eur J Med Chem 206:112787

Marciano D, Llorente C, Maugeri DA, de la Fuente C, Opperdoes F, Cazzulo JJ, Nowicki C (2008) Biochemical characterization of stage-specific isoforms of aspartate aminotransferases from *Trypanosoma cruzi* and *Trypanosoma brucei*. Mol Biochem Parasitol 161(1):12–20

Martínez-Reyes I, Chandel NS (2020) Mitochondrial TCA cycle metabolites control physiology and disease. Nat Commun 11(1):102

Mattos EC, Canuto G, Manchola NC, Magalhães RDM, Crozier TWM, Lamont DJ, Tavares MFM, Colli W, Ferguson MAJ, Alves MJM (2019) Reprogramming of Trypanosoma cruzi metabolism triggered by parasite interaction with the host cell extracellular matrix. PLoS Negl Trop Dis 13(2):e0007103

Maugeri DA, Cannata JJB, Cazzulo J-J (2011) Glucose metabolism in Trypanosoma cruzi. Essays Biochem 51:15–30

Morales J, Mogi T, Mineki S, Takashima E, Mineki R, Hirawake H, Sakamoto K, Omura S, Kita K (2009) Novel mitochondrial complex II isolated from Trypanosoma cruzi is composed of 12 peptides including a heterodimeric Ip subunit. J Biol Chem 284(11):7255–7263

Mounkoro P, Michel T, Golinelli-Cohen M-P, Blandin S, Davioud-Charvet E, Meunier B (2021) A role for the succinate dehydrogenase in the mode of action of the redox-active antimalarial drug, plasmodione. Free Radic Biol Med 162:533–541

Negreiros RS, Lander N, Chiurillo MA, Vercesi AE, Docampo R (2021) Mitochondrial pyruvate carrier subunits are essential for pyruvate-driven respiration, infectivity, and intracellular replication of Trypanosoma cruzi. MBio 12(2):10–1128

Olaoye I, Awotula A, Oso B, Agboola O, Akhigbe G, Olaoye T (2022) Disminución de glutatión reducido en ratas Wistar machos tras la exposición a naftaleno. Biotecnología Aplicada 39(1):1201–1210

Oyedotun KS, Lemire BD (2004) The quaternary structure of the Saccharomyces cerevisiae succinate dehydrogenase: homology modeling, cofactor docking, and molecular dynamics simulation studies. J Biol Chem 279(10):9424–9431

Pecsi I, Hards K, Ekanayaka N, Berney M, Hartman T, Jacobs WR Jr, Cook GM (2014) Essentiality of succinate dehydrogenase in Mycobacterium smegmatis and its role in the generation of the membrane potential under hypoxia. MBio 5(4):10–1128

Pinheiro MP, Iulek J, Nonato MC (2008) Crystal structure of Trypanosoma cruzi dihydroorotate dehydrogenase from Y strain. Biochem Biophys Res Commun 369(3):812–817

Al Rasheed MRH, Tarjan G (2018) Succinate dehydrogenase complex: an updated review. Arch Pathol Lab Med 142(12):1564–1570

Reis RAG, Ferreira P, Medina M, Nonato MC (2016) The mechanistic study of Leishmania major dihydro-orotate dehydrogenase based on steady-and pre-steady-state kinetic analysis. Biochem J 473(5):651–660

Relman DA, Choffnes ER (2011) The causes and impacts of neglected tropical and zoonotic diseases: opportunities for integrated intervention strategies. National Academies Press

Rubbo H, Denicola A, Radi R (1994) Peroxynitrite inactivates thiol-containing enzymes of Trypanosoma cruzi energetic metabolism and inhibits cell respiration. Arch Biochem Biophys 308(1):96–102

Rustin P, Munnich A, Rötig A (2002) Succinate dehydrogenase and human diseases: new insights into a well-known enzyme. Eur J Hum Genet 10(5):289–291

Rutter J, Winge DR, Schiffman JD (2010) Succinate dehydrogenase–assembly, regulation and role in human disease. Mitochondrion 10(4):393–401

Sayed AA, Cook SK, Williams DL (2006) Redox balance mechanisms in Schistosoma mansoni rely on peroxiredoxins and albumin and implicate peroxiredoxins as novel drug targets. J Biol Chem 281(25):17001–17010

Sharma P, Maklashina E, Cecchini G, Iverson TM (2019) Maturation of the respiratory complex II flavoprotein. Curr Opin Struct Biol 59:38–46

Silva TM, Peloso EF, Vitor SC, Ribeiro LHG, Gadelha FR (2011) O 2 consumption rates along the growth curve: new insights into Trypanosoma cruzi mitochondrial respiratory chain. J Bioenerg Biomembr 43:409–417

Silva MF, Douglas K, Sandalli S, Maclean AE, Sheiner L (2023) Functional and biochemical characterisation of the Toxoplasma gondii succinate dehydrogenase complex. PLoS Pathog 19(12):e1011867

Sun F, Huo X, Zhai Y, Wang A, Xu J, Su D, Bartlam M, Rao Z (2005) Crystal structure of mitochondrial respiratory membrane protein complex II. Cell 121(7):1043–1057

Takashima E, Inaoka DK, Osanai A, Nara T, Odaka M, Aoki T, Inaka K, Harada S, Kita K (2002) Characterization of the dihydroorotate dehydrogenase as a soluble fumarate reductase in Trypanosoma cruzi. Mol Biochem Parasitol 122(2):189–200

Van Hellemond JJ, Van Remoortere A, Tielens AGM (1997) Schistosoma mansoni sporocysts contain rhodoquinone and produce succinate by fumarate reduction. Parasitology 115(2):177–182

Wang Q, Li M, Zeng N, Zhou Y, Yan J (2023) Succinate dehydrogenase complex subunit C: Role in cellular physiology and disease. Exp Biol Med 248(3):263–270

Wilson JL, Mayr HK, Weichhart T (2019) Metabolic programming of macrophages: implications in the pathogenesis of granulomatous disease. Front Immunol 10:2265

Xiao Y, Chen X, Wang Z, Quan J, Zhao X, Tang H, Wu H, Di Q, Wu Z, Chen W (2022) Succinate is a natural suppressor of antiviral immune response by targeting MAVS. Front Immunol 13:816378

Yamey G, Batson A, Kilmarx PH, Yotebieng M (2018) Funding innovation in neglected diseases. BMJ

Yan L-J, Sumien N, Thangthaeng N, Forster MJ (2013) Reversible inactivation of dihydrolipoamide dehydrogenase by mitochondrial hydrogen peroxide. Free Radic Res 47(2):123–133

Yankovskaya V, Horsefield R, Tornroth S, Luna-Chavez C, Miyoshi H, Léger C, Byrne B, Cecchini G, Iwata S (2003) Architecture of succinate dehydrogenase and reactive oxygen species generation. Science 299(5607):700–704

Yu J, Wang S, Li W, Dai G, Xu S, He Z, Peng X, Zhou S, Liu X (2007) Cloning, expression and protective immunity evaluation of the full-length cDNA encoding succinate dehydrogenase iron-sulfur protein of Schistosoma japonicum. Sci China Ser C Life Sci 50:221–227

Zagoto JN, Bracht A, Pagadigorria CLS, Ishii-Iwamoto EL, Cortez DAG, Yamamoto NS (2006) Effects of the Kielmeyera coriacea extract on energy metabolism in the rat liver. J Ethnopharmacol 105(1–2):47–54

Dehghan Manshadi M, Setoodeh P, Zare H (2023) Systematic analysis of microorganisms' metabolism for selective targeting. Sci Rep14(1):16446

Zhou X, Gao Y, Wang W, Yang X, Yang X, Liu F, Tang Y, Lam SM, Shui G, Yu L, Tian C, Guddat LW, Wang Q, Rao Z, Gong H (2021) Architecture of the mycobacterial succinate dehydrogenase with a membrane-embedded Rieske FeS cluster. Proc Natl Acad Sci 118(15):e2022308118

Zhu J, Liu J, Yan C, Wang D, Pan W (2023) Trained immunity: a cutting edge approach for designing novel vaccines against parasitic diseases? Front Immunol 14

Santos VL d A, Gonsalves A dA, Guimarães DG, Simplicio SS, de Oliveira HP, Ramos LPS, Tang Y, Lam SM, Shui G, Yu L, Tian C, Guddat LW, Wang Q, Rao Z, Gong H (2023) Naphth [1, 2-d] imidazoles bioactive from β-Lapachone: fluorescent probes and cytotoxic agents to cancer cells. Molecules 28(7):3008

Succinate Dehydrogenase in Mitochondrial Disorders

5

Siddhant Tripathi, Yashika Sharma, and Dileep Kumar

Abstract

Neurodegenerative diseases including Alzheimer's, Parkinson's, and Huntington's disease are becoming more prevalent as people age. Reducing clinical symptoms and shielding neurons from degradation are the primary goals of studies to develop novel treatment approaches. Numerous studies have shown that mitochondria are essential to high-energy tissues like the brain. A disturbance in the physiology or function of the mitochondria may expose neurons to stress and deterioration. Succinate dehydrogenase (SDH) is the enzyme that joins the electron transport chain and the tricarboxylic cycle. As a result, SDH failure may affect the cell's ability to produce ATP, maintain energy homeostasis, and function of the mitochondria. Beyond the production of lipids, SDH may regulate the generation of excitotoxicity in neurodegenerative diseases via direct and indirect mechanisms. Furthermore, there is a correlation between the start of neurodegenerative diseases and mutation in SDH. SDH may therefore act as a crucial regulator in neuroprotection. This review will include current research on SDH activity and associated pathways that may be crucial for neuronal survival. We will also talk about every potential SDH candidate as a neuroprotective agent.

Keywords

Succinate dehydrogenase · Neurodegenerative disorders · Mitochondria · Neuroprotection · Lipid synthesis · Excitotoxicity

S. Tripathi · Y. Sharma
Poona College of Pharmacy, Bharati Vidyapeeth (Deemed to be) University, Pune, Maharashtra, India

D. Kumar (✉)
Department of Pharmaceutical Chemistry, Manipal College of Pharmaceutical Sciences, Manipal Academy of Higher Education, Manipal, Karnataka, India

V. M. Patil et al. (eds.), *Deciphering The Role of Succinate Dehydrogenase in Drug Discovery*, https://doi.org/10.1007/978-981-96-6831-1_5

Abbreviations

AD	Alzheimer's disease
CRM	Cross-reacting material
Fp	Flavoprotein
HD	Huntington's disease
HIF	Hypoxia-inducible factor
HMG-CoA	3-Hydroxy-3-methylglutaryl-CoA
Htt	Huntingtin protein
MDH	Malate dehydrogenase
MPTP	1-Methyl 4-phenyl 1,2,3,6 tetrahydropyridine
mtDNA	Mitochondrial DNA
NMDA	N-methyl-D-aspartate
NO	Nitric oxide
NOS	Nitric oxide synthetase
OAA	Oxaloacetate
PD	Parkinson disease
PDHC	Pyruvate dehydrogenase complex
PINK1	PTEN-induced putative kinase 1
ROS	Reactive oxygen species
SDH	Succinate dehydrogenase
SNEM	Subacute necrotizing encephalomyelopathy
SOD	Superoxide dismutase

5.1 Introduction

The respiratory chain's ubiquinone (UQ) pool receives electrons from the mitochondrial succinate dehydrogenase (SDH) complex, which also catalyzes the Krebs cycle's oxidation of succinate to fumarate (Fig. 5.1) (Cecchini et al. 2002; Siedow 1982). The structure and genes of the four nuclearly encoded subunits that make up SDH have mostly remained constant throughout history. The other respiratory complexes, on the other hand, have either acquired more subunits or, in some cases, may not exist at all, such as complex I in the yeast *Saccharomyces cerevisiae*. SDH is defined as the rigorous union of subunits A and B. The other two, subunits C and D, are also known as the anchoring subunits because they bind a b-type cytochrome. However, they are also necessary for the transfer of electrons from succinate to the ubiquinone pool.

In humans, succinate dehydrogenase defects are quite uncommon (Rustin and Rötig 2002). However, individuals with mutations in one of the four genes producing the SDH subunit have been shown to present with markedly different clinical presentations. The typical mitochondrial encephalopathy known as Leigh syndrome is caused by mutations in the flavoprotein component SDHA, which were the first nuclear gene alterations to be recognized as causing a mitochondrial RC deficiency in the mid-1990s (Bourgeron et al. 1995; Parfait et al. 2000). Recent research has

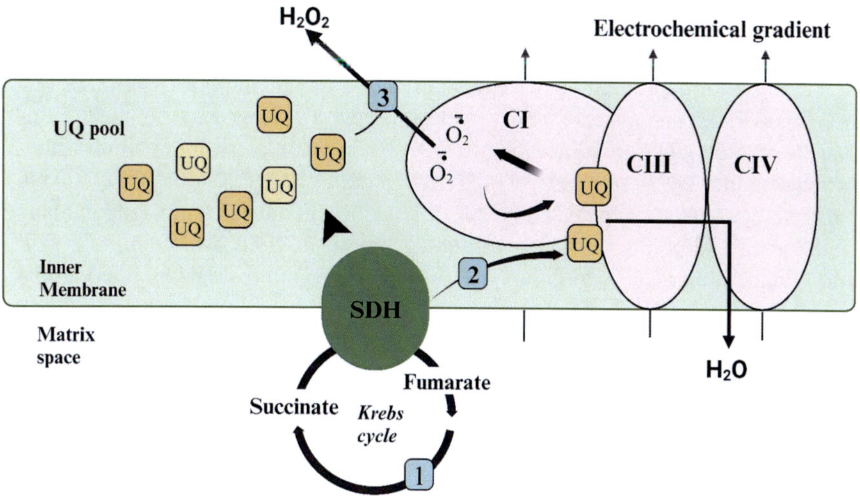

Fig. 5.1 The several roles that the mitochondrial succinate dehydrogenase plays. In the Krebs cycle (1), succinate is oxidized into fumarate by the enzyme succinate dehydrogenase. The electrons generated from this process are then sent to the respiratory chain complex III, where they decrease oxygen as well as make water (2). To facilitate the creation of ATP, this creates an electrochemical gradient across the inner membrane of the mitochondria, redirecting electrons may also reduce the amount of ubiquinone (UQ) in the system while offering the reducing equivalents required for reducing superoxide anions through the respiratory chain or other sources (3). A total absence of succinate dehydrogenase activity would impede electron transport to the quinone pool and respiratory chain complex III, causing significant oxidative stress that has been linked to the development of human tumors. UQ stands for ubiquinone; CI, CIII, and CIV are the different respiratory chain complexes

shown that mutations in subunits B, C, and D may result in pheochromocytomas, which are vascularized tumors in the head and neck that are usually benign. Paragangliomas are another name for these disorders (Astuti et al. 2001; Baysal et al. 2000; Niemann and Müller 2000). Situated near the intersection along the common carotid artery, the carotid body serves as a highly vascularized microscopic organ that gauges blood oxygen saturation levels. It is also known as the most often occurring tumor location. As a result of these three genes' somatic mutations occurring in tumors, they need to be considered tumor-suppressor genes (Baysal et al. 2001).

As of now, SDH is only known to catalyze a single process that involves all four of its subunits. Decreases in SDH activity should always result from significant mutations across all of the SDH genes. Therefore, the remarkable behavioral variations linked to mutations in each of the four subunits provide perplexing issues. A closer look at the well-known SDH characteristics might shed some light on this problem.

5.2 Neurodegeneration of Mitochondria

The mitochondrion's role in neurodegenerative diseases including Alzheimer's disease (AD), Parkinson disease (PD), and Huntington's disease (HD) makes this organelle a potential target for treatment. With 16,569 base pairs, mitochondria contain their circular DNA, mtDNA encodes certain rRNA and tRNA that are unique to the respiratory system (Lin and Beal 2006). The mitochondria are involved in many physiological and cellular processes, such as apoptosis, lipid oxidation, ATP generation, intracellular Ca^{2+} regulation, and the cell's capacity for reduction oxidation. Each of these roles has a bearing on neurodegenerative disorders (Martin 2010; Nicholls 2002).

The brain consumes 20% of oxygen while making up just 2% of the body weight. Reactive oxygen species (ROS) damages proteins, lipids, and DNA in neurons, leading to neurodegeneration. ROS include superoxide anion radical ($O_2\cdot-$), hydroxyl radical ($\cdot OH$), and nonradical oxidants such as H_2O_2. According to Petlicki and van de Ven, superoxide has the most oxidation capability in a cell (Petlicki and van de Ven 1998). The primary source of superoxide production is mitochondria. The primary process for producing superoxide is the oxygen reaction with electrons that escape from the mitochondrial respiratory chain (Loschen et al. 1974). According to Fukai and Ushio-Fukai, superoxide dismutase (SOD) catalyzes the conversion of superoxide radicals to hydrogen peroxide (Fukai and Ushio-Fukai, 2011). The mitochondrial matrix contains the Mn-dependent isoform (Mn-SOD, SOD2) (Weisiger and Fridovich 1973). The respiratory chain's complexes I and III are the primary sources of ROS production (Mailloux 2015).

Nitric oxide synthetase (NOS) catalyzes the process that converts L-arginine to NO and L-citrulline, which produces nitric oxide (NO) (Knowles and Moncada 1994). In addition to acting as a second messenger, NO additionally binds to cytochrome c oxidase, therefore lowering the oxygen-binding affinity of the enzyme. The cytochrome c oxidase-NO binding process influences electron flow and ATP synthesis (Moncada and Bolaños 2006). In addition, superoxide and hydrogen peroxide levels are raised when NO inhibits the respiratory chain. According to Poderos et al., superoxide undergoes a conversion from hydrogen peroxide to peroxynitrite (ONOO-) at high NO levels (Poderoso et al. 1998). Accordingly, an abundance of NO is a contributing factor to neurological disorders (Dawson and Dawson 1996). NO may disrupt mitochondrial activity and affect how the respiratory chain operates. The brain mitochondria exhibit a decrease in activity in complex IV when incubated with ONOO (Bolaños et al. 1995). When complex IV loses its activity, cardiolipin, the most prevalent phospholipid in the inner membrane of the mitochondria, peroxidizes (Soussi et al. 1990). In addition, ONOO exposure to the brain mitochondria reduces complex II activity (Brookes et al. 1998). Through the nitrosylation process, NO inhibits the normal function and character of proteins. NO targets sulfhydryls, the primary group of the active site of many enzymes, altering the activity of proteins (Radi et al. 1991). The channels at the outer and inner membrane contact sites of the mitochondria are known as mitochondrial permeability transition pores or mPTPs. Ca^{2+} hemostasis in nonpathological conditions is their responsibility. However, proapoptotic proteins are released into the cytoplasm when oxidative stress, ATP depletion, and excess Ca^{2+} activate these channels (Brenner and Grimm 2006). According to Chavez et al., one of the primary chemicals that might cause mPTP opening under pathological circumstances is ONOO- (Chávez et al. 1997). Through the nitrosylation of proteins, opening of the mPTP, as well as disruption of mitochondrial functions, particularly the electron transport chain, PD and HD peroxynitrite play a critical role in AD (Asiimwe et al. 2016; Jamwal and Kumar 2017; Jiménez-Jiménez et al. 2016).

Massive ionic flux in neurodegenerative diseases excites neurons, causing mitochondrial malfunction and death. These disorders cause an overabundance of glutamate, the primary excitatory neurotransmitter. N-methyl-D-aspartate (NMDA) as well as α-amino-3-hydroxy-5-methyl-4-isoxazolepropionic acid (AMPA) receptors are the two receptor types through which glutamate operates (Hara and Snyder 2007). According to Deng et al., amyloid beta (Aβ) in AD enhances excitatory behavior by indirectly affecting NMDA receptors (Deng et al. 2014). Furthermore, Aβ causes a reduction in glutamate reabsorption from the synaptic cleft in AD (Scimemi et al. 2013). Previous research revealed that NMDA receptor agonists exacerbate HD-like striatal neurodegeneration (Coyle and Schwarcz 1976). Different HD mice models have higher active NMDA receptors (Shehadeh et al. 2006). According to Zhang et al., PD is associated with glutamate buildup in the synaptic cleft and excessive NMDA receptor activation (Zhang et al. 2016). Glutamate-induced NMDA activation increases Ca^{2+} influx into the cell in addition to excitating postsynaptic neurons (Zipfel et al. 2000). To buffer cytoplasmic Ca^{2+}, mitochondria largely use uniporters, which open when Ca^{2+} concentrations are high (Rizzuto et al. 2000). The electrochemical gradient determines how much Ca^{2+} enters the mitochondrial matrix. During oxidative phosphorylation, the respiratory chain produces mitochondrial membrane potential ($\Delta\Psi$) (Sparagna et al. 1995). Preventing mitochondria from taking in Ca^{2+} shields neural cells from damage and deterioration (Stout et al. 1998). According to Georgi et al., mitochondrial Ca^{2+} excess causes an increase in ROS generation, a reduction in or elimination of $\Delta\Psi$, the opening of the PTP, disruptions in mitochondrial action, as well as ultimately ATP depletion (Giorgi et al. 2012). Research has shown a potential correlation between neurodegeneration and Ca^{2+} hemostasis. For instance, in the AD model, to penetrate the plasma membrane as well as improve Ca^{2+} influx entering the cytoplasm, Aβ serves as an ion channel (Demuro et al. 2005). Aβ oligomer also results in mitochondrial Ca^{2+} excess. According to Brustovetsky et al., too much Ca^{2+} in the mitochondria increases the creation of ROS, reduces the synthesis of ATP, and releases cytochrome c, which causes apoptosis (Brustovetsky et al. 2003). Similar to AD, PD models exhibit increased Ca^{2+} influx into the neuron due to α-synuclein buildup (Furukawa et al. 2006). One mitochondrial protein that demonstrates the pathophysiology of Parkinson's disease (PD) is mutant PTEN-induced putative kinase 1 (PINK1). According to some data, mitochondria in dopaminergic neurons of PINK1-deficient mice are responsive to Ca^{2+} (Akundi et al. 2011). The striatal neurodegeneration is a hallmark of HD. The primary cause of HD is an abnormal growth of polyglutamine (Poly Q) in the huntingtin protein (Htt). A mutant version of huntingtin (mHtt) increases the susceptibility of neurons to Ca^{2+} overloading, which causes neuronal degeneration and death, according to Choo et al. (2004). The cytoplasmic Ca^{2+} buffering function of mitochondria is essential, and their overabundance of Ca^{2+} not only impairs mitochondrial function but also sets off other signaling cascades that result in dementia. The unique mitochondrial DNA (mtDNA) encodes certain respiratory chain proteins as well as particular rRNA and tRNA for the mitochondria. According to Ehringer and Giguère and Legros et al., human mtDNA is a circular, double-stranded molecule of about 16.6 base pairs. It codes for 13 respiratory chain polypeptides, 22 tRNA, as well as 2 rRNA. (Eichner and Giguère 2011; Legros et al. 2004). Of the 45 subunits that make up complex I (NADH CoQ dehydrogenase), seven are mtDNA-encoded (Davis and Williams 2012). Specifically, 1-methyl 4-phenyl 1,2,3,6 tetrahydropyridine (MPTP), or its active metabolite, inhibits complex I in dopaminergic neurons. One-methyl 4-phenylpyridinium (MPP^+) degenerates dopaminergic neurons (Langston and Ballard 1983; Shi et al. 2017). Prior research has shown that inactive complex I is present in the platelets of PD and HD patients (Haas et al. 1995; Silva et al. 2013). PD patients' dopaminergic neurons had mtDNA deletion in addition to neurotoxins such as MPP^+ (Kraytsberg et al. 2006).

Early and definitive AD brain specimens are characterized by damaged mtDNA and poor expression of complex I subunits (Manczak et al. 2004; Yan et al. 2013). The four subunits of complex II, known as succinate dehydrogenase (SDH) and ubiquinone oxidoreductase (SQR), are all encoded by nuclear DNA. Research on individuals with AD, PD, and HD revealed a mitochondrial complex II malfunction and decreased activity (Damiano et al. 2013; Hattori et al. 1991; Long et al. 2012). One of complex III's eleven components, ubiquinol (cytochrome c oxidoreductase), is mtDNA-encoded. In models of AD, Aβ inhibits this complex (Devi and Anandatheerthavarada 2010). According to recent research, striatal atrophy in HD may result from mutations in complex III subunits (Gu et al. 1996), of the 12 subunits that make up complex IV (cytochrome c oxidase), three come from mtDNA. As the primary misfolded protein in Parkinson's disease (PD), α-synuclein may affect complex IV function and activity (Martin et al. 2006). Furthermore, prior research revealed that PD patients' platelets had abnormalities in intricate IV functions (Benecke et al. 1993). Aβ suppresses the HD model and cytochrome c oxidase activity in AD patients (Fig. 5.2) (Casley et al. 2002; Tabrizi et al. 2000).

In addition to generating ROS, anomalies in the electron transport chain's activity also indicate ATP depletion. Given that neurons have significant energy demands, such as those for synaptic activity, which requires a lot of energy to operate, neurons rely on mitochondrial activities (Jodeiri Farshbaf et al. 2016).

To summarize, disorders related to neurodegeneration and malfunction of the mitochondria. A treatment plan should take into account certain tactics that raise mitochondrial efficiency and activity.

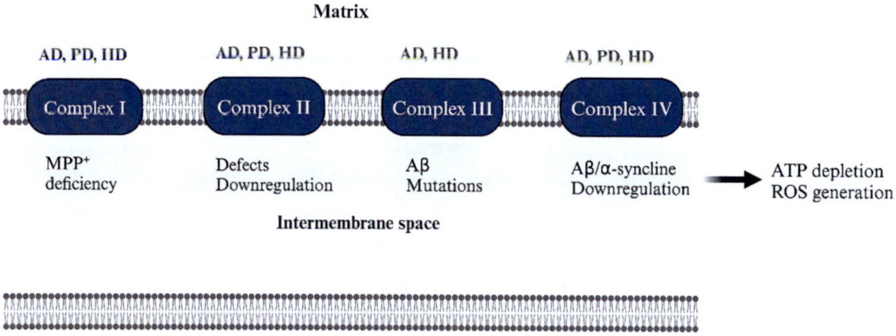

Fig. 5.2 Diverse forms of neurodegenerative diseases, including AD, PD, and HD, may result from abnormalities in the electron transport chain's activity. The production of ROS and the depletion of ATP are the main results of a defective oxidative phosphorylation system

5.3 Structure and Enzymology of SDH

SDH is a component of the respiratory electron transfer chain in the citric acid cycle. SDH oxidizes succinate to fumarate during the citric acid cycle. Fumarate reductase, a metabolic enzyme that catalyzes an opposite reaction throughout anaerobic respiration in bacteria, and SDH have structural similarities (Hägerhäll 1997). When produced during anaerobic growth, SDH can functionally replace fumarate reductase in *Escherichia coli*, while during aerobic respiration, fumarate reductase within *E. coli* can successfully substitute for SDH (Maklashina et al. 1998).

The nuclear genome encodes four subunits that make up eukaryotic SDH. SDH is the only respiratory complex that does not pump protons across the IM throughout its catalytic cycle as well as the only oxidative phosphorylation complex that lacks components encoded by the mitochondrial genome. Yankovskaya and Sun et al. have shown that the pig heart SDH is made up of a hydrophobic tail that is securely entrenched in the IM and projects a brief section into the soluble intermembrane gap, as well as a hydrophilic head that extends into the matrix compartment (Sun et al. 2005; Yankovskaya et al. 2003). Two subunits (SDH1 and SDH2 in yeast as well as SDHA and SDHB in mammals) make up the hydrophilic head, which is responsible for the catalytic core. In this review, we shall stick to the yeast terminology for consistency and ease of understanding.

These redox cofactors are present in the catalytic core SDH1 and SDH2 subunits and are involved in the electron transfer to ubiquinone. SDH1 has the succinate binding site and the covalently attached FAD cofactor. According to Hagerhall and Sun et al., the three Fe–S centers on SDH2 aid in the transfer of electrons to ubiquinone (Hägerhäll 1997; Sun et al. 2005). The Fe/S centers in SDH2 consist of a 3Fe–4S center, a nearby 4Fe–4S center, and a 2Fe–2S center near the FAD site (Sun et al. 2005; Yankovskaya et al. 2003). Furthermore, SDH2 creates the link that connects the complex's membrane anchor domain as well as the catalytic domain. The surface area of the packing interface between SDH1, SDH2, and SDH3 is comparable to every contact. It also means that the catalytic core cannot be found as a free dimer entity in the absence of the membrane anchor. In fact, in yeast lacking one of the membrane anchor subunits, both of the hydrophilic subunits, SDH1 and SDH2, are much less numerous (Hao et al. 2009). On the other hand, when the membrane domain subunits are absent, soluble succinate dehydrogenase that is active is what *E. coli* SDH does (Nakamura et al. 1996). Soluble enzymes do not have ubiquinone reductase activity; they only become active when they come into contact with artificial electron acceptors.

Two subunits make up the membrane domain: SDHC, SDHD in humans, and SDH3 and SDH4 in yeast. At the subunit interface of the membrane domain, there is a bound heme B moiety, SDH3 as well as SDH4 together provide each of the two axial His ligands. In SDH complexes from *E. coli* and mammals, two ubiquinone binding sites have been found (Sun et al. 2005; Yankovskaya et al. 2003). Residues from SDH2, SDH3, and SDH4 combine to produce the matrix side of the IM's high-affinity ubiquinone site (QP-proximal). The 3Fe–4S redox center's 7Å contains the QP site, which is the main ubiquinone site in yeast SDH (Oyedotun and Lemire 2001; Silkin et al. 2007). Near the IMS side of the IM is where you will find QD–distal, the second low-affinity ubiquinone site. Unlike the two-electron reduction of FAD, ubiquinone reduction happens via two successive single-electron stages. The Qp site, according to Yankovskaya, significantly stabilizes the partly reduced semi-quinone, facilitating the entire reduction to the ubiquinol (Yankovskaya et al. 2003). A conserved Tyr residue in the Qp pocket most likely facilitates the protonation of ubiquinol (Silkin et al. 2007).

Mammal, yeast, and *E. coli* SDHs all include the heme moiety linked to SDH3 and SDH4, although different SDH species differ in the quantity and redox characteristics of their heme moieties. This is in line with the finding that, in contrast to the highly conserved catalytic core domains, membrane domain subunits across SDHs and fumarate reductases exhibit more variability (Hägerhäll 1997). In certain SDH complexes, such as bovine SDH, succinate may diminish the membrane anchor heme but not in others (Oyedotun and Lemire 2001). According to Oyedotun, the mutation of both axial heme ligands produces a heme-free SDH complex that is capable of assembling as well as mediating succinate oxidation in yeast (Oyedotun and Lemire 2001). The double mutant exhibits just a little reduction in catalytic efficiency. Therefore, there is no essential function for the membrane domain heme in catalysis. Similar to this, when produced in an aerobic, *E. coli* fumarate reductase works correctly in succinate oxidation despite lacking heme in its membrane domain (Maklashina et al. 1998).

It is still unknown what the distal QD site and the conserved heme moiety in eukaryotic SDHs mean. The heme may facilitate electron transport to the distal QD site, but ubiquinone reduction on the QP site does not require it. Though there is not much evidence to support this theory, at the QD site, succinate-reduced heme-containing SDH complexes have the potential to produce ubiquinol. Since SDH is unable to pump protons, two Q sites fail to generate a Q cycle as seen in bc1 complex III.

When succinate attaches to an open state in SDH1, the SDH enzymatic activity starts. The binding of succinate causes domain closure, which advances the succinate toward the FAD's oxidized isoalloxazine ring. The covalent bonding of FAD to an active site His residue is necessary for succinate oxidation (Hägerhäll 1997). In the *E. coli* SDH, changing the His residue keeps bound FAD intact, but the mutant enzyme is unable to oxidize succinate. According to Tomasiak, the covalent link raises the FAD redox potential by 60 mV, allowing succinate oxidation to occur (Tomasiak et al. 2008). According to Hao, SDH is the main covalent flavoprotein in yeast. For succinate oxidation to occur, FAD must undergo two-electron reductions, as well as the resulting Fe/S centers constitute single electron carriers. This means that there must be two distinct electron transfer steps connecting $FADH_2$ to the 2Fe–2S center (Hao et al. 2009). The heme moiety and 3Fe–4S core of $FADH_2$ receive electron transfers quickly, resulting in the restoration of oxidized FAD, according to calculations based on the midpoint potentials of the SDH redox cofactors of *E. coli*. One possible explanation for the minimal ROS production from SDH might be the absence of partly reduced FAD (Yankovskaya et al. 2003). The dissociation of semiquinone may result in the formation of ROS. Nonetheless, according to Yankovskaya, the QP is useful in stabilizing semiquinone, which promotes the release of completely reduced ubiquinol (Yankovskaya et al. 2003).

5.4 Functions of SDH

The 70 kDa protein (621 aa) known as the SDHA subunit binds physiological regulators (oxaloacetate and ATP) as well as enzyme substrates (succinate and fumarate) via a covalently linked FAD prosthetic group.

When ATP binds to the enzyme, it is possible that malate, an inferior substrate of SDH, will displace oxaloacetate, a succinate competitive inhibitor (Gutman and Silman 1975). Initially, several iron-sulfur clusters receive the electrons supplied to the SDHA subunit. Before reaching the cyt b type (b558), -binding SDHB subunits (27 kDa, 280 aa; three clusters: [2Fe-2S], [4Fe-4S], and [3FeS].- binds to the SDHC and SDHD subunits (cybL and cybS, 15 and 12-13 kDa, respectively). (Beinert 2002). Two quinone binding sites help transport electrons to the quinone pool (Lemire and Oyedotun 2002). However, further research is necessary to completely comprehend the role of the additional iron-sulfur clusters linked to SDH as well as the b-type cytochrome (Beinert 2002; Oyedotun et al. 2004).

Under normal circumstances, the reverse reaction is scarcely noticeable in the majority of human cells and tissues, whereas the human enzyme easily oxidizes succinate to fumarate. A tunnel diode effect opposes the retrograde direction across a broad range of potentials, which might suggest that evolution has adapted toward aerobic metabolism (Sucheta et al. 1992). However, there is a conjecture that in hypoxic conditions, fumarate decreases sufficiently to result in a physiological impact, perhaps contributing to the production of ROS in the pulmonary vasculature (Paddenberg et al. 2003). Notably, the reverse process, which turns fumarate into succinate and drives fumarate-driven anaerobic respiration, is catalyzed by the enzymes found in a wide range of bacteria. The enzyme's in vivo activity, however, is unexpectedly unpredictable based just on the amino acid sequence (Lancaster 2002).

5.5 Unique Properties of SDH and Its Correlation Regarding the Ubiquinone Pool

Oxaloacetate, SDH's physiological inhibitor, does not provide electrons to the RC as the majority of other dehydrogenases do and dissociates in the active site of the enzyme following reduction of the RC as well as in the presence of ATP (Gutman 1978). Consequently, a large reduction of the UQ pool is nevertheless consistent with the reduced electron transport from succinate to oxygen in the presence of ATP. Moreover, UQ concentration within the mitochondria's inner membrane, which rapidly limits the majority of substrate oxidation after UQ extraction, makes succinate oxidation only mildly limiting (Geromel et al. 2001). This implies that whereas other dehydrogenases do not need UQ, the SDH may be able to "see" it. Therefore, the decrease in electron transport from the NADH dehydrogenase is the primary reason for the oxygen absorption measured using isolated mitochondria in the presence of a mixture of NADH and succinate being smaller than the total of the individual rates (Geromel et al. 1997). This proved yet another time how successfully the SDH fought to reduce the UQ pool. The unique capacity of the SDH to potentially initiate reverse electron flow via respiratory chain complex I, which would then lead to the generation of NADH, further demonstrates the enzyme's strong reducing activity. There have been reports of reverse electron flow in human cells and higher plants, from succinate to NADH (Eaton et al. 1994; Rustin and Lance 1991; Rustin et al. 1996). These findings all lend credence to the theory that SDH has a special function in maintaining the decrease in the mitochondrial UQ pool.

Apart from its function being an electron transporter that promotes the flow of electrons among several dehydrogenases as well as the cytochrome route in the respiratory chain, ubiquinone is also known to operate as a potent antioxidant in biological membranes (Ernster and Dallner 1995). It is present in much higher quantities than other RC electron carriers, such as the total of the dehydrogenases, maybe for this same reason. Then, depending on the dehydrogenases engaged, just a fraction of the UQ pool may be active in electron transport. Therefore, the oxidizing activity of complex III, the reducing behavior of the several dehydrogenases, as well as the kinetic balance within the pool ought should all be involved in the discernible redox state of the UQ pool. As a result, the UQ pool in the inner membrane of the mitochondria serves as an electron sink and, when decreased, an antioxidant reservoir. Nevertheless, UQ is a two-faced substance that may function as a prooxidant when partially reduced to the unstable form of ubisemiquinone or as an antioxidant when completely reduced to ubiquinol (Ernster and Dallner 1995). Semireduced quines, perhaps in conjunction with reduced cytochrome b, are the main source of superoxides produced by the RC in state 4 (Siedow 1982). Lastly, when the RC is malfunctioning, it might create an excessive number of superoxides that include other RC elements, such as complex I flavin radicals (Raha and Robinson 2000). To reduce oxygen toxicity in the mitochondria, it may thus be crucial to provide electrons for the complete reduction of UQ to UQH2. Because of its special redox characteristics, the SDH may thus be a crucial enzyme to regulate the redox state of the UQ pool under these circumstances.

5.6 SDH and Succinate: A Junction Where Cell Metabolism and the Mitochondria Meet

Oxidizing succinate connects SDH to the fast-cycling region of the Krebs cycle (Yudkoff et al. 1994), in which it participates in the total acetyl-coenzyme A (CoA) breakdown in the Krebs cycle. The n-butylmalonate- (or phenylsuccinate-) sensitive dicarboxylate carrier may easily import succinate into the mitochondrial matrix in exchange for malate or inorganic phosphate (Siedow 1982). The location of mitochondrial carriers differs across organs in some animals (Palmieri 1994). It is only partly understood how the mitochondrial carrier apparatus distributes itself among individuals, Nevertheless, it controls the accumulation of both intra- and extra-mitochondrial substrate and product. One possible explanation for tissue-specific susceptibility to anomalies in SDH might be the distribution of carriers. As a result, fumarate, a byproduct of the SDH reaction, often fails to cross the inner membrane of the mitochondria. Fumarate may slow down the SDH process in the mitochondria of skeletal muscle by building up intra-mitochondrially when fumarase is deficient (Bourgeron et al. 1994). A ketoacid-CoA transferase or a GTP-forming succinyl-CoA ligase may also produce succinate in the matrix when acetoacetate is present (Fig. 5.3) (Chayen 1993; Siedow 1982). This latter enzyme, which is found in the heart, kidney, and brain among other extra-hepatic organs, is in charge of ketolysis and consumes ketone bodies produced by the liver (acetoacetate and hydroxybutyrate) under fasting and lipolytic stress. Mevalonate may be produced by combining 3-hydroxy-3-methylglutaryl-CoA (HMG-CoA) synthase as well as HMG-CoA reductase processes with acetoacetyl-CoA, which is obtained from succinate and functions as an insulin secretagogue (MacDonald et al. 2005).

The highly variable enzyme activity across tissues as compared to other RC components suggests a tissue-specific role (Chretien et al. 1994). Compared to other mitochondrial enzymes, such as citrate synthase, SDH activity, and transcript levels are notably high in the mitochondria of the kidney and liver. SDH may provide more electrons to the quinone pool in these tissues than the cytochrome portion of the chain can hold, favoring the quinone pool's over-reduction (Rustin et al. 2002). The electron flow through the RC complex I may then reverse as a result of this over-reduction. It would therefore be possible to redirect electrons from SDH to create NADH in vivo. The high activity of fumarase that is present in these tissues may also facilitate the fast elimination of fumarate which is necessary for a sustained reverse electron flow. The hepatic mitochondria's synthesis of CoA and carnitine

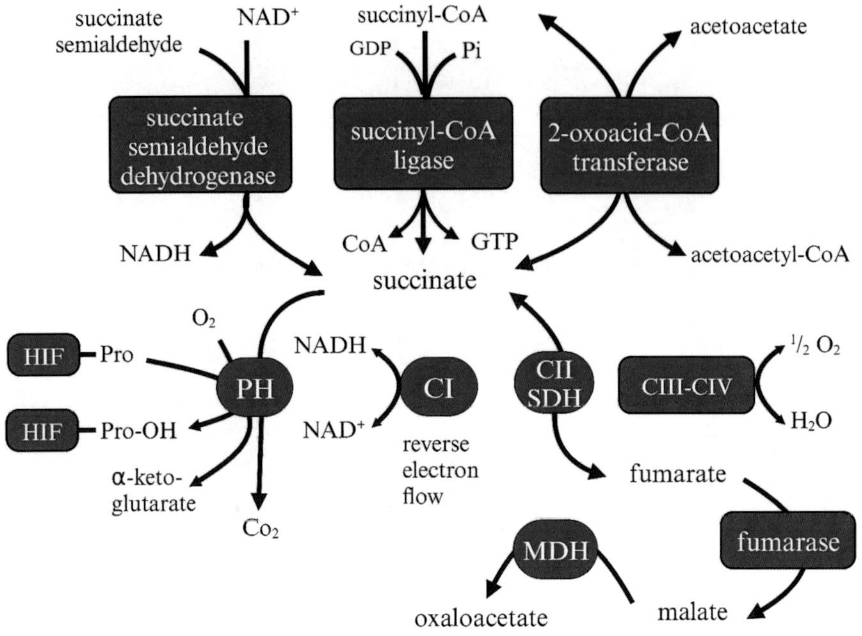

Fig. 5.3 Succinate metabolism

esters might profit from the redox control that the opposite electron flow across the NAD^+ pool imposes (Eaton et al. 1994).

In contrast to the many enzymes in the cell that metabolize other organic acids, very few extra-mitochondrial enzymes metabolize succinate. Since reduced SDH activity causes succinate to accumulate, it is doubtful that any other enzyme activity will be able to buffer it. Even minor SDH deficits might soon cause injury to the cell. Accordingly, 50% residual activity would be detrimental enough to result in late-onset neurodegenerative illness (Birch-Machin et al. 2000). However, the finding that nonsymptomatic members of families of patients with paraganglioma (PGL) may have a heterozygous harmful mutation in one of the SDH genes has cast doubt on this assertion.

The primary sources and enzymatic functions of succinate are inside the cell. The plan also includes the HIF1a prolyl hydroxylase (PH), whose response regulates the activation of the hypoxia-sensitive pathway seen in tumors lacking SDH; respiratory chain complexes I, III, and IV, respectively; MDH, malate dehydrogenase; and hypoxia-inducible factor (HIF).

5.7 Deficits in SDH: Decreased Electron Flow and/or Elevated Oxygen Toxicity

Hereditary deficits in SDH associated with mutations in SDHA correspond with rather significant residual activities, which range from 25% to 50% of control mean values (Bourgeron et al. 1995; Parfait et al. 2000). In contrast, individuals with significant defects of complex IV or I often have residual activity measurements of less than 5%. Patients with such SDH deficiencies, however, do not vary clinically from those with other RC complex problems since they exhibit typical Leigh syndrome (Munnich and Rustin 2001). Remarkably, in two cases with 50% residual SDH activity, a heterozygous mutation in SDHA has also been linked to late-onset optic atrophy (Birch-Machin et al. 2000). There was no mention of a tumor in any of these instances.

The residual activity of the SDHA gene is quite high in individuals with mutations, but in tumor tissues from patients with hereditary paraganglioma, SDH activity is hardly noticeable (Gimenez-Roqueplo et al. 2001). It is not surprising that the single allele expressed in the tumor has very harmful mutations that are the source of the complete loss of function. As previously mentioned, reduced ATP generation (a situation that also arises from severe CI or CIV deficiency) and the RC's inability to maintain the UQ pool's high reduction status due to the absence of SDH activity. This might thus result in the respiratory chain's diminished ability to combat free radicals and perhaps oxidative stress, which is known to easily set off the development of tumors.

5.8 SDH and Oxygen: Adding Worms

Caenorhabditis elegans worm observations corroborate the notion that the SDH has a specialized function in mitochondria's oxygen metabolism. Instead of discovering a class of respiratory chain mutants, it was unexpected to find that an oxygen-hypersensitivity worm had a single SDH mutation (Ishii et al. 1998). Although *C. elegans* has no history of tumor formation, the mutant worm has a shorter life expectancy. Any one of the SDH mutations causes complex II to entirely break down among humans. Then, complex II may be ignored as a potential source of extra superoxide generation since it is lacking. We propose to attribute the overproduction of superoxide, which is responsible for the development of tumors in humans, and the hypersensitivity to oxygen in mutant worms, to the SDH's reduced capacity to sufficiently decrease the UQ pool, an essential need for resistance to oxidative stress.

Thus, research on SDH-associated human illnesses indicates that the enzyme is distinct from other mitochondrial dehydrogenases due to its special redox characteristics, in addition to playing a crucial part in the Krebs cycle and the respiratory chain. SDH would be an essential antioxidant enzyme in the mitochondria that regulates the RC's ability to scavenge superoxides in conjunction with ubiquinone. Therefore, it should come as no surprise that mutations in this multifunctional enzyme are linked to a broad range of human disorders.

5.9 SDH Regulation

Both active site inhibition and post-translational phosphorylation and acetylation affect SDH's catalytic activity. It has been shown that the catalytic activity of mouse SDH1 is reduced by reversible acetylation at several Lys residues (Cimen et al. 2010). The main deacetylase regulating the amount of Sdh1 acetylation is SIRT 3. Similar to acetylation, the phosphorylation of the SDH1 subunit in mammalian cells seems to reduce its activity (Tomitsuka et al. 2009). In vitro, the Fgr tyrosine kinase may phosphorylate SDH1 at two tyrosine residues; however, it is unknown what the physiological implications of this change are (Salvi et al. 2007).

Oxaloacetate, a strong inhibitor, is one of the Krebs cycle intermediates that affects SDH catalytic activity. Succinate encourages oxaloacetate to separate from SDH, which activates the enzyme. The suppression of oxaloacetate might be a factor in the recognized regulation of SDH activity by mitochondrial metabolic state (Gutman and Silman 1975).

5.10 Disease Linked to Malfunctioning of SDH

5.10.1 Leigh Syndrome: SDHA Mutations

Leigh disease, also known as subacute necrotizing encephalomyelopathy (SNEM), is a neurological condition that progresses slowly and develops early (Finsterer 2008). Leigh syndrome patients have a distinctive neuropathology that includes dystonia, vomiting, seizures, lactic acidosis, ataxia, weakness, external ophthalmoplegia, developmental delay or psychomotor regression, and dystonia. The cause of the increasing neuropathy and related symptoms, which are often identified in the early stages of infancy, is either hereditary or spontaneous mitochondrial metabolic malfunction. Patients often present with bilateral lesions that are foci of necrosis along the brain, brain stem, or spinal cord. Where these increasingly necrotic lesions are located will determine the specific symptoms. Leigh syndrome has no known treatment, and after a diagnosis, people often pass away from the condition within a few months.

Among other possible causes of altered mitochondrial function, Leigh syndrome is a genetically diverse illness that results from changes as well as deficiencies within the pyruvate dehydrogenase complex (PDHC), electron transport chain complexes I–V, mutations in the SURF1 gene, and mitochondrial DNA (mtDNA). According to Rustin et al. and Parfait et al., complex II deficiency is very uncommon, making up just 2–4% of respiratory chain deficits (Parfait et al. 2000; Rustin et al. 1997).

It was Bourgeron and colleagues who initially described a mutation in the nuclear-encoded flavoprotein (Fp) subunit gene, also known as SDHA, after a pair of siblings having complex II deficient Leigh's syndrome appeared clinically (Bourgeron et al. 1995). These children's parents were first cousins who had the heterozygous SDHA mutation, which did not exist in 120 controls. The significance of this case report stemmed from the discovery that a nuclear gene mutation (SDHA) causing a mitochondrial respiratory chain deficit was occurring in humans for the first time (Bourgeron et al. 1995). Following this research, Parfait et al. reported on another patient with compound heterozygous mutations in SDHA and complex II deficient Leigh syndrome, some years later (Parfait et al. 2000). From that time on, two further case reports have described the Leigh syndrome-causing mutations in SDHA. These include a patient with compound heterozygous mutations and a homozygous Gly555Glu mutation (Horváth et al. 2006; Van Coster et al. 2003). Horvath et al. looked into six more individuals with isolated complex II deficiency and neurological symptoms of Leigh syndrome, but they were unable to find any SDHA mutations, which adds to the evidence supporting the disorder's genetic variability (Horváth et al. 2006). In a family exhibiting Leigh syndrome-like symptoms of late-onset ocular atrophy, ataxia, and myopathy, Birch-Machin et al. reported single gene SDHA mutations (Horváth et al. 2006). However, these symptoms tended to appear in the fourth decade of life. It is noteworthy that all of these patients had a limited deficit of complex II and SDH in the muscle mitochondria, which might perhaps account for the delayed onset of illness symptoms. Interestingly, none of these progressive neurodegenerative disorders associated with deficits in mitochondrial complex II have yet been linked to mutations in SDHB, -C, -D, or –5.

5.11 Mechanisms of Disease Related with SDH

5.11.1 Leigh Syndrome

Leigh syndrome and/or a progressive neurodegenerative disease have been linked to the mutations found in the nuclear-encoded flavoprotein (Fp) subunit gene (SDHA) of the SDH complex (Birch-Machin et al. 2000; Bourgeron et al. 1995; Horváth et al. 2006; Parfait et al. 2000; Van Coster et al. 2003). It is intriguing to note that, despite being a component of the same mitochondrial complex II that SDH5 constructs, which also contains SDHB, -C, and -D, no instances of these other SDH genes causing neurological problems have been documented. It is unclear what molecular mechanism underlies these deficits in respiratory chain enzymes. The study conducted by Bourgeron et al. revealed that SDHA exhibited heightened sensitivity to the inhibitory impact of oxaloacetate (OAA) (Bourgeron et al. 1995). The authors conjectured that the mutation might potentially modify the conformation or redox state of SDHA, hence potentially affecting enzyme catalysis and reaction to OAA. Both the Krebs cycle and oxidative phosphorylation malfunction in complex II deficits may be factors in the hypoglycemia associated with these illnesses. When compared to normal controls, Van Coster and colleagues found a substantial drop in cross-reacting material (CRM) for the SDHA as well as SDHB subunits. The authors speculate that this might be a factor in complex II's instability or insufficient assembly (Van Coster et al. 2003). Table 5.1 shows a summary of the diseases linked to SDH mutations.

Table 5.1 Synopsis of illnesses associated with SDH mutations

Diseases	PGL1	PGL3	PGL4	Leigh syndrome
SDH gene	SDHD	SDHC	SDHB	SDHA
Chromosomal location	11q23	1q21	1p35–36.1	5p15-3q29

PGL Paraganglioma

Why SDHA mutations, unlike those in SDHB, SDHC, SDHD, and SDH5, have not been linked to cancers in the heterozygous condition is yet unknown. With SDHB as well as SDHC on chromosome 1 along with SDHD as well as SDH5 on chromosome 11, perhaps it has to do with a neglected facet of chromosomal biology. Even though it is clear that chromosome 11 is sensitive to imprinting as well as that this is essential for PGL formation, when both SDHC as and SDHB mutations result in heterozygosity, chromosome 1 dynamics can be significant.

5.12 Conclusion and Future Perspectives

Numerous indicators point to SDH as the primary source of ROS production in mitochondria. The nuclear and mitochondrial genomes encode the majority of the subunits of the complexes in the electron transport chain, while the nuclear genome entirely encodes the SDH complex. Any alteration in SDH activity or assembly may affect the number of metabolites in TCA by tying the TCA cycle and oxidative phosphorylation together. According to studies, brain neurodegeneration is brought on by mutations in SDH subunits and the suppression of its action. SDH forms indirect connections with many signaling pathways, including mTOR/SREBP, which is essential for lipid synthesis. SDH both directly and indirectly controls mTOR/SREBP. For instance, mTOR prevents the SDH from functioning, which results in reduced fatty acid oxidation and lipid buildup. The overall findings of earlier research point to SDH as a promising target for treatments in aging and neurological disorders. In neurodegenerative diseases, increasing SDH activity may regulate lipid overloading, both NMDA-dependent signaling and excitotoxicity. The identification and characterization of the SDH protein have yielded significant new information about the mechanisms by which SDH protects neurons against excess lipid and ROS. Additionally, SDH may be a promising therapeutic target to slow the progression of neurodegeneration.

References

Akundi RS, Huang Z, Eason J, Pandya JD, Zhi L, Cass WA, Sullivan PG, Büeler H (2011) Increased mitochondrial calcium sensitivity and abnormal expression of innate immunity genes precede dopaminergic defects in Pink1-deficient mice. PLoS One 6(1):e16038. https://doi.org/10.1371/journal.pone.0016038

Asiimwe N, Yeo SG, Kim MS, Jung J, Jeong NY (2016) Nitric oxide: exploring the contextual link with Alzheimer's disease. Oxidative Med Cell Longev 2016:7205747. https://doi.org/10.1155/2016/7205747

Astuti D, Latif F, Dallol A, Dahia PL, Douglas F, George E, Sköldberg F, Husebye ES, Eng C, Maher ER (2001) Gene mutations in the succinate dehydrogenase subunit SDHB cause susceptibility to familial pheochromocytoma and to familial paraganglioma. Am J Hum Genet 69(1):49–54. https://doi.org/10.1086/321282

Baysal BE, Ferrell RE, Willett-Brozick JE, Lawrence EC, Myssiorek D, Bosch A, van der Mey A, Taschner PE, Rubinstein WS, Myers EN, Richard CW 3rd, Cornelisse CJ, Devilee P, Devlin B (2000) Mutations in SDHD, a mitochondrial complex II gene, in hereditary paraganglioma. Science 287(5454):848–851. https://doi.org/10.1126/science.287.5454.848

Baysal BE, Rubinstein WS, Taschner PE (2001) Phenotypic dichotomy in mitochondrial complex II genetic disorders. J Mol Med (Berl) 79(9):495–503. https://doi.org/10.1007/s001090100267

Beinert H (2002) Spectroscopy of succinate dehydrogenases, a historical perspective. Biochim Biophys Acta 1553(1–2):7–22. https://doi.org/10.1016/s0005-2728(01)00235-3

Benecke R, Strümper P, Weiss H (1993) Electron transfer complexes I and IV of platelets are abnormal in Parkinson's disease but normal in Parkinson-plus syndromes. Brain 116(Pt 6):1451–1463. https://doi.org/10.1093/brain/116.6.1451

Birch-Machin MA, Taylor RW, Cochran B, Ackrell BA, Turnbull DM (2000) Late-onset optic atrophy, ataxia, and myopathy associated with a mutation of a complex II gene. Ann Neurol 48(3):330–335

Bolaños JP, Heales SJ, Land JM, Clark JB (1995) Effect of peroxynitrite on the mitochondrial respiratory chain: differential susceptibility of neurones and astrocytes in primary culture. J Neurochem 64(5):1965–1972. https://doi.org/10.1046/j.1471-4159.1995.64051965.x

Bourgeron T, Chretien D, Poggi-Bach J, Doonan S, Rabier D, Letouzé P, Munnich A, Rötig A, Landrieu P, Rustin P (1994) Mutation of the fumarase gene in two siblings with progressive encephalopathy and fumarase deficiency. J Clin Invest 93(6):2514–2518. https://doi.org/10.1172/jci117261

Bourgeron T, Rustin P, Chretien D, Birch-Machin M, Bourgeois M, Viegas-Péquignot E, Munnich A, Rötig A (1995) Mutation of a nuclear succinate dehydrogenase gene results in mitochondrial respiratory chain deficiency. Nat Genet 11(2):144–149. https://doi.org/10.1038/ng1095-144

Brenner C, Grimm S (2006) The permeability transition pore complex in cancer cell death. Oncogene 25(34):4744–4756. https://doi.org/10.1038/sj.onc.1209609

Brookes PS, Land JM, Clark JB, Heales SJ (1998) Peroxynitrite and brain mitochondria: evidence for increased proton leak. J Neurochem 70(5):2195–2202. https://doi.org/10.1046/j.1471-4159.1998.70052195.x

Brustovetsky N, Dubinsky JM, Antonsson B, Jemmerson R (2003) Two pathways for tBID-induced cytochrome c release from rat brain mitochondria: BAK- versus BAX-dependence. J Neurochem 84(1):196–207. https://doi.org/10.1046/j.1471-4159.2003.01545.x

Casley CS, Canevari L, Land JM, Clark JB, Sharpe MA (2002) Beta-amyloid inhibits integrated mitochondrial respiration and key enzyme activities. J Neurochem 80(1):91–100. https://doi.org/10.1046/j.0022-3042.2001.00681.x

Cecchini G, Schröder I, Gunsalus RP, Maklashina E (2002) Succinate dehydrogenase and fumarate reductase from Escherichia coli. Biochim Biophys Acta 1553(1–2):140–157. https://doi.org/10.1016/s0005-2728(01)00238-9

Chávez E, Meléndez E, Zazueta C, Reyes-Vivas H, Perales SG (1997) Membrane permeability transition as induced by dysfunction of the electron transport chain. Biochem Mol Biol Int 41(5):961–968. https://doi.org/10.1080/15216549700202021

Chayen J (1993) The mitochondrion in health and disease. D. Tyler VCH Publishers: Weinheim, New York. xv + 557 pages, 225 DM (£84) (1992). Cell Biochem Funct 11(4): 291–291. https://doi.org/10.1002/cbf.290110411

Choo YS, Johnson GV, MacDonald M, Detloff PJ, Lesort M (2004) Mutant huntingtin directly increases susceptibility of mitochondria to the calcium-induced permeability transition and cytochrome c release. Hum Mol Genet 13(14):1407–1420. https://doi.org/10.1093/hmg/ddh162

Chretien D, Rustin P, Bourgeron T, Rötig A, Saudubray JM, Munnich A (1994) Reference charts for respiratory chain activities in human tissues. Clin Chim Acta 228(1):53–70. https://doi.org/10.1016/0009-8981(94)90056-6

Cimen H, Han MJ, Yang Y, Tong Q, Koc H, Koc EC (2010) Regulation of succinate dehydrogenase activity by SIRT3 in mammalian mitochondria. Biochemistry 49(2):304–311. https://doi.org/10.1021/bi901627u

Coyle JT, Schwarcz R (1976) Lesion of striatal neurones with kainic acid provides a model for Huntington's chorea. Nature 263(5574):244–246. https://doi.org/10.1038/263244a0

Damiano M, Diguet E, Malgorn C, D'Aurelio M, Galvan L, Petit F, Benhaim L, Guillermier M, Houitte D, Dufour N, Hantraye P, Canals JM, Alberch J, Delzescaux T, Déglon N, Beal MF, Brouillet E (2013) A role of mitochondrial complex II defects in genetic models of Huntington's disease expressing N-terminal fragments of mutant huntingtin. Hum Mol Genet 22(19):3869–3882. https://doi.org/10.1093/hmg/ddt242

Davis RE, Williams M (2012) Mitochondrial function and dysfunction: an update. J Pharmacol Exp Ther 342(3):598–607. https://doi.org/10.1124/jpet.112.192104

Dawson VL, Dawson TM (1996) Nitric oxide neurotoxicity. J Chem Neuroanat 10(3–4):179–190. https://doi.org/10.1016/0891-0618(96)00148-2

Demuro A, Mina E, Kayed R, Milton SC, Parker I, Glabe CG (2005) Calcium dysregulation and membrane disruption as a ubiquitous neurotoxic mechanism of soluble amyloid oligomers. J Biol Chem 280(17):17294–17300. https://doi.org/10.1074/jbc.M500997200

Deng Y, Xiong Z, Chen P, Wei J, Chen S, Yan Z (2014) B-Amyloid impairs the regulation of N-methyl-D-aspartate receptors by glycogen synthase kinase 3. Neurobiol Aging 35(3):449–459. https://doi.org/10.1016/j.neurobiolaging.2013.08.031

Devi L, Anandatheerthavarada HK (2010) Mitochondrial trafficking of APP and alpha synuclein: relevance to mitochondrial dysfunction in Alzheimer's and Parkinson's diseases. Biochim Biophys Acta 1802(1):11–19. https://doi.org/10.1016/j.bbadis.2009.07.007

Eaton S, Turnbull DM, Bartlett K (1994) Redox control of β-oxidation in rat liver mitochondria. Eur J Biochem 220(3):671–681. https://doi.org/10.1111/j.1432-1033.1994.tb18668.x

Eichner LJ, Giguère V (2011) Estrogen related receptors (ERRs): a new dawn in transcriptional control of mitochondrial gene networks. Mitochondrion 11(4):544–552. https://doi.org/10.1016/j.mito.2011.03.121

Ernster L, Dallner G (1995) Biochemical, physiological and medical aspects of ubiquinone function. Biochim Biophys Acta 1271(1):195–204. https://doi.org/10.1016/0925-4439(95)00028-3

Finsterer J (2008) Leigh and leigh-like syndrome in children and adults. Pediatr Neurol 39(4):223–235. https://doi.org/10.1016/j.pediatrneurol.2008.07.013

Fukai T, Ushio-Fukai M (2011) Superoxide dismutases: role in redox signaling, vascular function, and diseases. Antioxid Redox Signal 15(6):1583–1606. https://doi.org/10.1089/ars.2011.3999

Furukawa K, Matsuzaki-Kobayashi M, Hasegawa T, Kikuchi A, Sugeno N, Itoyama Y, Wang Y, Yao PJ, Bushlin I, Takeda A (2006) Plasma membrane ion permeability induced by mutant alpha-synuclein contributes to the degeneration of neural cells. J Neurochem 97(4):1071–1077. https://doi.org/10.1111/j.1471-4159.2006.03803.x

Geromel V, Parfait B, von Kleist-Retzow JC, Chretien D, Munnich A, Rötig A, Rustin P (1997) The consequences of a mild respiratory chain deficiency on substrate competitive oxidation in human mitochondria. Biochem Biophys Res Commun 236(3):643–646. https://doi.org/10.1006/bbrc.1997.7024

Geromel V, Kadhom N, Ceballos-Picot I, Chrétien D, Munnich A, Rötig A, Rustin P (2001) Human cultured skin fibroblasts survive profound inherited ubiquinone depletion. Free Radic Res 35(1):11–21. https://doi.org/10.1080/10715760100300551

Gimenez-Roqueplo AP, Favier J, Rustin P, Mourad JJ, Plouin PF, Corvol P, Rötig A, Jeunemaitre X (2001) The R22X mutation of the SDHD gene in hereditary paraganglioma abolishes the enzymatic activity of complex II in the mitochondrial respiratory chain and activates the hypoxia pathway. Am J Hum Genet 69(6):1186–1197. https://doi.org/10.1086/324413

Giorgi C, Agnoletto C, Bononi A, Bonora M, De Marchi E, Marchi S, Missiroli S, Patergnani S, Poletti F, Rimessi A, Suski JM, Wieckowski MR, Pinton P (2012) Mitochondrial calcium homeostasis as potential target for mitochondrial medicine. Mitochondrion 12(1):77–85. https://doi.org/10.1016/j.mito.2011.07.004

Gu M, Gash MT, Mann VM, Javoy-Agid F, Cooper JM, Schapira AH (1996) Mitochondrial defect in Huntington's disease caudate nucleus. Ann Neurol 39(3):385–389. https://doi.org/10.1002/ana.410390317

Gutman M (1978) Modulation of mitochondrial succinate dehydrogenase activity, mechanism and function. Mol Cell Biochem 20(1):41–60. https://doi.org/10.1007/BF00229453

Gutman M, Silman N (1975) The steady state activity of succinate dehydrogenase in the presence of opposing effectors.II. Reductive activation of succinate dehydrogenase in presence of oxaloacetate. Mol Cell Biochem 7(3):177–185. https://doi.org/10.1007/bf01731407

Haas RH, Nasirian F, Nakano K, Ward D, Pay M, Hill R, Shults CW (1995) Low platelet mitochondrial complex I and complex II/III activity in early untreated Parkinson's disease. Ann Neurol 37(6):714–722. https://doi.org/10.1002/ana.410370604

Hägerhäll C (1997) Succinate: quinone oxidoreductases. Variations on a conserved theme. Biochim Biophys Acta 1320(2):107–141. https://doi.org/10.1016/s0005-2728(97)00019-4

Hao HX, Khalimonchuk O, Schraders M, Dephoure N, Bayley JP, Kunst H, Devilee P, Cremers CW, Schiffman JD, Bentz BG, Gygi SP, Winge DR, Kremer H, Rutter J (2009) SDH5, a gene required for flavination of succinate dehydrogenase, is mutated in paraganglioma. Science 325(5944):1139–1142. https://doi.org/10.1126/science.1175689

Hara MR, Snyder SH (2007) Cell signaling and neuronal death. Annu Rev Pharmacol Toxicol 47:117–141. https://doi.org/10.1146/annurev.pharmtox.47.120505.105311

Hattori N, Tanaka M, Ozawa T, Mizuno Y (1991) Immunohistochemical studies on complexes I, II, III, and IV of mitochondria in Parkinson's disease. Ann Neurol 30(4):563–571. https://doi.org/10.1002/ana.410300409

Horváth R, Abicht A, Holinski-Feder E, Laner A, Gempel K, Prokisch H, Lochmüller H, Klopstock T, Jaksch M (2006) Leigh syndrome caused by mutations in the flavoprotein (Fp) subunit of succinate dehydrogenase (SDHA). J Neurol Neurosurg Psychiatry 77(1):74–76. https://doi.org/10.1136/jnnp.2005.067041

Ishii N, Fujii M, Hartman PS, Tsuda M, Yasuda K, Senoo-Matsuda N, Yanase S, Ayusawa D, Suzuki K (1998) A mutation in succinate dehydrogenase cytochrome b causes oxidative stress and ageing in nematodes. Nature 394(6694):694–697. https://doi.org/10.1038/29331

Jamwal S, Kumar P (2017) L-theanine, a component of green tea prevents 3-nitropropionic acid (3-NP)-induced striatal toxicity by modulating nitric oxide pathway. Mol Neurobiol 54(3):2327–2337. https://doi.org/10.1007/s12035-016-9822-5

Jiménez-Jiménez FJ, Alonso-Navarro H, Herrero MT, García-Martín E, Agúndez JA (2016) An update on the role of nitric oxide in the neurodegenerative processes of Parkinson's disease. Curr Med Chem 23(24):2666–2679. https://doi.org/10.2174/0929867323666160812151356

Jodeiri Farshbaf M, Ghaedi K, Megraw TL, Curtiss J, Shirani Faradonbeh M, Vaziri P, Nasr-Esfahani MH (2016) Does PGC1α/FNDC5/BDNF elicit the beneficial effects of exercise on neurodegenerative disorders? NeuroMolecular Med 18(1):1–15. https://doi.org/10.1007/s12017-015-8370-x

Knowles RG, Moncada S (1994) Nitric oxide synthases in mammals. Biochem J 298(Pt 2):249–258. https://doi.org/10.1042/bj2980249

Kraytsberg Y, Kudryavtseva E, McKee AC, Geula C, Kowall NW, Khrapko K (2006) Mitochondrial DNA deletions are abundant and cause functional impairment in aged human substantia Nigra neurons. Nat Genet 38(5):518–520. https://doi.org/10.1038/ng1778

Lancaster CR (2002) Succinate:quinone oxidoreductases: an overview. Biochim Biophys Acta 1553(1–2):1–6. https://doi.org/10.1016/s0005-2728(01)00240-7

Langston JW, Ballard PA Jr (1983) Parkinson's disease in a chemist working with 1-methyl-4-phenyl-1,2,5,6-tetrahydropyridine. N Engl J Med 309(5):310. https://doi.org/10.1056/nejm198308043090511

Legros F, Malka F, Frachon P, Lombès A, Rojo M (2004) Organization and dynamics of human mitochondrial DNA. J Cell Sci 117(Pt 13):2653–2662. https://doi.org/10.1242/jcs.01134

Lemire BD, Oyedotun KS (2002) The Saccharomyces cerevisiae mitochondrial succinate:ubiquinone oxidoreductase. Biochim Biophys Acta 1553(1–2):102–116. https://doi.org/10.1016/s0005-2728(01)00229-8

Lin MT, Beal MF (2006) Mitochondrial dysfunction and oxidative stress in neurodegenerative diseases. Nature 443(7113):787–795. https://doi.org/10.1038/nature05292

Long J, He P, Shen Y, Li R (2012) New evidence of mitochondria dysfunction in the female Alzheimer's disease brain: deficiency of estrogen receptor-β. J Alzheimers Dis 30(3):545–558. https://doi.org/10.3233/jad-2012-120283

Loschen G, Azzi A, Richter C, Flohé L (1974) Superoxide radicals as precursors of mitochondrial hydrogen peroxide. FEBS Lett 42(1):68–72. https://doi.org/10.1016/0014-5793(74)80281-4

MacDonald MJ, Fahien LA, Brown LJ, Hasan NM, Buss JD, Kendrick MA (2005) Perspective: emerging evidence for signaling roles of mitochondrial anaplerotic products in insulin secretion. Am J Physiol Endocrinol Metab 288(1):E1–E15. https://doi.org/10.1152/ajpendo.00218.2004

Mailloux RJ (2015) Teaching the fundamentals of electron transfer reactions in mitochondria and the production and detection of reactive oxygen species. Redox Biol 4:381–398. https://doi.org/10.1016/j.redox.2015.02.001

Maklashina E, Berthold DA, Cecchini G (1998) Anaerobic expression of Escherichia coli succinate dehydrogenase: functional replacement of fumarate reductase in the respiratory chain during anaerobic growth. J Bacteriol 180(22):5989–5996. https://doi.org/10.1128/jb.180.22.5989-5996.1998

Manczak M, Park BS, Jung Y, Reddy PH (2004) Differential expression of oxidative phosphorylation genes in patients with Alzheimer's disease: implications for early mitochondrial dysfunction and oxidative damage. NeuroMolecular Med 5(2):147–162. https://doi.org/10.1385/nmm:5:2:147

Martin LJ (2010) Mitochondrial and cell death mechanisms in neurodegenerative diseases. Pharmaceuticals (Basel) 3(4):839–915. https://doi.org/10.3390/ph3040839

Martin LJ, Pan Y, Price AC, Sterling W, Copeland NG, Jenkins NA, Price DL, Lee MK (2006) Parkinson's disease alpha-synuclein transgenic mice develop neuronal mitochondrial degeneration and cell death. J Neurosci 26(1):41–50. https://doi.org/10.1523/jneurosci.4308-05.2006

Moncada S, Bolaños JP (2006) Nitric oxide, cell bioenergetics and neurodegeneration. J Neurochem 97(6):1676–1689. https://doi.org/10.1111/j.1471-4159.2006.03988.x

Munnich A, Rustin P (2001) Clinical spectrum and diagnosis of mitochondrial disorders. Am J Med Genet 106(1):4–17. https://doi.org/10.1002/ajmg.1391

Nakamura K, Yamaki M, Sarada M, Nakayama S, Vibat CRT, Gennis RB, Nakayashiki T, Inokuchi H, Kojima S, Kita K (1996) Two hydrophobic subunits are essential for the Heme b ligation and functional assembly of complex II (succinate-ubiquinone oxidoreductase) from Escherichia coli(∗). J Biol Chem 271(1):521–527. https://doi.org/10.1074/jbc.271.1.521

Nicholls DG (2002) Mitochondrial function and dysfunction in the cell: its relevance to aging and aging-related disease. Int J Biochem Cell Biol 34(11):1372–1381. https://doi.org/10.1016/s1357-2725(02)00077-8

Niemann S, Müller U (2000) Mutations in SDHC cause autosomal dominant paraganglioma, type 3. Nat Genet 26(3):268–270. https://doi.org/10.1038/81551

Oyedotun KS, Lemire BD (2001) The Quinone-binding sites of the Saccharomyces cerevisiae succinate-ubiquinone oxidoreductase. J Biol Chem 276(20):16936–16943. https://doi.org/10.1074/jbc.M100184200

Oyedotun KS, Yau PF, Lemire BD (2004) Identification of the heme axial ligands in the cytochrome b562 of the Saccharomyces cerevisiae succinate dehydrogenase. J Biol Chem 279(10):9432–9439. https://doi.org/10.1074/jbc.M311877200

Paddenberg R, Ishaq B, Goldenberg A, Faulhammer P, Rose F, Weissmann N, Braun-Dullaeus RC, Kummer W (2003) Essential role of complex II of the respiratory chain in hypoxia-induced ROS generation in the pulmonary vasculature. Am J Physiol Lung Cell Mol Physiol 284(5):L710–L719. https://doi.org/10.1152/ajplung.00149.2002

Palmieri F (1994) Mitochondrial carrier proteins. FEBS Lett 346(1):48–54. https://doi.org/10.1016/0014-5793(94)00329-7

Parfait B, Chretien D, Rötig A, Marsac C, Munnich A, Rustin P (2000) Compound heterozygous mutations in the flavoprotein gene of the respiratory chain complex II in a patient with Leigh syndrome. Hum Genet 106(2):236–243. https://doi.org/10.1007/s004390051033

Petlicki J, van de Ven TGM (1998) The equilibrium between the oxidation of hydrogen peroxide by oxygen and the dismutation of peroxyl or superoxide radicals in aqueous solutions in contact with oxygen [10.1039/A804551H]. J Chem Soc Faraday Trans 94(18):2763–2767. https://doi.org/10.1039/A804551H

Poderoso JJ, Peralta JG, Lisdero CL, Carreras MC, Radisic M, Schöpfer F, Cadenas E, Boveris A (1998) Nitric oxide regulates oxygen uptake and hydrogen peroxide release by the isolated beating rat heart. Am J Phys 274(1):C112–C119. https://doi.org/10.1152/ajpcell.1998.274.1.C112

Radi R, Beckman JS, Bush KM, Freeman BA (1991) Peroxynitrite oxidation of sulfhydryls. The cytotoxic potential of superoxide and nitric oxide. J Biol Chem 266(7):4244–4250

Raha S, Robinson BH (2000) Mitochondria, oxygen free radicals, disease and ageing. Trends Biochem Sci 25(10):502–508. https://doi.org/10.1016/s0968-0004(00)01674-1

Rizzuto R, Bernardi P, Pozzan T (2000) Mitochondria as all-round players of the calcium game. J Physiol 529(Pt 1):37–47. https://doi.org/10.1111/j.1469-7793.2000.00037.x

Rustin P, Lance C (1991) Succinate-driven reverse electron transport in the respiratory chain of plant mitochondria. The effects of rotenone and adenylates in relation to malate and oxaloacetate metabolism. Biochem J 274(Pt 1):249–255. https://doi.org/10.1042/bj2740249

Rustin P, Rötig A. (2002) Inborn errors of complex II–unusual human mitochondrial diseases. Biochim Biophys Acta 1553(1–2):117–122. https://doi.org/10.1016/s0005-2728(01)00228-6

Rustin P, Parfait B, Chretien D, Bourgeron T, Djouadi F, Bastin J, Rötig A, Munnich A (1996) Fluxes of nicotinamide adenine dinucleotides through mitochondrial membranes in human cultured cells. J Biol Chem 271(25):14785–14790. https://doi.org/10.1074/jbc.271.25.14785

Rustin P, Bourgeron T, Parfait B, Chretien D, Munnich A, Rötig A (1997) Inborn errors of the Krebs cycle: a group of unusual mitochondrial diseases in human. Biochim Biophys Acta 1361(2):185–197. https://doi.org/10.1016/s0925-4439(97)00035-5

Rustin P, Munnich A, Rötig A (2002) Succinate dehydrogenase and human diseases: new insights into a well-known enzyme. Eur J Hum Genet 10(5):289–291. https://doi.org/10.1038/sj.ejhg.5200793

Salvi M, Morrice NA, Brunati AM, Toninello A (2007) Identification of the flavoprotein of succinate dehydrogenase and aconitase as *in vitro* mitochondrial substrates of Fgr tyrosine kinase. FEBS Lett 581(29):5579–5585. https://doi.org/10.1016/j.febslet.2007.11.005

Scimemi A, Meabon JS, Woltjer RL, Sullivan JM, Diamond JS, Cook DG (2013) Amyloid-β1-42 slows clearance of synaptically released glutamate by mislocalizing astrocytic GLT-1. J Neurosci 33(12):5312–5318. https://doi.org/10.1523/jneurosci.5274-12.2013

Shehadeh J, Fernandes HB, Zeron Mullins MM, Graham RK, Leavitt BR, Hayden MR, Raymond LA (2006) Striatal neuronal apoptosis is preferentially enhanced by NMDA receptor activation in YAC transgenic mouse model of Huntington disease. Neurobiol Dis 21(2):392–403. https://doi.org/10.1016/j.nbd.2005.08.001

Shi X, Li W, Liu H, Yin D, Zhao J (2017) The ROS/NF-κB/NR4A2 pathway is involved in H_2O_2 induced apoptosis of resident cardiac stem cells via autophagy. Oncotarget 8(44):77634–77648. Retrieved 2017/09, from http://europepmc.org/abstract/MED/29100414. http://www.oncotarget.com/index.php?journal=oncotarget&page=article&op=download&path%5B%5D=20747&path%5B%5D=66090. https://doi.org/10.18632/oncotarget.20747. https://europepmc.org/articles/PMC5652805. https://europepmc.org/articles/PMC5652805?pdf=render

Siedow JN (1982) Organelles: Mitochondria. Alexander Tzagoloff. Plenum, New York, 1982. xvi, 342 pp., illus. Cloth, $42.50; paper, $19.95. Cellular organelles. Science 217(4565):1134–1134. https://doi.org/10.1126/science.217.4565.1134

Silkin Y, Oyedotun KS, Lemire BD (2007) The role of SDH4p Tyr-89 in ubiquinone reduction by the Saccharomyces cerevisiae succinate dehydrogenase. Biochim Biophys Acta 1767(2):143–150. https://doi.org/10.1016/j.bbabio.2006.11.017

Silva AC, Almeida S, Laço M, Duarte AI, Domingues J, Oliveira CR, Januário C, Rego AC (2013) Mitochondrial respiratory chain complex activity and bioenergetic alterations in human platelets derived from pre-symptomatic and symptomatic Huntington's disease carriers. Mitochondrion 13(6):801–809. https://doi.org/10.1016/j.mito.2013.05.006

Soussi B, Idström JP, Scherstén T, Bylund-Fellenius AC (1990) Cytochrome c oxidase and cardiolipin alterations in response to skeletal muscle ischaemia and reperfusion. Acta Physiol Scand 138(2):107–114. https://doi.org/10.1111/j.1748-1716.1990.tb08822.x

Sparagna GC, Gunter KK, Sheu SS, Gunter TE (1995) Mitochondrial calcium uptake from physiological-type pulses of calcium. A description of the rapid uptake mode. J Biol Chem 270(46):27510–27515. https://doi.org/10.1074/jbc.270.46.27510

Stout AK, Raphael HM, Kanterewicz BI, Klann E, Reynolds IJ (1998) Glutamate-induced neuron death requires mitochondrial calcium uptake. Nat Neurosci 1(5):366–373. https://doi.org/10.1038/1577

Sucheta A, Ackrell BA, Cochran B, Armstrong FA (1992) Diode-like behaviour of a mitochondrial electron-transport enzyme. Nature 356(6367):361–362. https://doi.org/10.1038/356361a0

Sun F, Huo X, Zhai Y, Wang A, Xu J, Su D, Bartlam M, Rao Z (2005) Crystal structure of mitochondrial respiratory membrane protein complex II. Cell 121(7):1043–1057. https://doi.org/10.1016/j.cell.2005.05.025

Tabrizi SJ, Workman J, Hart PE, Mangiarini L, Mahal A, Bates G, Cooper JM, Schapira AH (2000) Mitochondrial dysfunction and free radical damage in the Huntington R6/2 transgenic mouse. Ann Neurol 47(1):80–86. https://doi.org/10.1002/1531-8249(200001)47:1<80::aid-ana13>3.3.co;2-b

Tomasiak TM, Maklashina E, Cecchini G, Iverson TM (2008) A threonine on the active site loop controls transition state formation in Escherichia coli respiratory complex II. J Biol Chem 283(22):15460–15468. https://doi.org/10.1074/jbc.M801372200

Tomitsuka E, Kita K, Esumi H (2009) Regulation of succinate-ubiquinone reductase and fumarate reductase activities in human complex II by phosphorylation of its flavoprotein subunit. Proc Jpn Acad Ser B Phys Biol Sci 85(7):258–265. https://doi.org/10.2183/pjab.85.258

Van Coster R, Seneca S, Smet J, Van Hecke R, Gerlo E, Devreese B, Van Beeumen J, Leroy JG, De Meirleir L, Lissens W (2003) Homozygous Gly555Glu mutation in the nuclear-encoded 70 kDa flavoprotein gene causes instability of the respiratory chain complex II. Am J Med Genet A 120a(1):13–18. https://doi.org/10.1002/ajmg.a.10202

Weisiger RA, Fridovich I (1973) Superoxide dismutase. Organelle specificity. J Biol Chem 248(10):3582–3592

Yan MH, Wang X, Zhu X (2013) Mitochondrial defects and oxidative stress in Alzheimer disease and Parkinson disease. Free Radic Biol Med 62:90–101. https://doi.org/10.1016/j.freeradbiomed.2012.11.014

Yankovskaya V, Horsefield R, Törnroth S, Luna-Chavez C, Miyoshi H, Léger C, Byrne B, Cecchini G, Iwata S (2003) Architecture of succinate dehydrogenase and reactive oxygen species generation. Science 299(5607):700–704. https://doi.org/10.1126/science.1079605

Yudkoff M, Nelson D, Daikhin Y, Erecińska M (1994) Tricarboxylic acid cycle in rat brain synaptosomes. Fluxes and interactions with aspartate aminotransferase and malate/aspartate shuttle. J Biol Chem 269(44):27414–27420

Zhang Y, Tan F, Xu P, Qu S (2016) Recent advance in the relationship between excitatory amino acid transporters and Parkinson's disease. Neural Plast 2016:8941327. https://doi.org/10.1155/2016/8941327

Zipfel GJ, Babcock DJ, Lee JM, Choi DW (2000) Neuronal apoptosis after CNS injury: the roles of glutamate and calcium. J Neurotrauma 17(10):857–869. https://doi.org/10.1089/neu.2000.17.857

Role of SDH in Cardiovascular Disorders

6

Vaishali Manikrao Patil, Neeraj Masand,
Praveen Thaggikuppe Krishnamurthy, and Satya P. Gupta

Abstract

Succinate dehydrogenase (SDH) is an enzyme linking the respiratory chain with the citric acid cycle. Cardiac-specific loss of SDH and other defects contributes toward molecular basis of the pathology of the cardiac disorders. Various cardiac disorders, namely ischemia, Barth syndrome, etc., have been well studied toward understanding the role of SDH in disease pathology.

In animal models, inhibition of SDH with malonate at the initiation of reperfusion has reduced generation of ROS and thus lowered the size of infarct. It also lowers the opening of mitochondrial permeability transition pore. SDH inhibitors such as diazoxide and 3-nitropropionic acid have also exhibited cardioprotective effect. Among cardiac arrest cases, brain damage is identified as the major cause of death. The mechanism related to succinate accumulation during ischemia by SDH is proposed. Thus, it is apparent to target succinate metabolism toward the development of potential therapeutics to treat cardiovascular diseases.

V. M. Patil (✉)
Charak School of Pharmacy, Chaudhary Charan Singh University (CCSU), Meerut, Uttar Pradesh, India

N. Masand
Department of Pharmacy, Lala Lajpat Rai Memorial Medical College, Meerut, Uttar Pradesh, India

P. T. Krishnamurthy
Department of Pharmacology, JSS College of Pharmacy, JSS Academy of Technical Education, Noida, Uttar Pradesh, India

S. P. Gupta
Ex-Professor, Birla Institute of Technology and Science (BITS), Pilani, Rajasthan, India

Meerut Institute of Engineering and Technology, Meerut, Uttar Pradesh, India

V. M. Patil et al. (eds.), *Deciphering The Role of Succinate Dehydrogenase in Drug Discovery*, https://doi.org/10.1007/978-981-96-6831-1_6

Keywords

Succinate dehydrogenase · SDH inhibitors · Cardioprotective effect · Ischemia · Barth syndrome

Abbreviations

AA5	Atpenin
BTHS	Barth Syndrome
CL	Cardiolipin
DMM	Dimethyl malonate
DSM	Disodium malonate
HIF	Hypoxia-inducible factor
IR	Ischemia–Reperfusion
MPTP	Mitochondrial permeability transition pore
RET	Reverse electron transport
ROS	Reactive oxygen species
ROSC	Return of spontaneous circulation
SDH	Succinate dehydrogenase
SDHB	Succinate dehydrogenase B
STEMI	Segment elevation myocardial infarction
SUCNR1	Succinate receptor 1
TCA	Tricarboxylic acid
3-NPA	3-Nitropropionic acid

6.1 Introduction

The AHA Heart Disease and Stroke Statistical Update released in 2024 reports data and the economic burden of diseases of the heart and circulatory system like cardiac arrest, congenital heart disease, cardiomyopathy, heart failure, atherosclerosis, hypertension, valvular disease, venous thromboembolism, peripheral artery disease, stroke, kidney disorders, etc. (Martin et al. 2024). The global burden of around 18 cardiovascular conditions suggests requirement to establish strategies toward prevention and treatment (Vaduganathan et al. 2022). To reduce the burden of this epidemiology and associated risk factors, preventive and treatment options have been widely studied and reviewed (Joseph et al. 2017; Leong et al. 2017). Therapeutic innovations have shown dramatic changes and focus is toward developing better understanding of the spectrum of cardiovascular pharmacology (Young and Raizner 2022). For the prevention of ischemia–reperfusion injury in myocardial infarction, effective therapeutics is not available (Heusch 2017, 2020; Heusch and Gersh 2017). Mechanism of mitochondria in the destructive events of ischemia–reperfusion injury is studied (Bøtker et al. 2020; Davidson et al. 2019, 2020; Ferdinandy et al. 2023; Zuurbier et al. 2020). Mitochondria not only generated ATP but also maintains ratio of ATP/ADP and consumption of oxygen for cardiomyocytes (Murashige et al. 2020; Hahn et al. 2023) and plays an important role in the pathology of ischemia. During ischemia, rapid consumption of oxygen is observed (Kalogeris et al. 2012; Heusch and Gersh 2017). Mitochondria has composite and complex effect in ischemia–reperfusion injury, i.e., it acts as key metabolic and signaling centers supporting pathophysiology (Davidson et al. 2020; Lesnefsky et al. 2017; Murphy and Steenbergen 2008; Picard and Shirihai 2022). Mitochondrial metabolites have an important role to interact with the cell, designing phenotype and shaping response to stimuli (Frezza 2017; Chandel 2021; Picard and Shirihai 2022). This is apparent in the case of ischemia and reperfusion, which defines shift response from homeostasis and the injury-related phenomenon (Hausenloy and Yellon 2013; Lesnefsky et al. 2017; Heusch 2020).

In the tricarboxylic acid cycle and glutamine metabolism, succinate is one of the major metabolic intermediates. Succinate also has role in regulating metabolism and immune homeostasis in different pathophysiologies. It plays an important role in signaling through its binding with the receptor. In the diagnosis and treatment of disorders related to the cardiovascular system, succinate has potential role as biomarker. Its definite function is not well established in the mode of action of cardiovascular therapeutics. As observed with the available studies, the disease progression cannot be controlled with available options, and thus, it requires newer interventions for its prevention and treatment. Thus, understanding the pivotal role of mitochondria in initiating the IR injury by generation of reactive oxygen species (ROS) is important to develop potential therapeutics. In addition, the reported studies support the involvement of succinate and suggest its significance toward the development of cardiovascular therapeutics. This chapter elaborates the role of succinate dehydrogenase in cardiovascular disorders as well as Barth syndrome.

6.2 Cardioprotective Role of SDH Inhibitors

In the case of ischemia, for succinate formation, SDH is a key enzyme and its oxidation upon reperfusion (Davidson et al. 2019). SDH is an important target to lower the concentration of succinate due to its protective effect (Chouchani et al. 2014). Cardioprotective effect of SDH inhibition has been evaluated in nondiabetic animal models. Mitochondrial dysfunction is reported with diabetes, and chances of ischemia are progressive with disease severity and duration (Bugger and Abel 2010; Ferdinandy et al. 2014; Jespersen et al. 2020). When compared to nondiabetic patients, myocardial infarction is found to be impaired among diabetic patients (Hafner et al. 1998). Alterations of the mitochondrial metabolism in the hearts of mature diabetic patients cause lack of cardioprotective effect through inhibition of SDH (Jespersen et al. 2020). To reduce reperfusion injury, malonate that acts as a competitive and selective inhibitor of SDH is an important target (Chouchani et al. 2014).

Role of SDH in cardiovascular diseases is illustrated in Figs. 6.1 and 6.2. In atherosclerosis, succinate binds to succinate receptor 1 (SUCNR1), which is expressed

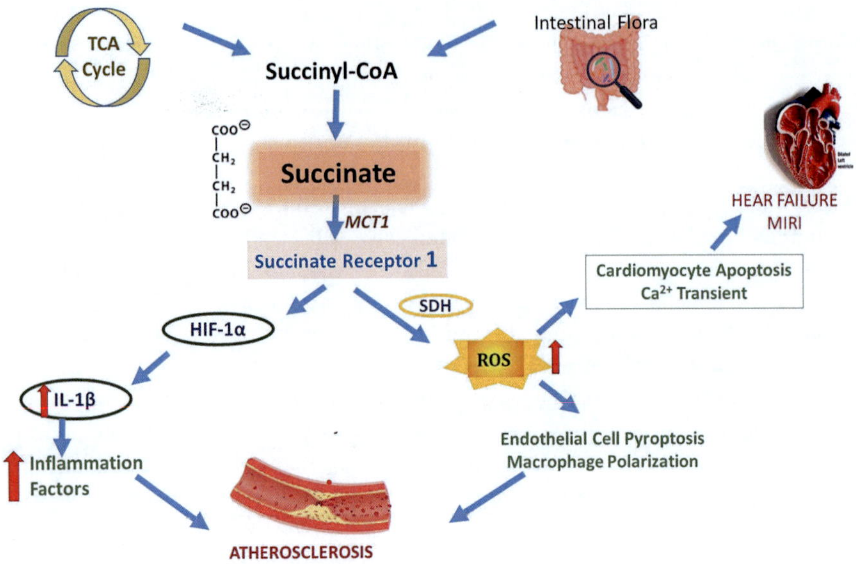

Fig. 6.1 Role of succinate in various cardiovascular diseases

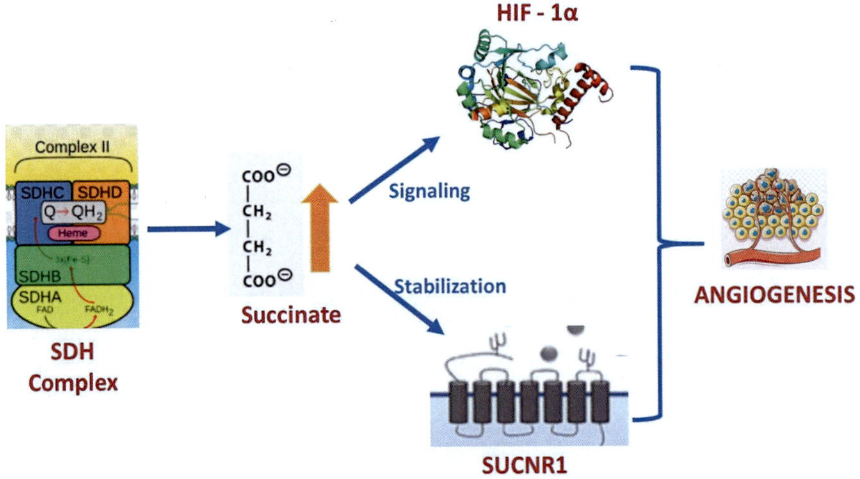

Fig. 6.2 SDH, HIF1α, and SUCNR1 are key targets modulating angiogenesis

in human umbilical vein endothelial cells and macrophages, activates hypoxia-inducible factor (HIF)-1α, promotes IL-1β expression, produces large amounts of inflammatory factors, as well as affects endothelial cell pyroptosis and macrophage polarization; alternatively, through a non-HIF-1α-dependent pathway, succinate upregulates succinate dehydrogenase B (SDHB) expression to promote ROS production, exacerbating the atherogenic inflammatory process of atherosclerosis. In heart failure and MIRI, succinate accumulates in ischemic tissues. After reperfusion, succinate is rapidly oxidized by SDH, generating excess ROS, which leads to cardiomyocyte apoptosis and calcium ion (Ca^{2+}) transients.

As shown in Fig. 6.2, in the conditions of altered SDH expression and activity, succinate levels increase. Accumulation of succinate causes angiogenesis *via* two mechanisms, *(i)* through HIF1α stabilization or *(ii)* by SUCNR1 signaling. Many pathological conditions are associated with this and will provide detailed role to ameliorate pathological angiogenesis (Atallah et al. 2022).

Dimethyl malonate, i.e., a prodrug of malonate, acts as a competitive inhibitor of SDH by binding to the succinate site (Drose et al. 2011). It helps to reduce succinate accumulation when administered prior to index ischemia while inhibiting succinate oxidation when administered during early reperfusion and limiting ROS generation and controlling reverse electron transport (Chouchani et al. 2014; Valls-Lacalle et al. 2018). In animal model of ischemia–reperfusion injury, intracoronary administration of dimethyl malonate at reperfusion has shown protective effect on the cardiac system (Valls-Lacalle et al. 2018). Administration of dimethyl malonate at reperfusion or before and throughout ischemia could alter the metabolism of succinate. During ischemia and reperfusion, it blocks accumulation and oxidation of succinate, respectively.

All the accumulated succinate in the heart does not get oxidized by mitochondria while a major part gets released into the circulation by percutaneous coronary intervention (Kohlhauer et al. 2018). It is further detected by the inflammation causing succinate receptor (SUCNR1). During ischemia, controlling succinate accumulation can be an important therapeutic approach for chronic heart failure. Malonate therapy has beneficial effect in heart attack, ischemia, kidney ischemia–reperfusion injury, etc. (Chouchani et al. 2014).

Cardioprotective effects from IR injury have been observed with dimethyl malonate in type 2 diabetic and nondiabetic animal models. Further, the effect was observed using human cardiac tissue using dimethyl malonate in preischemic SDH inhibition in type 2 diabetes and nondiabetes. It also aims to examine contractile force recovery and function of mitochondria. In the case of diabetes, higher dose of dimethyl malonate is required due to less sensitivity associated with compromised mitochondrial respiration capacity. In contrast to nondiabetic models, the mechanism is not associated with postischemic mitochondrial activity and ischemic preconditioning.

Dimethyl malonate is a synthetic compound with a limited clinical potential and it therapeutic range changes among diabetic and nondiabetic cases (Kula-Alwar et al. 2019; Jespersen et al. 2020). It is toxic to nonischemic tissue and causes cell death by inhibiting SDH (Fernandez-Gomez et al. 2005). Thus, it is required to access the role of SDH modulation in controlling the mitochondrial activity. Studies focusing on the role of dimethyl malonate to alter TCA cycle and inhibit mtROS generation have demonstrated the use to treat IR injury (Chouchani et al. 2014; Valls-Lacalle et al. 2016, 2018). Experiments have demonstrated the role of SDH modulation in reducing infarct size among diabetic and nondiabetic hearts, but the mechanism of dimethyl malonate toward differential response is unclear

(Kula-Alwar et al. 2019; Jespersen et al. 2020). At the onset of diabetes, higher dose of dimethyl malonate has detrimental action on the infarct size in heart. In mature diabetes models, SDH action is permanently upregulated, and effect was restored after 1 month of diabetes induction (Satav and Katyare 2004; Katyare and Satav 2005). Variation in cardiac SDH modulation with severity of diabetes needs to be analyzed. No difference was observed in SDH activity among the dimethyl malonate treated and nontreated groups of mature animal models of diabetes (Jespersen et al. 2020). Insulin deficiency and resistance have been found to be correlated with increased mitochondrial uncoupling and excess ROS generation (Ruegsegger et al. 2018). Animal and human studies have shown correlation between myocardial insulin resistance and lower mitochondrial respiration (Buchanan et al. 2005; Boudina et al. 2007; Montaigne et al. 2014; Ruegsegger et al. 2018). Thus, imbalance between ROS generation and antioxidant effect has role in mitochondrial dysfunction during diabetes (Bonnefont-Rousselot 2002; Pell et al. 2016; Lejay et al. 2016).

Cardioprotective effects of dimethyl malonate at two different doses (0.1 and 0.6 nM) in animal models of type 2 diabetes at three stages, i.e., pre, onset, and mature, have been evaluated (Lejay et al. 2016; Hjortbak et al. 2018; Tonnesen et al. 2021). Cardioprotective action of dimethyl malonate varies in all three stages, and it is found to be correlated with the inherent changes in the function of mitochondria. Response toward IR was also found varying in different stages of diabetes. Dimethyl malonate at high dose overrides the endogenous cardioprotective action, and this correlates the innate protective action of SDH (Guo et al. 2013). In mature diabetic hearts, the SDH action was observed as differential response and reduced with dose of dimethyl malonate. Accumulation of succinate in case of ischemia causes the reversal of SDH effect. Accumulated SDH overloads the working of electron transport chain and initiates the reverse electron transport through complex I and increased generation of ROS (Chouchani et al. 2014). Administration of dimethyl malonate before or during ischemia has shown to exhibit equal effect (Chouchani et al. 2014; Pell et al. 2016; Valls-Lacalle et al. 2016; Kula-Alwar et al. 2019). The study has demonstrated increased susceptibility for IR injury with mature diabetes as compared to observed at the onset of the diabetes (Tosaki et al. 1996; Povlsen et al. 2013; Hjortbak et al. 2018).

Alternatives to dimethyl malonate are oxaloacetate and malate, i.e., physiological SDH inhibitors. In animal models, oxaloacetate has shown neuroprotection from IR injury and needs to further evaluate its pharmacological effects (Knapp et al. 2015; Stepanova et al. 2016). The lack of resemblance to human physiology and systemic and humoral influences on results are the reported limitations. It also concludes that SDH modulation for cardioprotective effect is a challenging approach.

Diazoxide is an ATP sensitive potassium channel opener (K_{ATP}) and is specific toward purported mitochondrial K_{ATP} channel (Mizutani et al. 2005, 2006; Prasad et al. 2006; Al-Dadah et al. 2007; Maffit et al. 2012; Garlid et al. 1997). Proposed hypothesis for mitochondrial based role of diazoxide is elaborated in Fig. 6.3. Diazoxide activates the respiratory chain and provides more ATP or alters the permeability of outer mitochondrial membrane for ADP causing restoration of segregated adenine nucleotides, reduced overload of calcium, lowering stress, and thus

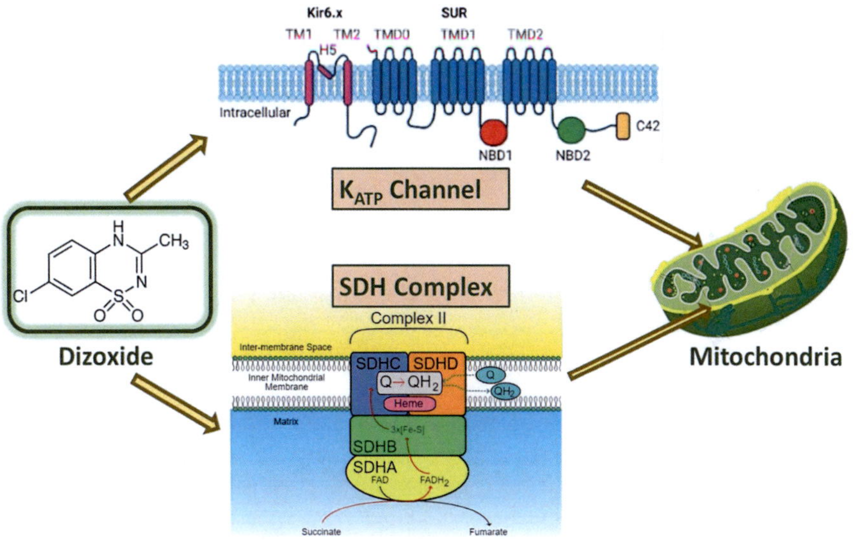

Fig. 6.3 K_{ATP} channel and SDH complex present on the mitochondria (inner membrane). Diazoxide acts by inhibiting SDH and acts as opener of K_{ATP} channel

reducing the cellular injury. It may act by maintaining potential of inner mitochondrial membrane causing the influx of potassium and efflux of hydrogen (Dos Santos et al. 2001; Das et al. 2003; McFalls et al. 2003; Rousou et al. 2004). The mechanism of diazoxide toward the inhibition of SDH is not K_{ATP} dependent. SDH causes oxidation of succinate to generate electrons, which are further utilized to produce ATP and with coenzyme Q it produces ROS (Vanden Hoek et al. 1997; Dzeja et al. 2003). Diazoxide acts as partial SDH inhibitor (Schafer et al. 1971; Hanley et al. 2002; Lim et al. 2002; Dzeja et al. 2003). SDH is suggested as part of purported mK_{ATP} channel, and the mitochondrial proteins associated with SDH show mK_{ATP} channel activity leading to generation of potassium current by improving the influx of potassium in the presence of diazoxide. Inhibitors of mK_{ATP} channel such as ATP, 5-hydroxydeconoate, and glibenclamide have shown ability to reverse the effect.

The cardioprotective effect of diazoxide was exhibited through SDH inhibition, and it was evaluated using glutathione in animal model (Anastacio et al. 2013). During stress, it maintains myocyte volume homeostasis and contractility (Mizutani et al. 2005, 2006; Prasad et al. 2006; Al-Dadah et al. 2007; Maffit et al. 2012). Indirect methods are required to study the mechanism of diazoxide because it is not feasible to confirm channel activity and measure ion flux across mitochondrial membrane. SUR1 subunit of K_{ATP} channel has role in providing cardioprotective action of diazoxide (Sellitto et al. 2010), and an inverse relationship between mK_{ATP} channel and SDH action has been proposed (Wojtovich and Brookes 2008). None to slight lowering of cellular volume was reported with the use of diazoxide alone (Mizutani et al. 2006; Prasad et al. 2006; Sellitto et al. 2010) and glutathione has shown cardioprotection in detrimental manner as reported earlier (Blaustein et al.

1989; Das et al. 1993; Yoshida et al. 1996; Dhalla et al. 2000; Aldakkak et al. 2011). This is helpful to produce synergistic cardioprotective effect with diazoxide, but it was not reported in this study, and thus, it suggests the role of glutathione to prevent beneficial inhibition of SDH by diazoxide. Glutathione, a potent antioxidant, has shown additional cardioprotective mechanism. It thus diverts the conclusions from the observed results. Figure 6.3 shows SDH complex and K_{ATP} channel present on the membrane of the mitochondria. Diazoxide has inhibitory effect on the SDH as well as opens K_{ATP} channel to preserve contractility during stress. It also causes myocyte volume homeostasis. In the situation of stress, the K_{ATP} and SDH are not reported. Diazoxide exhibits cardioprotection by preventing SDH and it was evaluated using glutathione.

Another SDH inhibitor, 3-nitropropionic acid (3-NPA), has also shown cardioprotective effect by reducing the generation of ROS (Ockaili et al. 2001; Busija et al. 2005). It is produced by fungi and has shown signs of Huntington's disease. It alters MDA levels, nitrite, GSH, and neuroinflammatory factors (Hosseini and Hosseinzadeh 2018). It shows irreversible and covalent binding with the active site of SDH causing its inactivation. In vivo studies have been reported to derive its mechanism (Scallet et al. 2003).

6.3 Role of SDH Inhibition During Reperfusion Injury

A large proportion of patients with ST segment elevation myocardial infarction (STEMI) have chances to develop large infarcts, postinfarction remodeling, and heart failure irrespective of the availability of reperfusion therapy. During reperfusion, cell death is observed during myocardial ischemia–reperfusion injury, and it is correlated with cardiomyocyte hypercontracture and mitochondrial permeability transition pore (MPTP). The cell death may be observed in the form of necrosis (Halestrap and Richardson 2015; Garcia-Dorado et al. 2012). Other contributing factors to MPTP include higher concentration ROS (Halestrap and Richardson 2015). Preclinical studies have discussed the role of ROS scavengers in the reduction of infarct size (Garciarena et al. 2011), while negative results obtained for the use of ROS scavengers like superoxide dismutase/edaravone (Dirksen et al. 2007). Translation of these treatments for clinical application is not successful. Attempts were also focused on use of alternate approaches like altering ROS generation from its main sources, namely mitochondrial complex I and III (Chen and Zweier 2014). Amobarbital, i.e., inhibitor of mitochondrial complex I, has lowered ROS production and infarct size when administered at the onset of reperfusion in isolated rat/mice hearts (Stewart et al. 2009; Xu et al. 2014). ROS may be lowered by the inhibition of mitochondrial complex II and succinate dehydrogenase (Drose et al. 2011; Chouchani et al. 2014). Some examples include atpenin A5, which has shown protective action against simulated ischemia–reperfusion in animal models and its mechanism involve mitochondrial ATP-dependent K^+ channels (Wojtovich and Brookes 2009). In another animal model, reversible SDH inhibition with malonate before ischemia has been evaluated (Chouchani et al. 2014). It has lowered ROS generation and infarct size, but it fails to differentiate the effects on ischemia from those on reperfusion and in STEMI cases.

Protective role against reperfusion injury by SDH inhibition with malonate at the onset of reperfusion was demonstrated in isolated mice hearts by Valls-Lacalle et al. (2016). Concentration-response curve was plotted to study the effects of disodium malonate on ischemia–reperfusion injury. Release of LDH and infarct size was measured and assessed the role of necrosis and apoptosis to the size of final infarct. ROS production in cardiac tissue was quantified. The study also focused on the analysis of myocardial metabolism during reperfusion and mitochondrial function. Activation of cytosolic signaling pathways during reperfusion was analyzed, and no significant alteration was observed in Akt, GSK3β, and STAT3 levels while ERK1/2 signal was lowered with or without malonate treatment. The mechanism for SDH inhibition with malonate during reperfusion to reduce infarct size involves the prevention of mitochondrial permeability transition.

6.4 Ischemia–Reperfusion

In reperfusion injury, higher levels of ROS have an important role (Chen and Zweier 2014; Valls-Lacalle et al. 2024). During myocardial infarction, ROS causes cell death through two mechanisms, *i.e.*, cardiomyocyte hypercontrature and opening of mitochondrial permeability transition pore during reperfusion (Hausenloy and Yellon 2013; Halestrap and Richardson 2015). Experimental studies have demonstrated the role of antioxidants in reducing infarct size when administered at the onset of reperfusion (Garciarena et al. 2011; Granger and Kvietys 2015). In ischemia, mitochondrial complexes I, II, and III have been reported to contribute toward ROS production (Quinlan et al. 2012; Siebels and Drose 2013), and significantly higher concentration of mitochondrial complex II is observed (Chouchani et al. 2014, 2016). Thus, therapeutic approaches targeting these mechanisms can contribute to reduce the infarct size by exhibiting protective action. The mechanism of action can be explained with the help of accumulated succinate, which is rapidly oxidized during initial reperfusion. It causes reverse electron transfer from mitochondrial complex II to complex I, and the latter reduces the ROS generation (Chouchani et al. 2014, 2016). This work aims to analyze the effects of reversible inhibition of mitochondrial complex II with a single dose of disodium malonate on postinfarction scar size and ventricular remodeling in an in vivo mouse model of transient coronary occlusion followed by 28 days of reperfusion.

In the cases of cardiac arrest, death that is caused due to brain damage and neurological and cognitive sequelae is reported among the survivors. Reversal of SDH causes the accumulation of ischemic succinate, and it is found to be correlated with mtROS generation (Chouchani et al. 2014). The mtROS is also generated by complex I in the case of a highly lowered CoQ pool and the protonmotive force support reverse electron transport (RET) (Chouchani et al. 2016). In the case of reperfusion, the accumulated succinate is reoxidized and the CoQ pool is maintained.

Accumulation of succinate can be prevented by dimethyl malonate, an SDH inhibitor. It also helps to restore the neurological function and lower the infarcted brain tissue volume. While in the case of out-of-hospital CA, CPR initiation causes the accumulation of succinate in brain tissue. Thus, in the case of CA a hypothesis was proposed and evaluated to reduce the oxidative stress and to diminish damage to the brain by preventing reoxidation of the pile of succinate (Xu et al. 2014). It was evaluated in an animal model to derive the mechanism involved. Use of dimethyl malonate as SDH inhibitor during CPR does not help to decrease death rate in the case of CA in rats and has significant role toward increased brain damage, while the use of dimethyl malonate during CPR has helped to improve return of spontaneous circulation (ROSC) after CA and also improved neurological functions. The study concluded following the role of dimethyl malonate:

- inhibition of neuronal apoptosis instead of systemic effect.
- inhibition of the MMP hyperpolarization and lowers cytochrome C production.
- acceleration of HIF-1α levels in the hippocampus.

During reperfusion, reversible inhibition of SDH with malonate has exhibited cardioprotective effect in acute myocardial infarction (Chouchani et al. 2014, 2016; Valls-Lacalle et al. 2016, 2018). Succinate accumulation in the reperfused myocardium causes reverse electron transfer to complex I, decreases generation of ROS, and lowers MPTP opening (Valls-Lacalle et al. 2016, 2018). Acute myocardial infarction leads to adverse left ventricular remodeling, and it causes heart failure and affects the quality of life among the affected individuals (Bhatt et al. 2017). The protective effect of malonate may not last longer and thus long-term effects of SDH inhibition on the area of scar and postinfarction left ventricular remodeling are observed. Size of necrotic area is useful to determine the adverse postinfarction left ventricular remodeling (Bhatt et al. 2017). Necrosis and apoptosis have an important role in cell death after myocardial infarction (Eefting et al. 2004; Oerlemans et al. 2013; Ihara et al. 2015). Therapeutic strategies used to lower infarct area cannot exhibit long-term protection while transient SDH inhibition can control it along with left ventricular remodeling is not supported.

The protective effect of single dose of malonate administration during reperfusion is not well explained. Effect of transient and reversible inhibition of SDH with single dose of disodium malonate on postinfarction scar size and ventricular remodeling in animal model was analyzed (Valls-Lacalle et al. 2024). It was administered at the onset of reperfusion and has shown long-term protective action and reduced scar size after myocardial infarction. Echocardiography has shown improved functioning of left ventricle and improved NF-κB and Smad2/3 expression.

Targeting succinate oxidation is another therapeutic approach to prevent IR injury by attenuating downstream superoxide production and signaling cascade. It is found clinically useful to remove vessel occlusion by administrating at the site of action.

Some of the advantages with malonate therapy for targeting RET and cardioprotective action are *(i)* favorable pharmacokinetics; *(ii)* lesser toxicity; *(iii)* low dose

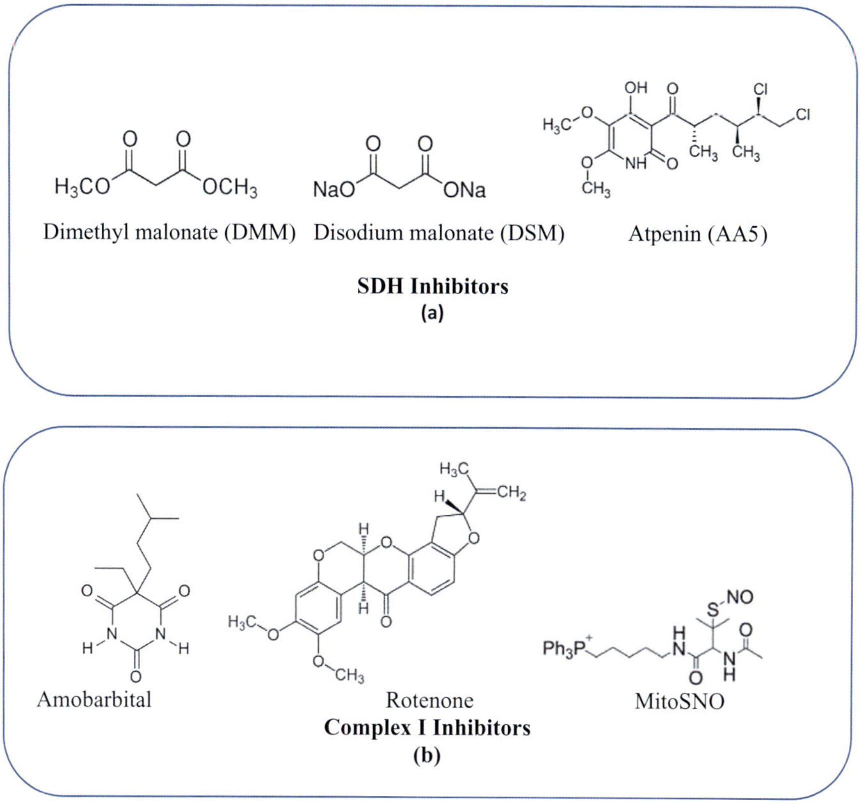

Fig. 6.4 Experimentally evaluated cardioprotective (**a**) SDH inhibitors and (**b**) complex I inhibitors that may impact RET

of acidic formulation; *(iv)* rapid renal excretion; and *(v)* no off-target effects (Samarzija et al. 1981). The noncompetitive inhibitor Atpenin has shown selective SDH inhibition at low dose (Wojtovich and Brookes 2009; Milliken et al. 2022). It has been evaluated using *in vivo* model. Complex I is also a potential target due to reperfusion-based superoxide formation from RET (Chouchani et al. 2013, 2014). Complex I can be catalytically activated and is a highly targeted approach (Babot et al. 2014; Burger et al. 2022). MitoPerSulf is a mitochondria-targeted hydrogen sulfide donor and reduces superoxide by inhibiting complex IV (Miljkovic et al. 2022). Experimentally evaluated cardioprotective SDH inhibitors and complex I inhibitors that may impact RET are presented in Fig. 6.4.

6.5 Atherosclerosis

Succinate has been identified as an extracellular driver of inflammation. It is an important metabolite accumulated during arterial inflammation associated with atherosclerosis. In macrophages, intracellular succinate stabilizes hypoxia-inducible factor 1α and boosts proinflammatory IL-1β generation during normoxia (He et al. 2004; Koppenol et al. 2011; Talreja et al. 2019). It can substitute the activity of prolyl hydroxylase domain enzyme, and thus, extracellular succinate can activate succinate receptor 1-expressing macrophages leading to the initiation of immune response (Eltzschig and Carmeliet 2011). Studies confirm higher levels of succinate in the immune disorders, namely Crohn's disease and rheumatoid arthritis (Littlewood-Evans et al. 2016; Chen et al. 2019). Vascular walls can release mediators of inflammation, which magnifies inflammatory effect by reacting with macrophages (Gimbrone Jr and García-Cardeña 2016; Jensen and Mehta 2016; Zhu et al. 2018). Endothelial and macrophage correlation is not well studied. Xu et al. (2022) evaluated the pathological characteristics related to atherosclerosis. Succinate secretions are activated by local deposit of macrophage causing inflammation through succinate/Hif-1α/IL-1β signaling axis. Studies have found selective antagonist of Sucnr1, NF-56-EJ40. It can affect succinate/IL-1β signal axis in endothelium and macrophages to control inflammation.

6.6 Barth Syndrome (BTHS)

Barth syndrome is an inherited disorder and is linked with X-chromosome. BTHS is related to cardiomyopathy, skeletal myopathy, retarded growth, and neutropenia (Barth et al. 1983). It is caused due to the lack of a mitochondrial acyltransferase enzyme, i.e., tafazzin (Neuwald 1997). Role of tafazzin in the synthesis of cardiolipin (CL) and in the development of cardiomyopathy is well reported. In case of BTHS patients, a defect in tafazzin changes the concentration, distribution, and peroxidation status of CL and is found to be correlated with cardiac diseases like aging heart, ischemia, reperfusion injury, heart failure, diabetic heart, etc. (Paradies et al. 2004; Han et al. 2005; He and Han 2014; Mulligan et al. 2014). CL has an important role in the functions of mitochondria such as maintaining the morphology, formation of cristae on inner membrane, biogenesis of Fe-S, metabolism, and apoptosis (Xu et al. 2005; Guiseppe 2014). It has been further attempted to elaborate with the help of cellular BTHS models and need further studies (Brandner et al. 2005; Xu et al. 2005; Mckenzie et al. 2006; Dudek et al. 2013; Gonzalvez et al., 2013). Defects in mitochondria of the heart tissues have been analyzed using BTHS mouse model and BTHS patient-induced pluripotent stem cell-derived cardiomyocytes (Acehan et al. 2011; Soustek et al. 2011).

6.7 Conclusion

In conclusion, modulation of SDH activity results in variable infarct size reduction depending not only on the presence but also on the stage of diabetes. Even though changes in cardioprotective efficacy were not associated with demonstrable alterations of mitochondrial respiratory capacity, modulation of SDH activity may be a challenging cardioprotective approach. SDH defect is also apparent in cell-derived cardiomyocytes, which form the molecular basis of cardiac pathology in BTHS. Essential role of mitochondrial succinate metabolism opens up exciting new therapeutic approaches for cardiovascular complications.

References

Acehan D, Vaz F, Houtkooper RH, James J, Moore V, Tokunaga C, Kulik W, Wansapura J, Toth MJ, Strauss A et al (2011) Cardiac and skeletal muscle defects in a mouse model of human barth syndrome. J Biol Chem 286:899–908

Al-Dadah AS, Voeller RK, Schuessler RB, Damiano RJ, Lawton JS (2007) Maintenance of myocyte volume homeostasis during stress by diazoxide is cardioprotective. Ann Thorac Surg 84:857–863

Aldakkak M, Stowe DF, Heisner JS, Riess ML, Camara AKS (2011) Adding ROS quenchers to cold Kþ cardioplegia reduces superoxide emission during 2-hour global cold cardiac ischemia. J Cardiovasc Pharmacol Ther 17:93–101

Anastacio MM, Kanter EM, Makepeace C, Keith AD, Zhang H, Schuessler RB, Nichols CG, Lawton JS (2013) Cardioprotective mechanism of diazoxide involves the inhibition of succinate dehydrogenase. Ann Thorac Surg 95(6):2042–2050

Atallah R, Olschewski A, Heinemann A (2022) Succinate at the Crossroad of Metabolism and Angiogenesis: Roles of SDH, HIF1α and SUCNR1. Biomedicines 10(12):3089

Babot M, Birch A, Labarbuta P, Galkin A (2014) Characterisation of the active/de-active transition of mitochondrial complex I. Biochimica Biophysica Acta 1837:1083–1092

Barth PG, Scholte HR, Berden JA, Van der Klei-Van Moorsel JM, Luyt-Houwen IE, Van 't Veer-Korthof ET, Van der Harten JJ, Sobotka-Plojhar MA (1983) An X-linked mitochondrial disease affecting cardiac muscle, skeletal muscle and neutrophil leucocytes. J Neurol Sci 62(1-3):327–355

Bhatt AS, Ambrosy AP, Velazquez EJ (2017) Adverse remodeling and reverse remodeling after myocardial Infarction. Curr Cardiol Rep 19:71

Blaustein A, Deneke SM, Stolz RI, Baxter D, Healey N, Fanburg BL (1989) Myocardial glutathione depletion impairs recovery after short periods of ischemia. Circulation 80:1449–1457

Bonnefont-Rousselot D (2002) Glucose and reactive oxygen species. Curr Opin Clin Nutr Metab Care 5(5):561–568

Bøtker HE, Cabrera-Fuentes HA, Ruiz-Meana M, Heusch G, Ovize M (2020) Translational issues for mitoprotective agents as adjunct to reperfusion therapy in patients with ST-segment elevation myocardial infarction. J Cell Mol Med 24:2717–2729

Boudina S, Sena S, Theobald H, Sheng X, Wright JJ, Hu XX, Aziz S, Johnson JI, Bugger H, Zaha VG, Abel ED (2007) Mitochondrial energetics in the heart in obesity-related diabetes: direct evidence for increased uncoupled respiration and activation of uncoupling proteins. Diabetes 56(10):2457–2466

Brandner K, Mick D, Frazier A, Taylor R, Meisinger C, Rehling P (2005) Taz1, an outer mitochondrial membrane protein, affects stability and assembly of inner membrane protein complexes: implications for Barth Syndrome. Mol Biol Cell 16:5202–5214

Buchanan J, Mazumder PK, Hu P, Chakrabarti G, Roberts MW, Yun UJ, Cooksey RC, Litwin SE, Abel ED (2005) Reduced cardiac efficiency and altered substrate metabolism precedes the onset of hyperglycemia and contractile dysfunction in two mouse models of insulin resistance and obesity. Endocrinology 146(12):5341–5349

Bugger H, Abel ED (2010) Mitochondria in the diabetic heart. Cardiovasc Res 88:229–240

Burger N, James AM, Mulvey JF, Hoogewijs K, Ding S, Fearnley IM, Loureiro-López M, Norman AAI, Arndt S, Mottahedin A, Sauchanka O, Hartley RC, Krieg T, Murphy MP (2022) ND3 Cys39 in complex I is exposed during mitochondrial respiration. Cell Chem Biol 29:636–649

Busija DW, Katakam P, Rajapakse NC, Kis B, Grover G, Domoki F, Bari F (2005) Effects of ATP sensitive potassium channel activators diazoxide and BMS191095 on membrane potential and reactive oxygen species production in isolated piglet mitochondria. Brain Res Bull 66:85–90
Chandel NS (2021) Signaling and metabolism. Cold Spring Harb Perspect Biol 13:a040600
Chen YR, Zweier JL (2014) Cardiac Mitochondria and Reactive Oxygen Species Generation. Circ Res 114:524–537
Chen H, Pan T, Liu P, Wang P, Xu S (2019) Baihu Jia Guizhi decoction improves rheumatoid arthritis inflammation by regulating succinate/Sucnr1 metabolic signaling pathway. Evid Based Complement Alternat Med 2019:3258572
Chouchani ET, Methner C, Nadtochiy SM, Logan A, Pell VR, Ding S, James AM, Cochemé HM, Reinhold J, Lilley KS, Partridge L, Fearnley IM, Robinson AJ, Hartley RC, Smith RAJ, Krieg T, Brookes PS, Murphy MP (2013) Cardioprotection by S-nitrosation of a cysteine switch on mitochondrial complex I. Nat Med 19:753–759
Chouchani ET, Pell VR, Gaude E, Aksentijevic D, Sundier SY, Robb EL, Logan A, Nadtochiy SM, Ord EN, Smith AC, Eyassu F, Shirley R, Hu CH, Dare AJ, James AM, Rogatti S, Hartley RC, Eaton S, Costa AS, Brookes PS, Davidson SM, Duchen MR, Saeb-Parsy K, Shattock MJ, Robinson AJ, Work LM, Frezza C, Krieg T, Murphy MP (2014) Ischaemic accumulation of succinate controls reperfusion injury through mitochondrial ROS. Nature 515:431–435
Chouchani ET, Pell VR, James AM, Work LM, Saeb-Parsy K, Frezza C, Krieg T, Murphy MP (2016) A unifying mechanism for mitochondrial superoxide production during ischemia-reperfusion injury. Cell Metab 23:254–263
Das DK, Engelman RM, Kimura Y (1993) Molecular adaptation of cellular defenses following preconditioning of the heart by repeated ischaemia. Cardiovasc Res 27:578–584
Das M, Parker JE, Halestrap AP (2003) Matrix volume measurements challenge the existence of diazoxide/glibenclamide sensitive KATP channels in rat mitochondria. J Physiol 547(Pt 3):893–902
Davidson SM, Ferdinandy P, Andreadou I, Bøtker HE, Heusch G, Ibáñez B, Ovize M, Schulz R, Yellon DM, Hausenloy DJ, Garcia-Dorado D (2019) Multitarget strategies to reduce myocardial ischemia/reperfusion injury. J Am Coll Cardiol 73:89–99
Davidson SM, Adameová A, Barile L, Cabrera-Fuentes HA, Lazou A, Pagliaro P, Stensløkken K-O, Garcia-Dorado D (2020) Mitochondrial and mitochondrial-independent pathways of myocardial cell death during ischaemia and reperfusion injury. J Cell Mol Med 24:3795–3806
Dhalla NS, Elmoselhi AB, Hata T, Makino N (2000) Status of myocardial antioxidants in ischemia-reperfusion injury. Cardiovasc Res 47:446–456
Dirksen MT, Laarman GJ, Simoons ML, Duncker DJ (2007) Reperfusion injury in humans: a review of clinical trials on reperfusion injury inhibitory strategies. Cardiovasc Res 74:343–355
Dos Santos P, Kowaltowski AJ, Laclau MN, Seetharaman S, Paucek P, Boudina S, Thambo JB, Tariosse L, Garlid KD (2001) Mechanisms by which opening the mitochondrial ATP-sensitive K(+) channel protects the ischemic heart. Am J Physiol Heart Circ Physiol 283:H284–H295
Drose S, Bleier L, Brandt U (2011) A common mechanism links differently acting complex II inhibitors to cardioprotection: modulation of mitochondrial reactive oxygen species production. Mol Pharmacol 79:814–822
Dudek J, Cheng I-F, Balleininger M, Vaz FM, Streckfuss-Bömeke K, Hübscher D, Vukotic M, Wanders RJA, Rehling P, Guan K (2013) Cardiolipin deficiency affects respiratory chain function and organization in an induced pluripotent stem cell model of Barth syndrome. Stem Cell Res 11:806–819
Dzeja PP, Bast P, Ozcan C, Valverde A, Holmuhamedov EL, Van Wylen DG, Terzic A (2003) Targeting nucleotiderequiring enzymes: implications for diazoxide-induced cardioprotection. Am J Physiol Heart Circ Physiol 254:H1048–H1056
Eefting F, Rensing B, Wigman J, Pannekoek WJ, Liu WM, Cramer MJ, Lips DJ, Doevendans PA (2004) Role of apoptosis in reperfusion injury. Cardiovasc Res 61:414–426
Eltzschig HK, Carmeliet P (2011) Hypoxia and Inflammation. N Engl J Med 364(7):656–665
Ferdinandy P, Hausenloy DJ, Heusch G, Baxter GF, Schulz R (2014) Interaction of risk factors, comorbidities, and comedications with ischemia/reperfusion injury and cardioprotection by preconditioning, postconditioning, and remote conditioning. Pharmacol Rev 66:1142–1174

Ferdinandy P, Andreadou I, Baxter GF, Bøtker HE, Davidson SM, Dobrev D, Gersh BJ, Heusch G, Lecour S, Ruiz-Meana M, Zuurbier CJ, Hausenloy DJ, Schulz R (2023) Interaction of cardiovascular nonmodifiable risk factors, comorbidities and comedications with ischemia/reperfusion injury and cardioprotection by pharmacological treatments and ischemic conditioning. Pharmacol Rev 75:159–216

Fernandez-Gomez FJ, Galindo MF, Gomez-Lazaro M, Yuste VJ, Comella JX, Aguirre N, Jordan J (2005) Malonate induces cell death via mitochondrial potential collapse and delayed swelling through an ROS-dependent pathway. Br J Pharmacol 144(4):528–537

Frezza C (2017) Mitochondrial metabolites: undercover signal ling molecules. Interface Focus 7:20160100

Garcia-Dorado D, Ruiz-Meana M, Inserte J, Rodriguez-Sinovas A, Piper HM (2012) Calciummediated cell death during myocardial reperfusion. Cardiovasc Res 94:168–180

Garciarena CD, Fantinelli JC, Caldiz CI, Chiappe DC, Ennis IL, Perez NG, Cingolani HE, Mosca SM (2011) Myocardial reperfusion injury: reactive oxygen species vs. NHE-1 reactivation. Cell Physiol Biochem 27:13–22

Garlid KD, Paucek P, Yarov-Yarovoy V, Murray HN, Darbenzio RB, D'Alonzo AJ, Lodge NJ, Smith MA, Grover GJ (1997) Cardioprotective effect of diazoxide and its interactions with mitochondrial ATP-sensitive K(+) channels: possible mechanism of cardioprotection. Circ Res 81:1072–1082

Gimbrone MA Jr, García-Cardeña G (2016) Endothelial Cell Dysfunction and the Pathobiology of Atherosclerosis. Circ Res 118(4):620–636

Gonzalvez F, D'Aurelio M, Boutant M, Moustapha A, Puech J-P, Landes T, Arnauné-Pelloquin L, Vial G, Taleux N, Slomianny C et al (2013) Barth syndrome: cellular compensation of mitochondrial dysfunction and apoptosis inhibition due to changes in cardiolipin remodeling linked to tafazzin (TAZ) gene mutation. Biochim Biophys Acta 1832: 1194–1206

Granger DN, Kvietys PR (2015) Reperfusion injury and reactive oxygen species: the evolution of a concept. Redox Biol 6:524–551

Guiseppe P (2014) Cardiolipin and mitochondrial function in health and disease. Antioxid Redox Signal 20: 1926–1955

Guo R, Liu B, Zhou S, Zhang B, Xu Y (2013) The protective effect of fasudil on the structure and function of cardiac mitochondria from rats with type 2 diabetes induced by streptozotocin with a high-fat diet is mediated by the attenuation of oxidative stress. Biomed Res Int 2013:430791

Hafner SM, Lehto S, Ronnemaa T, Pyorala K, Laakso M (1998) Mortality from coronary heart disease in subjects with type 2 diabetes and in nondiabetic subjects with and without prior myocardial infarction. N Engl J Med 339:229–234

Hahn VS, Petucci C, Kim M-S, Bedi KC, Wang H, Mishra S, Koleini N, Yoo EJ, Margulies KB, Arany Z, Kelly DP, Kass DA, Sharma K (2023) Myocardial metabolomics of human heart failure with preserved ejection fraction. Circulation 147:1147–1161

Han X, Yang J, Cheng H, Yang K, Abendschein DR, Gross RW (2005) Shotgun lipidomics identifies cardiolipin depletion in diabetic myocardium linking altered substrate utilization with mitochondrial dysfunction. Biochemistry 44:16684–16694

Halestrap AP, Richardson AP (2015) The mitochondrial permeability transition: a current perspective on its identity and role in Ischaemia/reperfusion injury. J Mol Cell Cardiol 78:129–141

Hanley PJ, Mickel M, Loffler M, Brandt U, Daut J (2002) K(ATP) channel-independent targets of diazoxide and 5- hydroxydecanoate in the heart. J Physiol 542(Pt 3):735–741

Hausenloy DJ, Yellon DM (2013) Myocardial ischemia-reperfusion injury: a neglected therapeutic target. J Clin Invest 123:92–100

He Q, Han X (2014) Cardiolipin remodeling in diabetic heart. Chem Phys Lipids 179:75–81

He W, Miao FJ, Lin DC, Schwandner RT, Wang Z, Gao J et al (2004) Citric acid cycle intermediates as ligands for Orphan G-protein-coupled receptors. Nature 429(6988):188–193

Heusch G (2017) Critical issues for the translation of cardio protection. Circ Res 120:1477–1486

Heusch G (2020) Myocardial ischaemia–reperfusion injury and cardioprotection in perspective. Nat Rev Cardiol 17:773–789

Heusch G, Gersh BJ (2017) The pathophysiology of acute myocardial infarction and strategies of protection beyond reperfusion: a continual challenge. Eur Heart J 38(11):774–784

Hjortbak MV, Hjort J, Povlsen JA, Jensen RV, Stottrup NB, Laursen MR, Jespersen NR, Lofgren B, Botker HE (2018) Infuence of diabetes mellitus duration on the efcacy of ischemic preconditioning in a zucker diabetic fatty rat model. PLoS One 13(2):e0192981

Hosseini A, Hosseinzadeh H (2018) Antidotal or protective effects of Curcuma longa (turmeric) and its active ingredient, curcumin, against natural and chemical toxicities: a review. Biomed Pharmacother 99:411–421

Ihara M, Asanuma H, Yamazaki S, Kato H, Asano Y, Shinozaki Y, Mori H, Minamino T, Asakura M, Sugimachi M et al (2015) An Interaction between glucagon-like Peptide-1 and adenosine contributes to cardioprotection of a dipeptidyl Peptidase 4 Inhibitor from myocardial ischemia-reperfusion injury. Am J Physiol Heart Circ Physiol 308:H1287–H1297

Jensen HA, Mehta JL (2016) Endothelial cell dysfunction as a novel therapeutic target in atherosclerosis. Expert Rev Cardiovasc Ther 14(9):1021–1033

Jespersen NR, Hjortbak MV, Lassen TR, Stottrup NB, Johnsen J, Tonnesen PT, Larsen S, Kimose HH, Botker HE (2020) Cardioprotective effect of succinate dehydrogenase inhibition in rat hearts and human myocardium with and without diabetes mellitus. Sci Rep 10(1):10344

Joseph P, Leong D, McKee M, Anand SS, Schwalm JD, Teo K, Mente A, Yusuf S (2017) Reducing the global burden of cardiovascular disease, part 1: the epidemiology and risk factors. Circ Res 121(6):677–694

Kalogeris T, Baines CP, Krenz M, Korthuis RJ (2012) Cell biology of ischemia/reperfusion injury. Int Rev Cell Mol Biol 298:229–317

Katyare SS, Satav JG (2005) Effect of streptozotocin-induced diabetes on oxidative energy metabolism in rat kidney mitochondria a comparative study of early and late effects. Diabetes Obes Metab 7(5):555–562

Knapp L, Gellert L, Kocsis K, Kis Z, Farkas T, Vecsei L, Toldi J (2015) Neuroprotective effect of oxaloacetate in a focal brain ischemic model in the rat. Cell Mol Neurobiol 35(1):17–22

Kohlhauer M, Dawkins S, Costa ASH, Lee R, Young T, Pell VR, Choudhury RP, Banning AP, Kharbanda RK, Saeb-Parsy K et al (2018) Metabolo mic profiling in acute ST-segment–elevation myocardial infarction identifies succinate as an early marker of human ischemia–reperfusion injury. J Am Heart Assoc 7:e007546

Koppenol WH, Bounds PL, Dang CV (2011) Otto Warburg's contributions to current concepts of cancer metabolism. Nat Rev Cancer 11(5):325–337

Kula-Alwar D, Prag HA, Krieg T (2019) Targeting succinate metabolism in ischemia/reperfusion injury. Circulation 140(24):1968–1970

Lejay A, Fang F, John R, Van JA, Barr M, Thaveau F, Chakfe N, Geny B, Scholey JW (2016) Ischemia reperfusion injury, ischemic conditioning and diabetes mellitus. J Mol Cell Cardiol 91:11–22

Leong DP, Joseph PG, McKee M, Anand SS, Teo KK, Schwalm JD, Yusuf S (2017) Reducing the global burden of cardiovascular disease, part 2: prevention and treatment of cardiovascular disease. Circ Res 121(6):695–710

Lesnefsky EJ, Chen Q, Tandler B, Hoppel CL (2017) Mitochon drial dysfunction and myocardial ischemia–reperfusion: implications for novel therapies. Annu Rev Pharmacol Toxicol 57:535–565

Lim KHH, Javadov SA, Das M, Clarke SJ, Suleiman MS, Halestrap AP (2002) The effects of ischaemic preconditioning, diazoxide and 5-hydroxydecanoate on rat heart mitochondrial volume and respiration. J Physiol 545(Pt 3):961–974

Littlewood-Evans A, Sarret S, Apfel V, Loesle P, Dawson J, Zhang J et al (2016) GPR91 senses extracellular succinate released from inflammatory macrophages and exacerbates rheumatoid arthritis. J Exp Med 213(9):1655–1662

Maffit SK, Sellitto AD, Al-Dadah AS, Schuessler RB, Damiano R Jr, Lawton JS (2012) Diazoxide maintains human myocyte volume homeostasis during stress. J Am Heart Assoc 1:1–8

Martin SS, Aday AW, Almarzooq ZI, Anderson CAM, Arora P, Avery CL, Baker-Smith CM, Barone Gibbs B, Beaton AZ, Boehme AK, Commodore-Mensah Y, Currie ME, Elkind MSV,

Evenson KR, Generoso G, Heard DG, Hiremath S, Johansen MC, Kalani R, Kazi DS, Ko D, Liu J, Magnani JW, Michos ED, Mussolino ME, Navaneethan SD, Parikh NI, Perman SM, Poudel R, Rezk-Hanna M, Roth GA, Shah NS, St-Onge MP, Thacker EL, Tsao CW, Urbut SM, Van Spall HGC, Voeks JH, Wang NY, Wong ND, Wong SS, Yaffe K, Palaniappan LP, American Heart Association Council on Epidemiology and Prevention Statistics Committee and Stroke Statistics Subcommittee (2024) 2024 heart disease and stroke statistics: a report of US and global data from the American Heart Association. Circulation 149(8):e347–e913. Erratum in: Circulation 2024 May 7;149(19):e1164

McFalls EO, Liem D, Schoonderwoerd K, Lamers J, Sluiter W, Duncker D (2003) Mitochondrial function: the heart of myocardial preservation. J Lab Clin Med 142:141–149

Mckenzie M, Lazarou M, Thorburn DR, Ryan MT (2006) Mitochondrial respiratory chain supercomplexes are destabilized in Barth Syndrome patients. J Mol Biol 361:462–469

Miljkovic JLJ, Burger N, Gawel JM, Mulvey JF, Norman AAI, Nishimura T, Tsujihata Y, Logan A, Sauchanka O, Caldwell ST, Morris JL, Prime TA, Warrington S, Prudent J, Bates GR, Aksentijević D, Prag HA, James AM, Krieg T, Hartley RC, Murphy MP (2022) Rapid and selective generation of H2S within mitochondria protects against cardiac ischemia−reperfusion injury. Redox Biol 55:102429

Milliken AS, Nadtochiy SM, Brookes PS (2022) Inhibiting succinate release worsens cardiac reperfusion injury by enhancing mitochondrial reactive oxygen species generation. J Am Heart Assoc 11:e026135

Mizutani S, Prasad SM, Sellitto AD, Schuessler RB, Damiano RJ, Lawton JS (2005) Myocyte volume and function in response to osmotic stress: observations in the presence of an adenosine triphosphate-sensitive potassium channel opener. Circulation 112(9 Suppl):I219–I223

Mizutani S, Al-Dadah AS, Bloch JB, Prasad SM, Diodato MD, Schuessler RB, Damiano RJ Jr, Lawton JS (2006) Hyperkalemic cardioplegia-induced myocyte swelling and contractile dysfunction: prevention by diazoxide. Ann Thorac Surg 81:154–159

Montaigne D, Marechal X, Coisne A, Debry N, Modine T, Fayad G, Potelle C, El Arid JM, Mouton S, Sebti Y, Duez H, Preau S, Remy-Jouet I, Zerimech F, Koussa M, Richard V, Neviere R, Edme JL, Lefebvre P, Staels B (2014) Myocardial contractile dysfunction is associated with impaired mitochondrial function and dynamics in type 2 diabetic but not in obese patients. Circulation 130(7):554–564

Mulligan CM, Le CH, Demooy AB, Nelson CB, Chicco AJ (2014) Inhibition of Delta-6 Desaturase Reverses Cardiolipin Remodeling and Prevents Contractile Dysfunction in the Aged Mouse Heart Without Altering Mitochondrial Respiratory Function. J Gerontol A Biol Sci Med Sci 69:799–809

Murashige D, Jang C, Neinast M, Edwards JJ, Cowan A, Hyman MC, Rabinowitz JD, Frankel DS, Arany Z (2020) Comprehensive quantification of fuel use by the failing and nonfailing human heart. Science 370:364–368

Murphy E, Steenbergen C (2008) Mechanisms underlying acute protection from cardiac ischemia−reperfusion injury. Physiol Rev 88:581–609

Neuwald AF, Landsman D (1997) GCN5-related histone N-acetyltransferases belong to a diverse superfamily that includes the yeast SPT10 protein. Trends Biochem Sci 22(5):154–155

Ockaili RA, Bhargave P, Kukreja RC (2001) Chemical preconditioning with 3-nitropropionic acid in hearts: role of mitochondrial KATP channel. Am J Physiol Heart Circ Physiol 280:H2406–H2411

Oerlemans MI, Koudstaal S, Chamuleau SA, de Kleijn DP, Doevendans PA, Sluijter JP (2013) Targeting cell death in the reperfused heart: pharmacological approaches for cardioprotection. Int J Cardiol 165:410–422

Paradies G, Petrosillo G, Pistolese M, Di Venosa N, Federici A, Ruggiero FM (2004) Decrease in mitochondrial complex I activity in ischemic/reperfused rat heart: involvement of reactive oxygen species and cardiolipin. Circ Res 94:53–59

Pell VR, Chouchani ET, Murphy MP, Brookes PS, Krieg T (2016) Moving forwards by blocking back-fow: the yin and yang of mi therapy. Circ Res 118(5):898–906

Picard M, Shirihai OS (2022) Mitochondrial signal transduction. Cell Metab 34:1620–1653

Povlsen JA, Lofgren B, Dalgas C, Birkler RI, Johannsen M, Stottrup NB, Botker HE (2013) Protection against myocardial ischemia-reperfusion injury at onset of type 2 diabetes in zucker diabetic fatty rats is associated with altered glucose oxidation. PLoS One 8(5):e64093

Prasad SM, Al-Dadah AS, Byrd GD, Flagg TP, Gomes J, Damiano RJ Jr, Nichols CG, Lawton JS (2006) Role of the sarcolemmal adenosine triphosphate sensitive potassium channel in hyperkalemic cardioplegia-induced myocyte swelling and reduced contractility. Ann Thorac Surg 81:148–153

Quinlan CL, Orr AL, Perevoshchikova IV, Treberg JR, Ackrell BA, Brand MD (2012) Mitochondrial complex II can generate reactive oxygen species at high rates in both the forward and reverse reactions. J Biol Chem 287:27255–27264

Rousou AJ, Ericsson M, Federman M, Levitsky S, McCully JD (2004) Opening of mitochondrial KATP channels enhances cardioprotection through the modulation of mitochondrial matrix volume, calcium accumulation, and respiration. Am J Physiol Heart Circ Physiol 287:H1967–H1976

Ruegsegger GN, Creo AL, Cortes TM, Dasari S, Nair KS (2018) Altered mitochondrial function in insulin-deficient and insulinresistant states. J Clin Invest 128(9):3671–3681

Samarzija I, Molnar V, Frömter E (1981) THE stoichiometry OF Na+ coupled anion absorption across the brushborder membrane of rat renal proximal tubule. In: Kidney and body fluids. Elsevier, pp 419–423

Satav JG, Katyare SS (2004) Effect of streptozotocin-induced diabetes on oxidative energy metabolism in rat liver mitochondria – a comparative study of early and late effects. Indian J Clin Biochem 19(2):23–31

Scallet AC, Haley RL, Scallet DM, Duhart HM, Binienda ZK (2003) 3-nitropropionic acid inhibition of succinate dehydrogenase (complex II) activity in cultured Chinese hamster ovary cells: antagonism by L-carnitine. Ann N Y Acad Sci 993:305–312

Schafer G, Portenhauser R, Trolp R (1971) Inhibition of mitochondrial metabolism by the diabetogenic thiadiazine diazoxide. I. Action on succinate dehydrogenase and TCA-cycle oxidations. Biochem Pharmacol 20:1271–1280

Sellitto AD, Maffit SK, Al-Dadah AS, Zhang H, Schuessler RB, Nichols CG, Lawton JS (2010) Diazoxide maintenance of myocyte volume and contractility during stress: evidence for a non-sarcolemmal K(ATP) channel location. J Thorac Cardiovasc Surg 140(5):1153–1159

Siebels I, Drose S (2013) Q-Site inhibitor induced ROS production of mitochondrial complex II is attenuated by TCA cycle Dicarboxylates. Biochim Biophys Acta 1827:1156–1164

Soustek MS, Falk DJ, Mah CS, Toth MJ, Schlame M, Lewin AS, Byrne BJ (2011) Characterization of a transgenic short hairpin RNA-induced murine model of Tafazzin deficiency. Hum Gene Ther 22:865–871

Stepanova A, Shurubor Y, Valsecchi F, Manfredi G, Galkin A (2016) Differential susceptibility of mitochondrial complex II to inhibition by oxaloacetate in brain and heart. Biochim Biophys Acta 1857(9):1561–1568

Stewart S, Lesnefsky EJ, Chen Q (2009) Reversible blockade of electron transport with amobarbital at the onset of reperfusion attenuates cardiac injury. Transl Res 153:224–231

Talreja J, Talwar H, Bauerfeld C, Grossman LI, Zhang K, Tranchida P, Samavati L (2019) Hif-1α Regulates IL-1β and IL-17 in Sarcoidosis. elife 8:e44519

Tonnesen PT, Hjortbak MV, Lassen TR, Seefeldt JM, Bøtker HE, Jespersen NR (2021) Myocardial salvage by succinate dehydrogenase inhibition in ischemia-reperfusion injury depends on diabetes stage in rats. Mol Cell Biochem 476(7):2675–2684

Tosaki A, Engelman DT, Engelman RM, Das DK (1996) The evolution of diabetic response to ischemia/reperfusion and preconditioning in isolated working rat hearts. Cardiovasc Res 31(4):526–536

Vaduganathan M, Mensah GA, Turco JV, Fuster V, Roth GA (2022) The global burden of cardiovascular diseases and risk: a compass for future health. JACC 80(25):2361–2371

Valls-Lacalle L, Barba I, Miro-Casas E, Alburquerque-Bejar JJ, Ruiz-Meana M, Fuertes-Agudo M, Rodriguez-Sinovas A, Garcia-Dorado D (2016) Succinate dehydrogenase inhibition with

malonate during reperfusion reduces infarct size by preventing mitochondrial permeability transition. Cardiovasc Res 109(3):374–384
Valls-Lacalle L, Barba I, Miro-Casas E, Ruiz-Meana M, Rodriguez-Sinovas A, Garcia-Dorado D (2018) Selective inhibition of succinate dehydrogenase in reperfused myocardium with intracoronary malonate reduces infarct size. Sci Rep 8(1):2442
Valls-Lacalle L, Consegal M, Ganse FG, Yáñez-Bisbe L, Pastor J, Ruiz-Meana M, Inserte J, Benito B, Ferreira-González I, Rodríguez-Sinovas A (2024) Long-Term Protective Effects of Succinate Dehydrogenase Inhibition during Reperfusion with Malonate on Post-Infarction Left Ventricular Scar and Remodeling in Mice. Int J Mol Sci 25(8):4366
Vanden Hoek TL, Shao Z, Li C, Schumacker PT, Becker LB (1997) Mitochondrial electron transport can become a significant source of oxidative injury in cardiomyocytes. J Mol Cell Cardiol 29:2441–2450
Wojtovich AP, Brookes PS (2008) The endogenous mitochondrial complex II inhibitor malonate regulates mitochondrial ATP sensitive potassium channels: implications for ischemic preconditioning. Biochim Biophys Acta 1777:882–889
Wojtovich AP, Brookes PS (2009) The complex II inhibitor atpenin A5 protects against cardiac ischemia–reperfusion injury via activation of mitochondrial KATP channels. Basic Res Cardiol 104:121–129
Xu Y, Sutachan JJ, Plesken H, Kelley RI, Schlame M (2005) Characterization of lymphoblast mitochondria from patients with Barth syndrome. Lab Invest 85:823–830
Xu A, Szczepanek K, Maceyka MW, Ross T, Bowler E, Hu Y, Kenny B, Mehfoud C, Desai PN, Baumgarten CM, Chen Q, Lesnefsky EJ (2014) Transient complex I inhibition at the onset of reperfusion by extracellular acidification decreases cardiac injury. Am J Physiol Cell Physiol 306:C1142–C1153
Xu J, Zheng Y, Zhao Y, Zhang Y, Li H, Zhang A, Wang X, Wang W, Hou Y, Wang J (2022) Succinate/IL-1β Signaling Axis Promotes the Inflammatory Progression of Endothelial and Exacerbates Atherosclerosis. Front Immunol 13:817572
Yoshida T, Watanabe M, Engelman DT, Engelman RM, Schley JA, Maulik N, Ho YS, Oberley TD, Das DK (1996) Transgenic mice overexpressing glutathione peroxidase are resistant to myocardial ischemia reperfusion injury. J Mol Cell Cardiol 28:1759–1767
Young JB, Raizner AE (2022) Cardiovascular pharmacology: contemporary physiologic and scientific approaches to cardiovascular disease prevention and treatment. Methodist Debakey Cardiovasc J 18(5):1–4
Zhu Y, Xian X, Wang Z, Bi Y, Chen Q, Han X, Tang D, Chen R (2018) Research progress on the relationship between atherosclerosis and inflammation. Biomol Ther 8(3):80
Zuurbier CJ, Bertrand L, Beauloye CR, Andreadou I, Ruiz Meana M, Jespersen NR, Kula-Alwar D, Prag HA, Eric Botker H, Dambrova M, Montessuit C, Kaambre T, Liepinsh E, Brookes PS, Krieg T (2020) Cardiac metabolism as a driver and therapeutic target of myocardial infarction. J Cell Mol Med 24:5937–5954

Implication of Succinate Dehydrogenase in Cancer

7

Viswapriya Viswalingam and Dileep Kumar

Abstract

Succinate dehydrogenase (SDH) has long been recognized as a mitochondrial enzyme crucial in both the citric acid cycle and the electron transport chain (ETC). However, recent research has shed light on an additional role of its substrate, succinate, beyond mere metabolism, particularly in conditions like tumorigenesis. Consequently, SDH is now acknowledged as a tumor suppressor, while succinate is identified as an oncometabolite. This chapter delves into contemporary discoveries concerning changes in SDH function that led to succinate buildup. These alterations encompass SDH mutations, mRNA expression control, posttranslational modifications, and natural inhibitors of SDH. Moreover, the chapter scrutinizes succinate's impact on cancer development, exploring how it induces changes in epigenetics and metabolism and influences the transition of epithelium to mesenchyme, cell migration, and promotes angiogenesis. This chapter will also emphasize the diagnostic potential of SDH as markers for cancers displaying modified SDH expression.

Keywords

Cancer · HIF · Succinate · SDH · Tumors · Warburg effect

V. Viswalingam
Department of Pharmaceutical Chemistry, Poona College of Pharmacy, Bharati Vidyapeeth (Deemed to be University), Pune, Maharashtra, India

D. Kumar (✉)
Department of Pharmaceutical Chemistry, Manipal College of Pharmaceutical Sciences, Manipal Academy of Higher Education, Manipal, Karnataka, India

V. M. Patil et al. (eds.), *Deciphering The Role of Succinate Dehydrogenase in Drug Discovery*, https://doi.org/10.1007/978-981-96-6831-1_7

Abbreviations

APCs	Antigen-presenting cells
ETC	Electron transport chain
GISTs	Gastrointestinal stromal tumors
GPR91	G protein-coupled receptor-91
HCC	Hepatocellular carcinoma
HIF-1	Hypoxia-inducible factor 1
HIF-1α	Hypoxia-inducible factor 1-alpha
HPGL/PCC	Hereditary paraganglioma/pheochromocytoma syndrome
IFN-γ	Interferon-gamma
IHC	Immunohistochemistry
mtROS	Mitochondrial reactive oxygen species
NAM	Nicotinamide
NAMPT	Nicotinamide phosphoribosyltransferase
NMN	Nicotinamide mononucleotide (NMN),
OXPHOS	Oxidative phosphorylation
PHDs	Prolyl hydroxylase domain-containing enzymes
PPP	Pentose phosphate pathway
SDH	Succinate dehydrogenase
SDHB	Succinate dehydrogenase subunit B
SUCNR-1	Succinate receptor-1
TCA	Tricarboxylic acid
TET	Ten-eleven translocation
TLR4	Toll-like receptor 4
TNF-α	Tumor necrosis factor-alpha
TTFA	Thenoyltrifluoroacetone
α-KG	Alpha-ketoglutarate

7.1 Introduction

Succinate dehydrogenase (SDH), also referred to as complex II of mitochondria, is an enzyme within mitochondria that possesses a unique ability to engage in both the citric acid cycle and the electron transport chain (ETC) (Ciccarone et al. 2019; Moreno et al. 2020; Rutter et al. 2010). In the citric acid cycle, it facilitates the oxidation of succinate to fumarate, while in the electron transport chain, it aids in the reduction of ubiquinone to ubiquinol. SDH consists of six subunits—SDHA, SDHB, SDHC, SDHD, SDHAF1, and SDHAF2—with the latter two encoding accessory elements (Bezawork-Geleta et al. 2017; Goetzman et al. 2023). This complex is situated in the inner mitochondrial membrane, exhibiting two distinct domains: one facing the matrix and the other integrated into the membrane. These domains carry out separate chemical reactions, generating fumarate (soluble mitochondrial matrix) and ubiquinol (inner membrane) (Alberts et al. 2002; Kühlbrandt 2015; Tretter et al. 2016). Germline mutations in SDH have been detected in various types of cancers, leading to abnormal accumulation of succinate in tumor cytosol and extracellular fluids of cells. These findings, which affirm its involvement in tumor genesis, have categorized SDH as a tumor suppressor and succinate as an oncometabolite (Arefin et al. 2021).

Succinate, initially isolated from amber in 1546 by a German chemist, swiftly became a focal point in biochemistry and bioenergetics studies (Tretter et al. 2016). Researchers concentrated their efforts on exploring its metabolic function. SDH, an enzyme complex comprising SDHA, SDHB, SDHC, and SDHD subunits, plays a crucial role in converting succinate to fumarate in the tricarboxylic acid (TCA) cycle (Cooper and Clark 1994; De Vivo et al. 2015). Any spontaneous mutation affecting the subunits of SDH, whether inherited or acquired, can lead to a resultant accumulation of succinate in the cytosol (Kantorovich et al. 2010).

Recent publications have shed light on previously overlooked roles of succinate beyond metabolism, particularly in signal transduction, reactive oxygen species (ROS) generation, subsequent activation and stabilization of hypoxia-inducible factor-1 (HIF-1), stimulation of G protein-coupled receptor-91 (GPR91), and downstream signaling pathways closely associated with inflammatory and carcinogenic progression (Jung et al. 2008; Moldogazieva et al. 2020). Mutations in SDH have been identified in various cancers, including hereditary and sporadic, familial paraganglioma, renal carcinoma, thyroid cancer, ovarian cancer, neuroblastoma, gastrointestinal stromal tumor (GIST), and even testicular seminoma. Furthermore, increased succinate levels have been observed in cancer patients (Aghamir et al. 2019; Nazar et al. 2019).

Excessive amounts of succinate leak into the cytoplasm and are released into the extracellular space. The buildup of succinate both inside cells and externally promotes cancer growth through distinct mechanisms (Kuo et al. 2022; Wu 2023). Inside cells, succinate, resembling α-ketoglutarate (α-KG), competes with α-KG-dependent enzymes (also known as 2-oxoglutarate-dependent dioxygenases [2OGDD]), inhibiting their function. These enzymes, involved in protein hydroxylation, demethylation of histone and DNA, collagen synthesis, and energy metabolism, are crucial for various cellular processes. Dysregulation of α-KG-dependent enzymes, such as prolyl hydroxylase domain (PHD)-containing enzymes, the ten-eleven translocation (TET) family of 5′-methylcytosine hydroxylases, and DNA/histone demethylases, occurs in cancer (Losman et al. 2020; Martínez-Reyes and Chandel 2020; Xu et al. 2017). The accumulation of cytosolic succinate due to SDH mutation inhibits these enzymes, contributing to tumorigenesis (Moosavi et al. 2020). Notably, cytosolic succinate stabilizes hypoxia-inducible factor-1-alpha (HIF-1α) by blocking proline–hydroxylase activity, preventing HIF-1α degradation through the ubiquitin–proteasome system. HIF-1α regulates gene transcription linked to glycolysis, inflammation, and angiogenesis (Zimna and Kurpisz 2015).

Recent research has revealed the cancer-promoting effects of signaling through the succinate receptor-1 (SUCNR1). These include inducing epithelial–mesenchymal transition (EMT), migration, metastasis of lung cancer cells, and immunosuppression. SUCNR1 has also been implicated in tumor angiogenesis (Wu et al. 2020). The downstream effects of SUCNR1 activation depend on cell type and G protein coupling. In kidney cells, SUCNR1 couples to Gq and Gi to activate extracellular signal-regulated kinase (ERK), generate inositol 1,4,5-trisphosphate (IP3), increase calcium, and decrease cyclic adenosine monophosphate (cAMP). In cardiomyocytes, it increases cAMP via Gs coupling. SUCNR1 is expressed in adrenal and carotid body chromaffin cells, although its role in these and chromaffin cell-derived tumors is unclear (Ariza et al. 2012; Sanchez et al. 2022). Succinate treatment and SDHB silencing can induce SUCNR1 expression in human hepatoma cells, suggesting a positive feedback loop from succinate accumulation to increased receptor levels. Consistently, elevated SUCNR1 is seen in SDHB-related pheochromocytomas and SDHD head and neck paragangliomas. The combination of high succinate and SUCNR1 may uniquely contribute to tumor formation and progression in SDHx-mutated cancers (Armstrong et al. 2022a, b). Since selective small molecule SUCNR1 inhibitors exist, targeting this receptor represents a promising therapeutic strategy for these tumors (Krzak et al. 2021; You et al. 2023).

Conversely, extracellular succinate exerts its effects by activating succinate receptor-1 (SUCNR-1; or GRP91), a plasma membrane G protein-coupled receptor. It has been found to significantly impact inflammation, tissue fibrosis, and the regulation of renin–angiotensin levels and hypertension (Li et al. 2020; Hu et al. 2017). Succinate buildup has been identified in tumors with SDH germline mutations, and cytosolic succinate acts as a signaling molecule, stabilizing HIF-1α and altering tumor activity (Atallah et al. 2022; Pollard et al. 2005). However, until recently, it remained uncertain whether cancer cells secrete succinate into the extracellular environment and whether extracellular succinate possesses biological activities that affect the behavior and progression of cancer. Recent studies have begun to uncover the secretion of succinate by cancer cells and the pivotal role of extracellular succinate in promoting metastasis.

7.2 How Does Succinate Manage the Warburg Effect Targeted Cancer Cells?

Roughly a century back, Otto Warburg made a significant discovery that cancer cells tend to opt for fermentative metabolism over oxidative metabolism, despite the latter being more efficient in energy production per glucose molecule (Liberti and Locasale 2016; Schiliro and Firestein 2021). Despite the presence of oxygen, cancer cells favor an additional fermentative process known as aerobic glycolysis or the Warburg effect. This alternative metabolic pathway is chiefly marked by heightened glycolytic rates, enabling cancer cells to yield greater amounts of total adenosine triphosphate (ATP) and generate lactate (Vander Heiden et al. 2009). Additionally, it triggers the activation of pro-tumoral signaling pathways and the creation of molecules supporting cancer progression. Among all these molecules, succinate, an intermediate in the Krebs cycle, sees an elevation in concentration within cancer cells and is recognized as an oncometabolite (Ryan et al. 2019). Various pro-tumoral actions have been linked to succinate, and its subsequent role in multiple cancer types has been elucidated. Despite its significant role in metabolism and cancer, the potential of succinate as a target in cancer prevention and treatment has largely remained unexplored (Kes et al. 2020; Schmidt et al. 2021). Many prior anticancer therapies following the Warburg effect have focused on other intermediates (Hillmann et al. 2017).

Given the intricate nature of the Warburg effect, which involves various causes and benefits for transformed cells, it presents an apparent target for novel therapeutic approaches (Jaworska et al. 2023; Li et al. 2022). However, the challenge lies in identifying the specific targets to disrupt this cancerous process. Recent studies indicate that the Warburg effect might not solely serve to generate ATP quickly and channel intermediates into biosynthetic pathways; rather, it may result from inadequate NAD^+ recovery compared to ATP production rates, leading cells to favor fermentation for NAD^+ regeneration (Burns and Manda 2017; Cassim et al. 2020). The reverse Warburg effect, initiated by ROS production, could itself be a therapeutic target by reducing oxidative stress in cancer cells (Arfin et al. 2021; Jaworska et al. 2023). Targeting enzymes like PFKFB3 and PFKFB4, regulating glucose flow between the pentose phosphate pathway (PPP) and glycolysis, may prove crucial if their catalytic function is well understood in specific cancers (Kotowski et al. 2021; Shi et al. 2017). Additionally, lactate, acting as an oncometabolite, plays a role in migration, invasion, immunosuppression, and angiogenesis. Inhibiting lactate dehydrogenase (LDH) could hinder NAD^+ regeneration and lactate production, impacting the reverse Warburg effect (Frisardi et al. 2023; Pérez-Tomás and Pérez-Guillén 2020; Wang et al. 2021). Utilizing lactate-consuming enzymes like lactate oxidase, especially in targeted nanoparticles, could be a more effective and tumor-selective intervention in reprogrammed cancer cell metabolism (Jin et al. 2022; Li and Cui 2023). Numerous inhibitors of proteins involved in the Warburg effect are currently under investigation for cancer treatment, some showing inhibitory effects on tumor growth in clinical trials (Angulo-Elizari et al. 2023; Siddiqui et al. 2018). However, challenges related to drug delivery, specificity, toxicity, cancer cell plasticity, and heterogeneity limit their effectiveness as monotherapy, prompting consideration for their potential as part of multiagent regimens (Navya et al. 2019; Paresishvili and Kakabadze 2023; Tefas et al. 2021). While many Warburg effect inhibitors have shown promise in preclinical research, few have progressed to clinical trials, including compounds like 2-deoxy glucose, lonidamine, indisulam, and AZD3965 (Bonatelli et al. 2019; Raez et al. 2012). Yet, none of these, alone or in combination, have yielded satisfactory outcomes for routine clinical use. As research and therapeutic possibilities advance, findings from cancer models may not universally apply to large patient cohorts but may hold significance in tailoring individualized approaches to patients (Lopez and Banerji 2017; Mokhtari et al. 2017; Zhong et al. 2021).

Targeting cancer cell metabolism has been attempted but with limited success, presenting an opportunity for alternative approaches like targeting succinate accumulation (Teicher et al. 2012). Inhibiting succinate production not only disrupts its oncometabolite functions that are exacerbated in Warburg metabolism but also reduces TCA cycle activity and ROS generation, which play dual roles in cancer (Sullivan and Chandel 2014). One option is targeting the succinate receptor-1

(SUCNR1) to block extracellular pro-tumoral succinate signaling that promotes angiogenesis, metastasis, and M2 macrophage activation. However, SUCNR1 is involved in several biological processes, so toxicity may result (Li et al. 2023; Xu et al. 2022; You et al. 2023). Sirtuin-3 (SIRT3) activation is another approach since SIRT3 increases SDH activity, reducing succinate accumulation, inhibiting the Warburg effect, regulating protein acetylation, and suppressing ROS (Lambona et al. 2024; Zhang et al. 2020a). SIRT3 appears to act as a tumor suppressor in glycolytic cancers, but it may also play a pro-tumoral role depending on the metabolic state. Natural compounds like honokiol and silybin can increase SIRT3 expression and show anticancer effects in models, especially as adjuvants with chemotherapy, although the role of succinate reduction is unclear (Ouyang et al. 2022; Torrens-Mas et al. 2017). Inhibiting Tumour Necrosis Factor-Receptor-Associated Protein (TRAP-1) is also promising since it increases succinate by inhibiting SDH, but it acts as a molecular chaperone with antiapoptotic and homeostatic functions and is not always overexpressed in tumors (Guzzo et al. 2014; Sciacovelli et al. 2013). Small-molecule heat shock protein (HSP) inhibitors like gamitrinib can target mitochondrial TRAP-1 but need specificity to avoid cytosolic effects (Wei et al. 2022). In summary, succinate-targeted therapies present opportunities but require further research on optimal applications selective for cancer cells, elucidating the roles of downstream effects like succinate accumulation, combining with standard chemotherapy.

The Warburg effect involves a metabolic switch to aerobic glycolysis in cancer cells, with enhanced glucose uptake mediated by increased glucose transporter (GLUT) expression (Shin and Koo 2021; To and Cho 2020). This is accompanied by upregulated glycolytic enzymes and diversion of pyruvate to lactate instead of acetyl coenzyme A (acetyl-CoA), driven by lactate dehydrogenase A (LDHA) and 3-phosphoinositide-dependent protein kinase 1 (PDK1) (Gray et al. 2014). Lactate accumulation leads to extracellular acidification through monocarboxylate transporters (MCTs), providing a malignant microenvironment. Lactate also feeds oxidative metabolism in other cells (the reverse Warburg effect) and binds GPR81 to activate pro-tumoral signaling (Li et al. 2022; Payen et al. 2020; Wang et al. 2021). Anaplerotic reactions regenerate TCA cycle intermediates. Accumulation of metabolites like lactate, 2-hydroxyglutarate, fumarate, and succinate promotes cancer progression as oncometabolites (Anderson et al. 2018). The hexosamine pathway generating Uridine diphosphate N-acetylglucosamine (UDP-GlcNAc) is also upregulated. ROS production is increased, so the pentose phosphate pathway is enhanced to generate protective nicotinamide adenine dinucleotide phosphate (NADPH) (Alhayaza et al. 2020; Chiaradonna et al. 2018). Targeting the Warburg effect has been attempted, including glycolytic enzyme inhibitors, GLUT and MCT inhibitors, LDHA inhibitors, HIF inhibitors, ketogenic diets, and glutaminolysis inhibitors (Icard et al. 2021; Kozal et al. 2021; Zhao and Li 2021). However, cancer cell plasticity enables compensation through alternate pathways. Specificity and toxicity remain challenges, resistance develops, and slow-proliferating cancer stem cells are resistant due to their different metabolism (Pérez-González et al. 2023; Shen and Clairambault 2020). While targeting the Warburg effect and associated oncometabolites remains a promising anticancer approach, further research is needed to overcome the current limitations and develop clinically relevant applications.

7.3 Succinate: The Inside Out Oncometabolite

In the past century, Otto Warburg's research uncovered the preference of cancer cells for fermentative metabolism, despite its less efficient energy production per glucose molecule compared to oxidative metabolism (Bose et al. 2021; Liberti and Locasale 2016). Even in oxygen-rich conditions, cancer cells favor aerobic glycolysis, also known as the Warburg effect, leading to increased glycolytic rates, elevated total ATP production, and lactate generation (Liao et al. 2023; Vander Heiden et al. 2009). This metabolic shift is associated with the activation of pro-tumoral signaling pathways and the production of molecules that promote cancer progression, including succinate, an intermediate in the TCA cycle (Benito-Cuesta et al. 2021; Schiliro and Firestein 2021). Succinate accumulation occurs due to SDH inhibition under hypoxia, high H^+ concentration over mitochondrial inner membranes, SDH hypermethylation, and SDH subunit mutations or when the TCA cycle reverses in low-oxygen states (Atallah et al. 2022; Castellano 2024). Additionally, TRAP1 inhibition of SDH contributes to succinate elevation. While in normal tissues, succinate is utilized intracellularly for ATP generation, its pathological accumulation leads to detection in extracellular spaces, functioning as a signaling molecule (Peruzzo et al. 2016; Tretter et al. 2016). SDH mutations are linked to various cancers, such as hereditary paraganglioma, pheochromocytoma, neuroblastomas, and gastrointestinal, colon, renal, and ovarian cancers, highlighting succinate's involvement in their pathogenesis (Table 7.1).

Succinate exerts a significant impact on angiogenesis through various pathways. It contributes to HIF-1α accumulation by inhibiting PHDs, enhances mitochondrial reactive oxygen species (mtROS) levels that similarly inhibit PHDs, stimulates angiogenesis in SDH-deficient cells, and inhibits ten–eleven translocation (TET) proteins, promoting DNA demethylation and increased transcription of HIF-1α and HIF-2α (Chen et al., 2018; Infantino et al. 2021). Succinate also activates the Phosphoinositide 3-kinase (PI3K) pathway, modulates nitric oxide (NO) expression, affects angiopoietins, induces Vascular Endothelial Growth Factor (VEGF) production, activates the ERK pathway, and promotes chemotactic motility, tube-like structure formation, and blood vessel formation (Jiang and Liu 2009; Liu et al. 2023).

Furthermore, succinate accumulation impacts immune cells within the tumor microenvironment. GPR91, associated with succinate receptor 1, correlates with immune cell infiltration in ovarian cancer, including T cells, natural killer (NK) cells, macrophages, dendritic cells, and myeloid-derived suppressor cells (Najm et al. 2022; Zhang et al. 2020b). It also affects markers of exhausted T cells and regulators of inhibitory pathways. Succinate promotes TAM polarization, induces interleukin (IL)-6 secretion, elicits anti-inflammatory responses, activates

Table 7.1 The involvement of SDH in paraganglioma and pheochromocytoma

SDH subunit	Protein	Gene	Clinical aspects	PGL	Function	Inheritance	Recurrence of multifocal disease	References
SDHA	70 kDa/664 aa	5p15	Necrotizing encephalopathy	Leigh	Substrate binding	Autosomal recessive	–	Salehiomran et al. (2013)
SDHB	27 kDa/280 aa	1p35-p36.1	PHEO/PGL	PGL4	Catalysis, electron	Autosomal dominant	Frequent	Jochmanova et al. (2017)
SDHC	15 kDa/169 aa	1q21	PHEO/PGL	PGL3	Anchoring, ubiquinone binding	Autosomal dominant	–	Kantorovich et al. (2010)
SDHD	12 kDa/159 aa	11q23	PHEO/PGL	PGL1	Anchoring, ubiquinone binding	Maternal imprinting	Frequent	Bardella et al. (2011)

macrophages through p38 MAPK, AKT, and AMPK, enhances dendritic cell-mediated T cell activation, and suppresses host defenses against cancer (Han et al. 2023; Huang et al. 2023).

Moreover, succinate contributes to cancer metastasis by inducing the transition of epithelium to mesenchymal through PI3K/AKT signaling, hypermethylation of histones and DNA cytosine, inhibiting histone lysine demethylases (KDMs) and TET proteins, modifying gene expression profiles, and promoting migration and invasion through epithelial-to-mesenchymal transition-dependent mechanisms (Miranda-Gonçalves et al. 2018; Wang and Han 2022). These succinate-induced effects underscore its multifaceted role in tumor progression across various pathways and cell types. These mutations, particularly those in SDHB, are often associated with aggressive forms of these tumors, their metastasis, and susceptibility to other cancer types like serous ovarian cancer, gastrointestinal stromal tumors (GISTs), and renal cell carcinoma (RCC) (Aghamir et al. 2019; Saxena et al. 2015).

Moreover, SDH mutations, especially those in SDHC and SDHD, have implications in different tumor types such as head and neck paragangliomas, Cowden syndrome, and cancers of the bladder, breast, lung, ovary, and nasopharynx (Nazar et al. 2019; Pasini and Stratakis 2009). However, the impact of SDHA mutations on cancer development remains less significant.

Mutations in *SDH* genes disrupt enzymatic activity, affecting the ETC and leading to increased production of superoxide anions (O^{2-}) within mitochondria. This elevated oxidative stress results in various cellular damages, including protein carbonylation, DNA damage, altered mitochondrial function, and reduced mitochondrial membrane potential (Kowalczyk et al. 2021; Patergnani et al. 2021).

One significant consequence of SDH mutations is the abnormal increase in succinate levels. Accumulated succinate functions as an oncometabolite, mimicking the structure of α-ketoglutarate (α-KG), interfering with α-KG-dependent dioxygenase enzymes, and leading to altered gene expression via epigenetic regulation (Xiao et al. 2012; Ye et al. 2018). This alteration results in a hypermethylated DNA profile and elevated histone modifications, impacting chromatin structure and gene expression patterns within cells.

Succinate's action inhibits prolyl hydroxylase domain (PHD)-containing enzymes, causing hypoxia-inducible factor-1-alpha (HIF-1α) stabilization, which promotes a cellular response akin to hypoxia, impacting various cellular processes (Selak et al. 2005; Yang et al. 2014).

Additionally, SDHB mutations, specifically, have been associated with more aggressive tumor behavior by promoting epithelial-to-mesenchymal transition (EMT) (Loriot et al. 2015; Røsland et al. 2019). This transition, marked by changes in gene expression related to S-adenosylmethionine (SAM) metabolism and histone modifications, results in a shift toward a more aggressive cancer phenotype (Audia and Campbell 2016; Yang et al. 2022b).

The intricate relationship between succinate accumulation, SDH mutations, and their multifaceted impact on cellular and tumorigenic processes underscores the importance of further research in understanding these mechanisms for potential therapeutic interventions in cancer treatment (Moreno et al. 2020; Zhang and Lang 2023).

The metabolic dynamics within cancer cells, particularly their heightened glucose and glutamine metabolism, serve as the fundamental drivers of tumor progression (Boroughs and DeBerardinis 2015; Cluntun et al. 2017). This intensive glucose metabolism, known as the Warburg effect, not only facilitates rapid ATP production but also initiates lactate generation via aerobic glycolysis, ensuring a constant supply of NAD^+ essential for sustaining high rates of glucose consumption and ATP production (Burns and Manda 2017; Vander Heiden et al. 2009). However, this relationship between NADH generated during glycolysis and the NAD+ formed during lactate production mandates additional pathways for NADH regeneration. Recent research by Vander Heiden's group highlights a surprising aspect: The demand for NAD^+ in cancer cells might surpass the need for ATP (Vander Heiden et al. 2009). Consequently, this preference for NAD^+ over ATP diminishes mitochondrial respiration, reducing the NAD^+/NADH ratio and fostering fermentation even in the presence of oxygen, thus promoting cell proliferation. Nicotinamide phosphoribosyltransferase (NAMPT), highly expressed in cancer cells, plays a pivotal role in maintaining NAD^+ levels by converting nicotinamide (NAM) to nicotinamide mononucleotide (NMN), crucial for replenishing the NAD^+ pool (Luengo et al. 2021; Yang and Sauve 2016). Additionally, lactate itself, aside from its role in NADH oxidation, acts as an oncometabolite, exerting significant influence on the tumor microenvironment (Vander Heiden et al. 2009). Beyond energy production, the rapid metabolism of glucose offers crucial precursors for biomass synthesis and generates NADPH, essential for anabolic processes and the synthesis of nucleotides, amino acids, and lipids (Alam et al. 2023; Schiliro and Firestein 2021). Intermediates from glycolysis serve diverse purposes, contributing to serine synthesis, the pentose phosphate pathway (PPP), and the maintenance of cellular redox balance through NADPH production (Patra and Hay 2014; Stincone et al. 2015). Notably, this metabolic flux into the PPP oxidative arm aids in protecting the cell against oxidative stress and subsequent DNA damage (Hayes et al. 2020). Furthermore, tumor-specific isoenzymes like PFKFB3 and PFKFB4 regulate glucose flux, directing it between glycolysis and the oxidative arm of the PPP, thereby influencing the metabolic fate of cancer cells (Bartrons et al. 2018).

7.4 SDH as an Oncotherapeutic Agent

Succinate is increasingly acknowledged for its involvement in immune responses, cancer-based processes, and its role in immunoinflammatory reactions. Various studies consistently demonstrate its ability to regulate tumor formation within specific microenvironments, establishing succinate as a key signal in tumorigenesis, alongside its well-known association with inflammation (Coussens and Werb 2002; Sever and Brugge 2015).

Generally, SDH complex contains four subunits encoded by nuclear genes: SDHA-D in mammals, SDH1-4 in yeast, and SDH1-8 in plants (Huang et al. 2019).

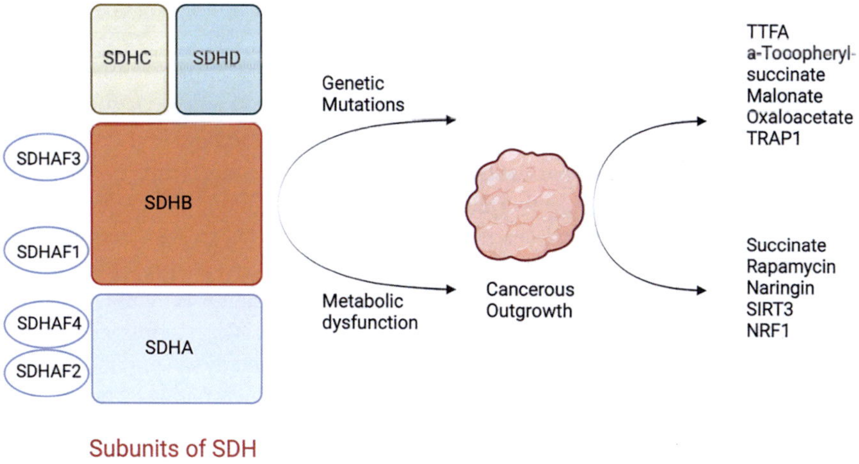

Fig. 7.1 Having provided Fe–S clusters to SDHB, SDHAF1 assisted in the maturation of subunits A and B. Working with dicarboxylate to stabilize the active site of SDHA, SDHAF2 also assisted in the maturation of subunits A and B. Containing heme and being responsible for ubiquinone reduction to ubiquinol, subunits C and D then housed the reaction's next steps

Assembly factors SDHAF1-4 (SDH5-8 in yeast) are also required for SDH function. SDHAF2 flavinates SDHA, stabilizing its active site to allow TCA cycle dicarboxylate handling. SDHAF1 provides iron–sulfur clusters to SDHB by initial binding followed by the recruitment of the cochaperone HscB (Fig. 7.1) (Bezawork-Geleta et al. 2017; Na et al. 2014). The SDHAP1-3 pseudogenes are part of the SDHA complex. Recently, SDHAP1 was found to upregulate EIF4G2 in ovarian cancer by reducing miR-4465 levels, suggesting that SDHAPs may regulate gene expression by sponging microRNAs (Fasoulakis et al. 2023; Zhao et al. 2020). Further study of SDHAP1-3 regulation beyond metabolic roles could provide an insight into SDH complex functions. Overall, the mammalian SDH complex contains four core enzymatic subunits (SDHA-D) and several assembly factors (SDHAF1-4) that enable its metabolic activities, while the related SDHAP1-3 pseudogenes may allow regulatory gene expression roles (Dwight et al. 2017; Moreno et al. 2020).

Mutations in the SDH gene and related succinate pathways contribute significantly to tumorigenesis (Nazar et al. 2019). These mutations accelerate tumor development and are detectable in tumor tissues and cells (Pasini and Stratakis 2009). SDH enzyme, critical in the TCA cycle, converts succinate to fumarate. Mutations in the SDH gene are found in various tumors, positioning SDH as an established tumor suppressor (Rutter et al. 2010; Takács-Vellai et al. 2021). However, these mutations can cause succinate accumulation, stabilizing HIF-1α and promoting tumor growth (Bao et al. 2021). Therefore, investigating succinate accumulation in the context of cancer becomes crucial as it contributes to diverse pathways fostering tumor development and immune responses.

The SDH activity and its regulation play a pivotal role in succinate buildup. Mutation-induced deficiencies in SDH subunits reduce its activity, leading to succinate accumulation and increased mitochondrial ROS generation (Mills et al. 2016). Furthermore, the SDH function relies on oxidized FAD and NAD^+ as cofactors. Under hypoxic conditions, SDH function is significantly impaired due to the existence of cofactors in reduced forms, causing succinate accumulation in the cancer–immunity cycle (Her and Maher 2015; Zhang and Lang 2023). Notably, succinate accumulation in M1 macrophages links to toll-like receptor 4 (TLR4) activation. Upon lipopolysaccharide (LPS) activation, TLR4 signaling disrupts the Krebs cycle (Lee et al. 2018; Zhang and Lang 2023), leading to succinate accumulation by inhibiting the PHD enzyme activity, thus stabilizing HIF-1α. This stabilization triggers the expression of genes encoding glycolytic enzymes and the proinflammatory cytokine IL-1β, affecting tumor cell growth, survival, and inflammation. Succinate's role in tumor development can either promote or hinder tumor growth. To date, SDH has primarily demonstrated its role in suppressing tumor growth and progression.

The relationship between cancer and the immune system involves the metabolite succinate playing a key regulatory role. Succinate can modulate the function of immune cells such as antigen-presenting cells (APCs) and T lymphocytes (Leone and Powell 2020; Wei et al. 2023). One effect of succinate is to increase the ability of dendritic cells (DCs) to present antigens. This leads to an enhanced adaptive immune response that can help suppress tumor growth. By acting on APCs and T cells, succinate links cancer and immunity through its impact on the antigen presentation process and downstream events (Coleman et al. 2020; Park et al. 2022). When antigen-specific T cells become activated after encountering antigenic stimuli, they release cytokines such as tumor necrosis factor-alpha (TNF-α) and interferon-gamma (IFN-γ) as part of the immune response (Arango Duque and Descoteaux 2014). These cytokines can inhibit the progression of cancer cells, which may prolong the survival of patients. By secreting TNF-α and IFN-γ, the activated T lymphocytes can mediate antitumor effects and impede the growth of malignant cells. The cytokine production induced upon T cell activation represents one way these immune cells can combat cancer and potentially improve patient outcomes (Dwivedi et al. 2022; Gonzalez et al. 2018).

Regarding tumorigenesis, Po-Lin Tseng et al. have shown an inverse correlation between hepatocellular carcinoma (HCC) malignancy and SDHB expression levels. Higher malignancy associates with lower SDHB expression, significantly tied to advanced tumor stage and unfavorable prognosis (Tseng et al. 2018). Experimental interventions manipulating SDHB expression in mouse models demonstrated the potential regulation of HCC growth and metastasis. Studies have shown that human HCC exhibits decreased expression of succinate dehydrogenase subunit B (SDHB). This reduction in SDHB shifts cellular metabolism away from aerobic respiration and toward glycolysis. This metabolic switch induces the Warburg effect in human HCC cells and promotes tumor malignancy. By lowering SDHB levels, human HCC tumors can activate the Warburg effect, a shift to glycolytic metabolism, which facilitates cancer cell malignancy and disease progression (Gnocchi et al. 2023;

Srivani et al. 2022; Tseng et al. 2018). Studies also indicate that reduced SDH activity, from silencing subunits, promotes HCC proliferation and metastasis by increasing ROS and activating nuclear factor-kappa B (NF-κB) signaling (Zhang and Lang 2023). These findings suggest that SDH could be a therapeutic target for HCC due to its tumor-suppressive role.

In addition to its link with HCC, SDH is also closely associated with hereditary paraganglioma/pheochromocytoma syndrome (HPGL/PCC), which is characterized by inherited loss-of-function mutations in SDH genes (Else et al. 1993; Hirose et al. 2021). Patients with HPGL/PCC have mutations in the SDHB, SDHC, and SDHD genes. Beyond HPGL/PCC, various neuroendocrine and non-neuroendocrine tumors, including gastrointestinal stromal tumors (GISTs), renal tumors, and thyroid tumors, show significant correlations with SDH gene mutations alongside HPGL/PCC. Research emphasizes the tumor suppressor role of SDH and the oncogenic function of succinate in driving cancer progression (Costa et al. 2015; Papathomas et al. 2014). Current studies focus on elucidating the genetic and molecular mechanisms of succinate accumulation due to SDH defects and their contribution to initiating tumor invasion and metastasis. Variants in SDH genes have also been found in thyroid C cell hyperplasia and papillary thyroid cancers, independent of SDH mutations (Kamani et al. 2022; Luzón-Toro et al. 2019). Nevertheless, the critical involvement of succinate and SDH in cancer development has led to the classification of succinate as an oncometabolite and SDH as a tumor suppressor.

As mentioned before, the oncometabolite function of succinate suggests that detecting these molecules in patient cancer samples could facilitate early detection, tumor screening, and comprehensive monitoring (Beyoğlu and Idle 2021; Wang et al. 2023a, b). Technologies like mass spectrometry and nuclear magnetic resonance provide promising methods for clinicians to identify succinate accumulation. Additionally, immunohistochemistry (IHC) offers an opportunity to evaluate oncometabolites as potential cancer biomarkers in patients (Kowalczyk et al. 2021; Letertre et al. 2021). The abnormal expression of SDH, particularly the loss of the B subunit (SDHB), is implicated in neuroendocrine tumor development as a prognostic factor, likely due to succinate buildup in cancerous tissues.

Impairment of SDH complex II can result in excessive ROS, triggering specific apoptotic cell death pathways in tumors. As a result, several compounds that influence SDH activity have been explored for their potential as anticancer agents (Bezawork-Geleta et al. 2017; Hadrava Vanova et al. 2020). These compounds selectively eliminate cancer cells by inducing ROS-mediated apoptosis while sparing normal cells (NavaneethaKrishnan et al. 2019; Nguyen and Pandey 2019). However, more research is still needed to fully elucidate the mechanisms involved and validate the efficacy and safety of these compounds as prospective cancer treatments. Understanding the various pathways leading to succinate accumulation presents exciting opportunities for regulating succinate levels and impacting tumorigenesis. Modulating succinate levels is a promising, innovative strategy for cancer therapy. By comprehending the molecular mechanisms governing succinate accumulation and its impact on tumor development, novel therapeutic approaches targeting succinate metabolism in malignant cells can be devised (Benito-Cuesta et al. 2021; Wigerup et al. 2016). These interventions could potentially improve patient outcomes and provide effective treatment options for various types of cancers. Further investigation is necessary to fully elucidate succinate's precise role in cancer and to develop targeted therapeutic strategies that efficiently modulate its concentrations.

The gastrointestinal stromal tumors (GISTs), the most prevalent mesenchymal neoplasms in the gastrointestinal tract, often involve mutations in *KIT* and *PDGFRA* genes. However, about 10–15% of GISTs, termed "wild type" (WT), lack these mutations and are categorized into two main subtypes based on SDH immunohistochemical status: SDH-competent and SDH-deficient tumors (Brčić et al. 2021; Miettinen and Lasota 2013). SDH-deficient GISTs, stemming from SDH enzyme complex mutations, are distinct and include most pediatric cases, some sporadic cases, and specific syndromic GISTs (Belinsky et al. 2013). Treatment with imatinib mesylate, a receptor tyrosine kinase inhibitor, has transformed GIST therapy, particularly for advanced cases, although WT GISTs, especially SDH-deficient tumors, exhibit less responsiveness to common targeted therapies like imatinib (Dudzisz-Śledź et al. 2022). Current investigations explore alternative treatment strategies, including antiangiogenic drugs and potential sensitivity to DNA alkylating agents like temozolomide, although further research into SDH-deficient GISTs is necessary for clearer therapeutic approaches.

Despite over 100 years of research on mitochondrial regulation and SDH, the precise molecular pathways involved in SDH activation and inhibition remain poorly understood (Bao et al. 2021; Kluckova et al. 2013). SDH is regulated at the genetic level and has three distinct binding sites that have been leveraged for applications ranging from fungicides to tumor cell regulation (Grünow et al. 2017; Xu et al. 2016). There are two main classes of SDH metabolic inhibitors based on their binding site: those that bind to the ubiquinone pocket at the proximal (QP) or distal (QD) sites and those that bind to the succinate pocket. Binding at the ubiquinone pocket affects ubiquinone reduction to ubiquinol, a part of the electron transport chain. Binding at the succinate pocket impacts SDH enzymatic activity in the TCA cycle (Guo and Lemire 2003; Moreno et al. 2020). Carboxin is a ubiquinone inhibitor used as a fungicide. Thenoyltrifluoroacetone (TTFA) is another ubiquinone SDH inhibitor originally thought to be a chelating agent. The analysis of 19 SDH inhibitors revealed common structural features—a carbonyl center, conserved amide function, and amine group (El-Fakharany et al. 2023). Yao et al. used these to design potential novel inhibitors via in silico library screening and pharmacophore mapping. A16c emerged as the most promising, with enhanced potency against fungal targets (Giordano et al. 2022; Kumar and Zhang 2015). SDH subunit expression is also genetically regulated. For example, SDHB is downregulated in maize leaves based on light levels, reflecting decreased TCA cycle energy needs during photosynthesis (Eprintsev et al. 2016; Gobelli et al. 2023). SDH inhibitors have long been used as fungicides, but there has been recent interest in them as cancer therapies. However, fungicidal SDH inhibitors have been challenging to adapt for human use due to cytotoxic similarities between fungi and human cells. While TTFA induces apoptosis in neuroblastoma cells, other compounds become promising (d'Hose et al. 2021; Xiong et al. 2015). Alpha-tocopherol succinate, similar to vitamin E, causes apoptosis in cancer cells by binding QP/QD sites. Malonate, a synthetic inhibitor, limits reperfusion injury and infarct size in animal hearts (Dong et al. 2008; Weber et al. 2002). TCA intermediates, oxaloacetate and malate, strongly inhibit SDH by binding its catalytic site. The protein TRAP1 is an SDH inhibitor overexpressed in cancers, which reduces SDH activity, leading to succinate accumulation and tumor cell survival (Cannino et al. 2018; Scagliola et al. 2020).

Having been observed in clear cell renal cell carcinoma (ccRCC), which is the most common kidney cancer subtype with high mortality and poor prognosis, SDH is unique in participating in both the tricarboxylic acid (TCA) cycle and oxidative phosphorylation (OXPHOS) (Yang et al. 2022a). Previous studies have found decreased SDH activity in ccRCC, but its role in tumor initiation and progression remains unclear. More research is needed to elucidate the contribution of impaired SDH function to the development and growth of ccRCC tumors, SDH dysfunction having been observed in ccRCC, the deadliest kidney cancer, but additional investigation into the molecular mechanisms linking SDH to ccRCC tumorigenesis is still needed (Aggarwal et al. 2021; Yang et al. 2022a). Pan-cancer analysis here revealed significantly downregulated SDH genes in ccRCC tissues, correlating with disease progression. Increased methylation and mutation rates of SDH gene promoters may explain the reduced SDH expression in ccRCC. Network analysis further demonstrated that SDH genes are connected with ferroptosis-related genes (Fang et al. 2021; Wang et al. 2023b). Functional experiments showed that inhibiting SDH suppressed oxidative phosphorylation (OXPHOS), decreased ferroptotic processes, and restored ferroptotic cell death. This was accompanied by reductions in mitochondrial reactive oxygen species (ROS), overall cellular ROS, and peroxide levels (Su et al. 2019; Zhang et al. 2023). The data demonstrated that SDH inhibition could modulate ferroptosis by lowering ROS production and oxidative activity. By suppressing SDH, OXPHOS and ferroptosis were inhibited, but this ultimately enhanced ferroptotic cell death, potentially through decreased mitochondrial ROS generation. The functional analyses revealed previously unknown roles for SDH in regulating ferroptosis and oxidative stress in cells. Together, this study provides new insights into the regulatory role of SDH in ccRCC development, identifying it as a potential target for advanced antitumor therapy through modulating ferroptosis (Cosialls et al. 2023) (Fig. 7.2).

Additionally, accumulated mitochondrial succinate can permeate the inner and outer membranes via SLC25A10 transporter and porins, respectively. In the cytosol, elevated succinate promotes protein succinylation and suppresses both prolyl hydroxylase domain enzymes and ten-eleven translocation methylcytosine dioxygenases. This results in hypoxia-inducible factor-1-alpha stabilization alongside a hypermethylated state. Succinate may also exit the cell through plasma membrane transporters. Extracellular succinate engagement with SUCNR1 receptors on neighboring malignant cells can potentiate angiogenesis through VEGF production and ERK/STAT3 signaling. Similarly, SUCNR1 binding on tumor-associated macrophages may elicit HIF and interleukin pathways that confer proangiogenic and proinflammatory phenotypes.

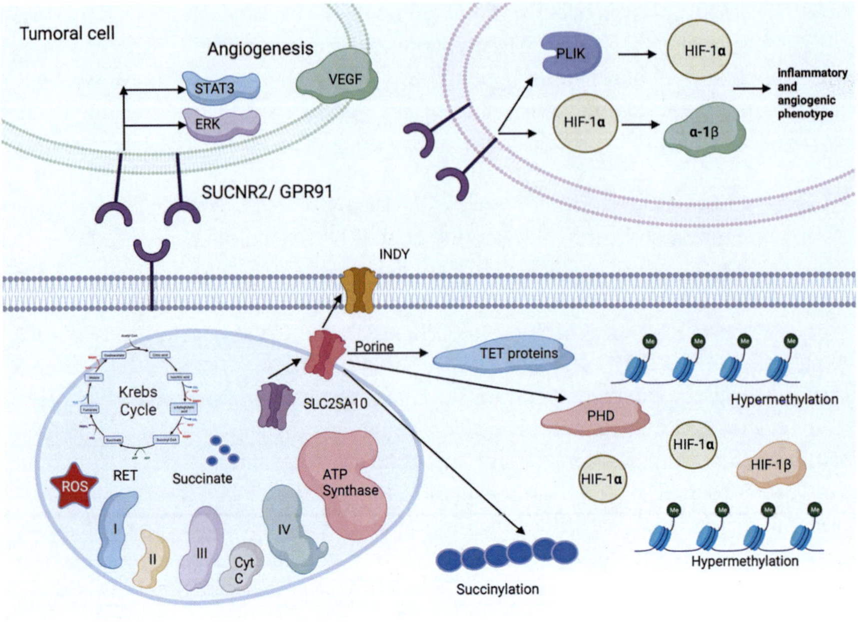

Fig. 7.2 The multifaceted consequences of excess succinate buildup within cancerous mitochondria. At the organelle level, fumarate accumulation due to heightened succinate saturation induces complex III inhibition, prompting reverse electron transport and consequent reactive oxygen species generation at complex I

7.5 Future Prospects

Various mechanisms linking dysfunction of the SDH enzyme complex to cancer have been proposed. One idea is that SDH mutations lead to increased reactive oxygen species and DNA damage, contributing to tumorigenesis (Bardella et al. 2011; Chang et al. 2015). However, the experimental knockdown of SDH did not alter reactive oxygen species levels in one study. Another prominent hypothesis is that the accumulation of succinate acts as an oncometabolite by inhibiting α-ketoglutarate-dependent dioxygenases, including prolyl hydroxylases that regulate HIF-1α stability (Her and Maher 2015; Tretter et al. 2016). Some studies have found only modest increases in succinate levels and no changes in HIF signaling with SDH dysfunction (Xu et al. 2022).

The Warburg effect is a metabolic phenomenon observed in cancer cells where they preferentially use glycolysis to generate energy, even in the presence of oxygen (aerobic glycolysis) (Liberti and Locasale 2016). This shift in metabolism provides advantages to cancer cells, facilitating rapid growth and proliferation. Succinate, a metabolite in the TCA cycle, has garnered attention for its role in this altered metabolic state. It is not just a bystander; it actively influences cancer progression by signaling pathways within cancer cells and impacting surrounding cells in the tumor microenvironment, including immune cells (Baghban et al. 2020).

Targeting succinate in cancer metabolism presents a promising avenue for novel anticancer therapies. This chapter outlines several potential strategies to interfere with succinate-related functions. One approach involves targeting the receptor of succinate (SUCN1), which plays a key role in succinate-mediated effects. Additionally, enhancing the degradation of succinate via enzymes like SIRT3 or TRAP-1 is a promising approach. These enzymes are involved in the breakdown of succinate, potentially reducing its levels within cancer cells. While these strategies hold promise, translating them into effective therapies presents challenges. Clinical application requires a deeper understanding of succinate's precise role in cancer progression and therapy outcomes. Studies focusing on augmenting SIRT3 expression have shown potential benefits, particularly in cancers reliant on glycolysis (Warburg phenotype). However, caution is needed, as SIRT3 overexpression might have adverse effects in tumors relying on oxidative phosphorylation.

The idea of combining these therapies with conventional chemotherapy agents is promising. It has been observed that these novel approaches might work better when used in conjunction with existing treatments rather than being standalone therapies. Moreover, exploring inhibitors specifically targeting TRAP-1 might open new avenues for treating certain types of cancers.

In order to reduce systemic toxicity, researchers are exploring localized delivery approaches. These approaches involve utilizing tailored compounds or antibodies that specifically target succinate within tumors while sparing normal cells. This targeted delivery system aims to reduce side effects while enhancing the treatment's efficacy (Chehelgerdi et al. 2023; Senapati et al. 2018).

More recently, research has shown that elevated succinate inhibits epigenome regulatory enzymes like TET hydroxylases and histone demethylases. This leads to widespread DNA and histone hypermethylation, silencing tumor suppressor genes (Huo et al. 2021; Wong et al. 2017). SDH-mutated tumors display this hypermethylation phenotype, which correlates with metastasis potential. Experiments support this model, as induced hypermethylation of histone H3K27 was sufficient to drive epithelial–mesenchymal transition in ovarian cancer cells (Janin et al. 2023; Wei et al. 2020). However, reversing H3K27 hypermethylation alone did not reverse the mesenchymal state, indicating complexity in maintaining this transition.

In addition to epigenetic effects, SDH dysfunction leads to metabolic reprogramming, increasing glucose flux through the pentose phosphate pathway and glutamine utilization in the TCA cycle, similar to the effects from fumarate hydratase deficiency (Nong et al. 2023; Ward and Thompson 2012). This may be mediated by the altered epigenome and transcriptome induced by SDH loss. Further research on how SDH mutations drive metabolic and epigenetic network reprogramming will provide insight into their role in cancer pathogenesis (Ge et al. 2022; Sun et al. 2022).

7.6 Conclusion

While traditionally recognized as part of the citric acid cycle, succinate has emerged beyond its role in energy metabolism, showcasing significance in gene expression and cell-to-cell signaling. Recent findings have highlighted succinate's accumulation as a pivotal factor in advancing carcinogenesis, labeling succinate as an oncometabolite and SDH as a tumor suppressor. This chapter delved into elucidating the genetic and molecular mechanisms governing succinate buildup, emphasizing its initiation of neoplasm invasion and metastasis. The potential of succinate as a diagnostic indicator for various tumors and of SDH as a potential target for anticancer therapies was outlined. The discussion underscored the necessity for comprehensive exploration of succinate/SDH in future research, recognizing their promising diagnostic and therapeutic implications.

The Warburg effect described how cancer cells preferentially use glycolysis for energy production even in the presence of oxygen. The metabolite succinate plays an active role in this altered metabolism, impacting cancer progression through signaling pathways and effects on the tumor microenvironment. Targeting succinate metabolism is a promising anticancer strategy, with potential approaches including inhibiting its receptor SUCNR1 or enhancing its degradation via enzymes like SIRT3 and TRAP-1. However, translating these ideas into effective therapies faces challenges, requiring a deeper understanding of succinate's precise contributions. Promising directions involve combining these novel therapies with conventional chemotherapy, developing targeted delivery systems to reduce systemic toxicity, and tailoring approaches based on a tumor's specific metabolic phenotype. Overall, interfering with succinate-mediated functions represents an exciting new frontier in cancer metabolism research and therapy development.

Furthermore, it is important to highlight that exploring succinate as a therapeutic target involves understanding the specific metabolic dependencies of different tumor types. Some cancers might respond favorably to treatments targeting succinate metabolism, while others might not. Therefore, careful patient selection and understanding the metabolic profiles of tumors are crucial in determining the effectiveness of these therapies.

Overall, while the idea of targeting succinate in cancer metabolism is promising, it requires extensive research and clinical trials to determine its efficacy, potential side effects, and optimal application in various cancer types. This approach could represent a significant advancement in the field of cancer treatment, potentially offering new avenues for more effective and targeted therapies.

References

Aggarwal RK, Luchtel RA, Machha V, Tischer A, Zou Y, Pradhan K, Ashai N, Ramachandra N, Albanese JM, Yang J, Wang X, Aluri S, Gordon S, Aboumohamed A, Gartrell BA, Hafizi S, Pullman J, Shenoy N (2021) Functional succinate dehydrogenase deficiency is a common adverse feature of clear cell renal cancer. Proc Natl Acad Sci USA 118(39):e2106947118

Aghamir SMK, Heshmat R, Ebrahimi M, Ketabchi SE, Dizaji SP, Khatami F (2019) The impact of succinate dehydrogenase gene (SDH) mutations in renal cell carcinoma (RCC): a systematic review. Onco Targets Ther 12:7929–7940

Alam S, Gu Y, Reichert P, Bähler J, Oliferenko S (2023) Optimization of energy production and central carbon metabolism in a non-respiring eukaryote. Curr Biol 33(11):2175–2186.e5. https://doi.org/10.1016/j.cub.2023.04.046

Alberts B, Johnson A, Lewis J, Raff M, Roberts K, Walter P (2002) The mitochondrion. In: Molecular biology of the cell, 4th edn. Garland Science. https://www.ncbi.nlm.nih.gov/books/NBK26894/

Alhayaza R, Haque E, Karbasiafshar C, Sellke FW, Abid MR (2020) The relationship between reactive oxygen species and endothelial cell metabolism. Front Chem 8:592688. https://doi.org/10.3389/fchem.2020.592688

Anderson NM, Mucka P, Kern JG, Feng H (2018) The emerging role and targetability of the TCA cycle in cancer metabolism. Protein Cell 9(2):216–237. https://doi.org/10.1007/s13238-017-0451-1

Angulo-Elizari E, Gaviria-Soteras L, Zubiri I, Ramos-Inza S, Sanmartin C, Plano D (2023) Unmasking the Warburg effect: unleashing the power of enzyme inhibitors for cancer therapy. Drugs Drug Candidates 2(3):3. https://doi.org/10.3390/ddc2030037

Arango Duque G, Descoteaux A (2014) Macrophage cytokines: involvement in immunity and infectious diseases. Front Immunol 5:491. https://doi.org/10.3389/fimmu.2014.00491

Arefin AME, Khatri NR, Kulkarni N, Egan PF (2021) Polymer 3D printing review: materials, process, and design strategies for medical applications. Polymers 13(9):9. https://doi.org/10.3390/polym13091499

Arfin S, Jha NK, Jha SK, Kesari KK, Ruokolainen J, Roychoudhury S, Rathi B, Kumar D (2021) Oxidative stress in cancer cell metabolism. Antioxidants 10(5):642

Ariza AC, Deen PMT, Robben JH (2012) The succinate receptor as a novel therapeutic target for oxidative and metabolic stress-related conditions. Front Endocrinol 3:22

Armstrong M, Mehrabi H, Naveed N (2022a) An overview of modern metal additive manufacturing technology. J Manuf Process 84:1001–1029

Armstrong N, Storey CM, Noll SE, Margulis K, Soe MH, Xu H, Yeh B, Fishbein L, Kebebew E, Howitt BE, Zare RN, Sage J, Annes JP (2022b) SDHB knockout and succinate accumulation are insufficient for tumorigenesis but dual SDHB/NF1 loss yields SDHx-like pheochromocytomas. Cell Rep 38(9):110453

Atallah R, Olschewski A, Heinemann A (2022) Succinate at the crossroad of metabolism and angiogenesis: roles of SDH, HIF1α and SUCNR1. Biomedicines 10(12):12

Audia JE, Campbell RM (2016) Histone modifications and cancer. Cold Spring Harb Perspect Biol 8(4):a019521

Baghban R, Roshangar L, Jahanban-Esfahlan R, Seidi K, Ebrahimi-Kalan A, Jaymand M, Kolahian S, Javaheri T, Zare P (2020) Tumor microenvironment complexity and therapeutic implications at a glance. Cell Commun Signal 18(1):59

Bao X, Zhang J, Huang G, Yan J, Xu C, Dou Z, Sun C, Zhang H (2021) The crosstalk between HIFs and mitochondrial dysfunctions in cancer development. Cell Death Dis 12(2):215

Bardella C, Pollard PJ, Tomlinson I (2011) SDH mutations in cancer. Biochim Biophys Acta 1807(11):1432–1443

Bartrons R, Simon-Molas H, Rodríguez-García A, Castaño E, Navarro-Sabaté À, Manzano A, Martinez-Outschoorn UE (2018) Fructose 2,6-Bisphosphate in cancer cell metabolism. Front Oncol 8:331

Belinsky MG, Rink L, von Mehren M (2013) Succinate dehydrogenase deficiency in pediatric and adult gastrointestinal stromal tumors. Front Oncol 3:117

Benito-Cuesta I, Ordóñez-Gutiérrez L, Wandosell F (2021) AMPK activation does not enhance autophagy in neurons in contrast to MTORC1 inhibition: different impact on β-amyloid clearance. Autophagy 17(3):656–671

Beyoğlu D, Idle JR (2021) Metabolic rewiring and the characterization of oncometabolites. Cancers 13(12):2900

Bezawork-Geleta A, Rohlena J, Dong L, Pacak K, Neuzil J (2017) Mitochondrial complex II: at the crossroads. Trends Biochem Sci 42(4):312–325

Bonatelli M, Silva ECA, Cárcano FM, Zaia MG, Lopes LF, Scapulatempo-Neto C, Pinheiro C (2019) The Warburg effect is associated with tumor aggressiveness in testicular germ cell tumors. Front Endocrinol 10

Boroughs LK, DeBerardinis RJ (2015) Metabolic pathways promoting cancer cell survival and growth. Nat Cell Biol 17(4):351–359

Bose S, Zhang C, Le A (2021) Glucose metabolism in cancer: the Warburg effect and beyond. In: The heterogeneity of cancer metabolism [internet], 2nd edn. Springer

Brčić I, Argyropoulos A, Liegl-Atzwanger B (2021) Update on molecular genetics of gastrointestinal stromal tumors. Diagnostics 11(2):194

Burns JS, Manda G (2017) Metabolic pathways of the Warburg effect in health and disease: perspectives of choice, chain or chance. Int J Mol Sci 18(12):2755

Cannino G, Ciscato F, Masgras I, Sánchez-Martín C, Rasola A (2018) Metabolic plasticity of tumor cell mitochondria. Front Oncol 8

Cassim S, Vučetić M, Ždralević M, Pouyssegur J (2020) Warburg and beyond: the power of mitochondrial metabolism to collaborate or replace fermentative glycolysis in cancer. Cancers 12(5):5

Castellano J (2024) Chapter 34 – molecular pathology of gastrointestinal stromal tumors. In: Coleman WB, Tsongalis GJ (eds) Diagnostic molecular pathology, 2nd edn. Academic Press, pp 553–559

Chang Y-L, Hsieh M-H, Chang W-W, Wang H-Y, Lin M-C, Wang C-P, Lou P-J, Teng S-C (2015) Instability of succinate dehydrogenase in SDHD polymorphism connects reactive oxygen species production to nuclear and mitochondrial genomic mutations in yeast. Antioxid Redox Signal 22(7):587–602

Chehelgerdi M, Chehelgerdi M, Allela OQB, Pecho RDC, Jayasankar N, Rao DP, Thamaraikani T, Vasanthan M, Viktor P, Lakshmaiya N, Saadh MJ, Amajd A, Abo-Zaid MA, Castillo-Acobo RY, Ismail AH, Amin AH, Akhavan-Sigari R (2023) Progressing nanotechnology to improve targeted cancer treatment: overcoming hurdles in its clinical implementation. Mol Cancer 22(1):169

Chen R, Lai UH, Zhu L, Singh A, Ahmed M, Forsyth NR (2018) Reactive oxygen species formation in the brain at different oxygen levels: the role of hypoxia inducible factors. Front Cell Dev Biol 6

Chiaradonna F, Ricciardiello F, Palorini R (2018) The nutrient-sensing Hexosamine biosynthetic pathway as the hub of cancer metabolic rewiring. Cells 7(6):53

Ciccarone F, Di Leo L, Ciriolo MR (2019) TCA cycle aberrations and cancer. In: Boffetta P, Hainaut P (eds) Encyclopedia of cancer, 3rd edn. Academic Press, pp 429–436

Cluntun AA, Lukey MJ, Cerione RA, Locasale JW (2017) Glutamine metabolism in cancer: understanding the heterogeneity. Trends Cancer 3(3):169–180

Coleman MF, Cozzo AJ, Pfeil AJ, Etigunta SK, Hursting SD (2020) Cell intrinsic and systemic metabolism in tumor immunity and immunotherapy. Cancers 12(4):4

Cooper JM, Clark JB (1994) 1 – the structural organization of the mitochondrial respiratory chain. In: Schapira AHV, DiMauro S (eds) Mitochondrial disorders in neurology. Butterworth-Heinemann, pp 1–30

Cosialls E, Pacreau E, Duruel C, Ceccacci S, Elhage R, Desterke C, Roger K, Guerrera C, Ducloux R, Souquere S, Pierron G, Nemazanyy I, Kelly M, Dalmas E, Chang Y, Goffin V, Mehrpour M, Hamaï A (2023) mTOR inhibition suppresses salinomycin-induced ferroptosis in breast cancer stem cells by ironing out mitochondrial dysfunctions. Cell Death Dis 14(11):744

Costa MHS, Ortiga-Carvalho TM, Violante AD, Vaisman M (2015) Pheochromocytomas and Paragangliomas: clinical and genetic approaches. Front Endocrinol Front Endocrinol (Lausanne). 6:126

Coussens LM, Werb Z (2002) Inflammation and cancer. Nature 420(6917):860–867

d'Hose D, Isenborghs P, Brusa D, Jordan BF, Gallez B (2021) The short-term exposure to SDHI fungicides Boscalid and Bixafen induces a mitochondrial dysfunction in selective human cell lines. Molecules 26(19):5842. https://doi.org/10.3390/molecules26195842
De Vivo DC, Paradas C, DiMauro S (2015) Chapter 41 – Mitochondrial Encephalomyopathies. In: Darras BT, Jones HR, Ryan MM, De Vivo DC (eds) Neuromuscular disorders of infancy, childhood, and adolescence, 2nd edn. Academic Press, pp 796–833
Dong L-F, Low P, Dyason JC, Wang X-F, Prochazka L, Witting PK, Freeman R, Swettenham E, Valis K, Liu J, Zobalova R, Turanek J, Spitz DR, Domann FE, Scheffler IE, Ralph SJ, Neuzil J (2008) A-Tocopheryl succinate induces apoptosis by targeting ubiquinone-binding sites in mitochondrial respiratory complex II. Oncogene 27(31):4324–4335
Dudzisz-Śledź M, Klimczak A, Bylina E, Rutkowski P (2022) Treatment of gastrointestinal stromal tumors (GISTs): a focus on younger patients. Cancers 14(12):2831
Dwight T, Na U, Kim E, Zhu Y, Richardson AL, Robinson BG, Tucker KM, Gill AJ, Benn DE, Clifton-Bligh RJ, Winge DR (2017) Analysis of SDHAF3 in familial and sporadic pheochromocytoma and paraganglioma. BMC Cancer 17(1):497
Dwivedi YK, Hughes L, Baabdullah AM, Ribeiro-Navarrete S, Giannakis M, Al-Debei MM, Dennehy D, Metri B, Buhalis D, Cheung CMK, Conboy K, Doyle R, Dubey R, Dutot V, Felix R, Goyal DP, Gustafsson A, Hinsch C, Jebabli I et al (2022) Metaverse beyond the hype: multidisciplinary perspectives on emerging challenges, opportunities, and agenda for research, practice and policy. Int J Inf Manag 66:102542
El-Fakharany ZS, Nissan YM, Sedky NK, Arafa RK, Abou-Seri SM (2023) New proapoptotic chemotherapeutic agents based on the quinolone-3-carboxamide scaffold acting by VEGFR-2 inhibition. Sci Rep 13:11346
Else T, Greenberg S, Fishbein L (1993) Hereditary Paraganglioma-Pheochromocytoma syndromes. In: Adam MP, Feldman J, Mirzaa GM, Pagon RA, Wallace SE, Bean LJ, Gripp KW, Amemiya A (eds) GeneReviews®. University of Washington, Seattle
Eprintsev A, Fedorin D, Karabutova L, Pokusina T (2016) Light regulation of succinate dehydrogenase subunit B gene SDH2-3 expression in maize leaves. Russ J Plant Physiol 63:505–510
Fang Z, Sun Q, Yang H, Zheng J (2021) SDHB suppresses the tumorigenesis and development of ccRCC by inhibiting glycolysis. Front Oncol 11:639408
Fasoulakis Z, Psarommati M-Z, Papapanagiotou A, Pergialiotis V, Koutras A, Douligeris A, Mortaki A, Mihail A, Theodora M, Stavros S, Karakalpakis D, Papamihail M, Kontomanolis EN, Daskalakis G, Antsaklis P (2023) MicroRNAs can influence ovarian cancer progression by dysregulating integrin activity. Cancers 15(18):4449
Frisardi V, Canovi S, Vaccaro S, Frazzi R (2023) The significance of microenvironmental and circulating lactate in breast cancer. Int J Mol Sci 24(20):15369
Ge T, Gu X, Jia R, Ge S, Chai P, Zhuang A, Fan X (2022) Crosstalk between metabolic reprogramming and epigenetics in cancer: updates on mechanisms and therapeutic opportunities. Cancer Commun 42(11):1049–1082
Giordano D, Biancaniello C, Argenio MA, Facchiano A (2022) Drug design by pharmacophore and virtual screening approach. Pharmaceuticals 15(5):646
Gnocchi D, Sabbà C, Massimi M, Mazzocca A (2023) Metabolism as a new avenue for hepatocellular carcinoma therapy. Int J Mol Sci 24(4):3710
Gobelli D, Serrano-Lorenzo P, Esteban-Amo MJ, Serna J, Pérez-García MT, Orduña A, Jourdain AA, Martín-Casanueva MÁ, de la Fuente Á, Simarro M (2023) The mitochondrial succinate dehydrogenase complex controls the STAT3-IL-10 pathway in inflammatory macrophages. iScience 26(8):107473
Goetzman E, Gong Z, Zhang B, Muzumdar R (2023) Complex II biology in aging, health, and disease. Antioxidants 12(7):7
Gonzalez H, Hagerling C, Werb Z (2018) Roles of the immune system in cancer: from tumor initiation to metastatic progression. Genes Dev 32(19–20):1267–1284
Li X, Xie L, Qu X, Zhao B, Fu W, Wu B, Wu J (2020) GPR91, a critical signaling mechanism in modulating pathophysiologic processes in chronic illnesses. FASEB J 34(10):13091-13105

Gray LR, Tompkins SC, Taylor EB (2014) Regulation of pyruvate metabolism and human disease. Cell Mol Life Sci 71(14):2577–2604
Grünow J, Rong C, Hischmann J, Zaoui K, Flechtenmacher C, Weber K-J, Plinkert P, Hess J (2017) Regulation of submaxillary gland androgen-regulated protein 3A via estrogen receptor 2 in radioresistant head and neck squamous cell carcinoma cells. J Exp Clin Cancer Res 36(1):25
Guo J, Lemire BD (2003) The ubiquinone-binding site of the Saccharomyces cerevisiae succinate-ubiquinone oxidoreductase is a source of superoxide. J Biol Chem 278(48):47629–47635
Guzzo G, Sciacovelli M, Bernardi P, Rasola A (2014) Inhibition of succinate dehydrogenase by the mitochondrial chaperone TRAP1 has anti-oxidant and anti-apoptotic effects on tumor cells. Oncotarget 5(23):11897–11908
Hadrava Vanova K, Kraus M, Neuzil J, Rohlena J (2020) Mitochondrial complex II and reactive oxygen species in disease and therapy. Redox Rep 25(1):26–32
Han J, Dong L, Wu M, Ma F (2023) Dynamic polarization of tumor-associated macrophages and their interaction with intratumoral T cells in an inflamed tumor microenvironment: from mechanistic insights to therapeutic opportunities. Front Immunol 14:1160340
Hayes JD, Dinkova-Kostova AT, Tew KD (2020) Oxidative stress in cancer. Cancer Cell 38(2):167–197
Her YF, Maher LJ (2015) Succinate dehydrogenase loss in familial paraganglioma: biochemistry, genetics, and epigenetics. Int J Endocrinol 2015:296167
Hillmann P, Schmitz D, Mestan J, D'Alonzo S, Fabbro D (2017) 5.01 – cancer biology and the principles of targeted cancer drug discovery. In: Chackalamannil S, Rotella D, Ward SE (eds) Comprehensive medicinal chemistry III. Elsevier, pp 1–38
Hirose R, Tsurutani Y, Sugisawa C, Inoue K, Suematsu S, Nagata M, Hasegawa N, Kakuta Y, Yonamine M, Takekoshi K, Kimura N, Saito J, Nishikawa T (2021) Hereditary pheochromocytoma/paraganglioma syndrome with a novel mutation in the succinate dehydrogenase subunit B gene in a Japanese family: two case reports. J Med Case Rep 15:282
Hu J, Li T, Du X, Wu Q, Le Y-Z (2017) G protein-coupled receptor 91 signaling in diabetic retinopathy and hypoxic retinal diseases. Vis Res 139:59–64
Huang S, Braun H-P, Gawryluk RMR, Millar AH (2019) Mitochondrial complex II of plants: subunit composition, assembly, and function in respiration and signaling. Plant J 98(3):405–417
Huang J, Wu Q, Geller DA, Yan Y (2023) Macrophage metabolism, phenotype, function, and therapy in hepatocellular carcinoma (HCC). J Transl Med 21:815
Huo M, Zhang J, Huang W, Wang Y (2021) Interplay among metabolism, epigenetic modifications, and gene expression in cancer. Front Cell Dev Biol 9:793428
Icard P, Loi M, Wu Z, Ginguay A, Lincet H, Robin E, Coquerel A, Berzan D, Fournel L, Alifano M (2021) Metabolic strategies for inhibiting cancer development. Adv Nutr 12(4):1461–1480
Infantino V, Santarsiero A, Convertini P, Todisco S, Iacobazzi V (2021) Cancer cell metabolism in hypoxia: role of HIF-1 as key regulator and therapeutic target. Int J Mol Sci 22(11):11
Janin M, Davalos V, Esteller M (2023) Cancer metastasis under the magnifying glass of epigenetics and epitranscriptomics. Cancer Metastasis Rev 42(4):1071–1112
Jaworska M, Szczudło J, Pietrzyk A, Shah J, Trojan SE, Ostrowska B, Kocemba-Pilarczyk KA (2023) The Warburg effect: a score for many instruments in the concert of cancer and cancer niche cells. Pharmacol Rep 75(4):876–890
Jiang B-H, Liu L-Z (2009) PI3K/PTEN signaling in angiogenesis and tumorigenesis. Adv Cancer Res 102:19–65
Jin M, Cao W, Chen B, Xiong M, Cao G (2022) Tumor-derived lactate creates a favorable niche for tumor via supplying energy source for tumor and modulating the tumor microenvironment. Front Cell Dev Biol 10:808859
Jochmanova I, Wolf KI, King KS, Nambuba J, Wesley R, Martucci V, Raygada M, Adams KT, Prodanov T, Fojo AT, Lazurova I, Pacak K (2017) SDHB-related Pheochromocytoma and Paraganglioma penetrance and genotype-phenotype correlations. J Cancer Res Clin Oncol 143(8):1421–1435

Jung S-N, Yang WK, Kim J, Kim HS, Kim EJ, Yun H, Park H, Kim SS, Choe W, Kang I, Ha J (2008) Reactive oxygen species stabilize hypoxia-inducible factor-1 alpha protein and stimulate transcriptional activity via AMP-activated protein kinase in DU145 human prostate cancer cells. Carcinogenesis 29(4):713–721

Kamani T, Charkhchi P, Zahedi A, Akbari MR (2022) Genetic susceptibility to hereditary non-medullary thyroid cancer. Hered Cancer Clin Pract 20(1):9

Kantorovich V, King KS, Pacak K (2010) SDH-related pheochromocytoma and paraganglioma best practice and research. Clin Endocrinol Metab 24(3):415–424

Kes MMG, Van den Bossche J, Griffioen AW, Huijbers EJM (2020) Oncometabolites lactate and succinate drive pro-angiogenic macrophage response in tumors. Biochim Biophys Acta Rev Cancer 1874(2):188427

Kluckova K, Bezawork-Geleta A, Rohlena J, Dong L, Neuzil J (2013) Mitochondrial complex II, a novel target for anti-cancer agents. Biochim Biophys Acta Bioenerg 1827(5):552–564

Kotowski K, Rosik J, Machaj F, Supplitt S, Wiczew D, Jabłońska K, Wiechec E, Ghavami S, Dzięgiel P (2021) Role of PFKFB3 and PFKFB4 in cancer: genetic basis, impact on disease development/progression, and potential as therapeutic targets. Cancers 13(4):909

Kowalczyk P, Sulejczak D, Kleczkowska P, Bukowska-Ośko I, Kucia M, Popiel M, Wietrak E, Kramkowski K, Wrzosek K, Kaczyńska K (2021) Mitochondrial oxidative stress – a causative factor and therapeutic target in many diseases. Int J Mol Sci 22(24):13384

Kozal K, Jóźwiak P, Krześlak A (2021) Contemporary perspectives on the Warburg effect inhibition in cancer therapy. Cancer Control J Moffitt Cancer Center 28:10732748211041243

Krzak G, Willis CM, Smith JA, Pluchino S, Peruzzotti-Jametti L (2021) Succinate receptor 1: An emerging regulator of myeloid cell function in inflammation. Trends Immunol 42(1):45–58

Kühlbrandt W (2015) Structure and function of mitochondrial membrane protein complexes. BMC Biol 13(1):89

Kumar A, Zhang KYJ (2015) Hierarchical virtual screening approaches in small molecule drug discovery. Methods 71:26–37

Kuo C-C, Wu J-Y, Wu KK (2022) Cancer-derived extracellular succinate: a driver of cancer metastasis. J Biomed Sci 29(1):93

Lambona C, Zwergel C, Valente S, Mai A (2024) SIRT3 activation a promise in drug development? New insights into SIRT3 biology and its implications on the drug discovery process. J Med Chem 67(3):1662–1689

Lee J-W, Nam H, Kim LE, Jeon Y, Min H, Ha S, Lee Y, Kim S-Y, Lee SJ, Kim E-K, Yu S-W (2018) TLR4 (toll-like receptor 4) activation suppresses autophagy through inhibition of FOXO3 and impairs phagocytic capacity of microglia. Autophagy 15(5):753–770

Leone RD, Powell JD (2020) Metabolism of immune cells in cancer. Nat Rev Cancer 20(9):516–531

Letertre MPM, Giraudeau P, de Tullio P (2021) Nuclear magnetic resonance spectroscopy in clinical metabolomics and personalized medicine: current challenges and perspectives. Front Mol Biosci 8:698337

Li Z, Cui J (2023) Targeting the lactic acid metabolic pathway for antitumor therapy. Mol Ther Oncolytics 31:100740

Li X, Yang Y, Zhang B, Lin X, Fu X, An Y, Zou Y, Wang J-X, Wang Z, Yu T (2022) Lactate metabolism in human health and disease. Signal Transduct Target Ther 7(1):1

Li M, Yang Y, Xiong L, Jiang P, Wang J, Li C (2023) Metabolism, metabolites, and macrophages in cancer. J Hematol Oncol 16(1):80

Liao M, Yao D, Wu L, Luo C, Wang Z, Zhang J, Liu B (2023) Targeting the Warburg effect: a revisited perspective from molecular mechanisms to traditional and innovative therapeutic strategies in cancer. Acta Pharm Sin B 14(3):953-1008

Liberti MV, Locasale JW (2016) The Warburg effect: how does it benefit cancer cells? Trends Biochem Sci 41(3):211–218

Liu Z-L, Chen H-H, Zheng L-L, Sun L-P, Shi L (2023) Angiogenic signaling pathways and anti-angiogenic therapy for cancer. Signal Transduct Target Ther 8(1):1

Lopez JS, Banerji U (2017) Combine and conquer: challenges for targeted therapy combinations in early phase trials. Nat Rev Clin Oncol 14(1):57–66

Loriot C, Domingues M, Berger A, Menara M, Ruel M, Morin A, Castro-Vega L-J, Letouzé É, Martinelli C, Bemelmans A-P, Larue L, Gimenez-Roqueplo A-P, Favier J (2015) Deciphering the molecular basis of invasiveness in Sdhb-deficient cells. Oncotarget 6(32):32955–32965

Losman J-A, Koivunen P, Kaelin WG (2020) 2-Oxoglutarate-dependent dioxygenases in cancer. Nat Rev Cancer 20(12):710–726

Luengo A, Li Z, Gui DY, Sullivan LB, Zagorulya M, Do BT, Ferreira R, Naamati A, Ali A, Lewis CA, Thomas CJ, Spranger S, Matheson NJ, Vander Heiden MG (2021) Increased demand for NAD^+ relative to ATP drives aerobic glycolysis. Mol Cell 81(4):691–707

Luzón-Toro B, Fernández RM, Villalba-Benito L, Torroglosa A, Antiñolo G, Borrego S (2019) Influencers on thyroid cancer onset: molecular genetic basis. Genes 10(11):11

Martínez-Reyes I, Chandel NS (2020) Mitochondrial TCA cycle metabolites control physiology and disease. Nat Commun 11(1):1

Matlac DM, Hadrava Vanova K, Bechmann N, Richter S, Folberth J, Ghayee HK, Ge G-B, Abunimer L, Wesley R, Aherrahrou R, Dona M, Martínez-Montes ÁM, Calsina B, Merino MJ, Schwaninger M, Deen PMT, Zhuang Z, Neuzil J, Pacak K, Lehnert H, Fliedner SMJ (2021) Succinate mediates tumorigenic effects via succinate receptor 1: potential for new targeted treatment strategies in succinate dehydrogenase deficient paragangliomas. Front Endocrinol 12:589451

Miettinen M, Lasota J (2013) Gastrointestinal stromal tumors. Gastroenterol Clin N Am 42(2):399–415

Mills EL, Kelly B, Logan A, Costa ASH, Varma M, Bryant CE, Tourlomousis P, Däbritz JHM, Gottlieb E, Latorre I, Corr SC, McManus G, Ryan D, Jacobs HT, Szibor M, Xavier RJ, Braun T, Frezza C, Murphy MP, O'Neill LA (2016) Succinate dehydrogenase supports metabolic repurposing of mitochondria to drive inflammatory macrophages. Cell 167(2):457–470

Miranda-Gonçalves V, Lameirinhas A, Henrique R, Jerónimo C (2018) Metabolism and epigenetic interplay in cancer: regulation and putative therapeutic targets. Front Genet 9:427

Mokhtari RB, Homayouni TS, Baluch N, Morgatskaya E, Kumar S, Das B, Yeger H (2017) Combination therapy in combating cancer. Oncotarget 8(23):38022–38043

Moldogazieva NT, Mokhosoev IM, Terentiev AA (2020) Metabolic heterogeneity of cancer cells: an interplay between HIF-1, GLUTs, and AMPK. Cancers 12(4):4

Moosavi B, Zhu X, Yang W-C, Yang G-F (2020) Molecular pathogenesis of tumorigenesis caused by succinate dehydrogenase defect. Eur J Cell Biol 99(1):151057

Moreno C, Santos RM, Burns R, Zhang WC (2020) Succinate dehydrogenase and ribonucleic acid networks in cancer and other diseases. Cancers 12(11):3237

Na U, Yu W, Cox J, Bricker DK, Brockmann K, Rutter J, Thummel CS, Winge DR (2014) The LYR factors SDHAF1 and SDHAF3 mediate maturation of the iron-sulfur subunit of succinate dehydrogenase. Cell Metab 20(2):253–266

Najm R, Hachim MY, Kandasamy RK (2022) Divulging a pleiotropic role of succinate receptor SUCNR1 in renal cell carcinoma microenvironment. Cancers 14(24):6064

NavaneethaKrishnan S, Rosales JL, Lee K-Y (2019) ROS-mediated cancer cell killing through dietary phytochemicals. Oxidative Med Cell Longev 2019:9051542

Navya PN, Kaphle A, Srinivas SP, Bhargava SK, Rotello VM, Daima HK (2019) Current trends and challenges in cancer management and therapy using designer nanomaterials. Nano Converg 6(1):23

Nazar E, Khatami F, Saffar H, Tavangar SM (2019) The emerging role of succinate dehyrogenase genes (SDHx) in tumorigenesis. Int J Hematol Oncol Stem Cell Res 13(2):72–82

Nguyen C, Pandey S (2019) Exploiting mitochondrial vulnerabilities to trigger apoptosis selectively in cancer cells. Cancers 11(7):916

Nong S, Han X, Xiang Y, Qian Y, Wei Y, Zhang T, Tian K, Shen K, Yang J, Ma X (2023) Metabolic reprogramming in cancer: mechanisms and therapeutics. MedComm 4(2):e218

Ouyang S, Zhang Q, Lou L, Zhu K, Li Z, Liu P, Zhang X (2022) The double-edged sword of SIRT3 in cancer and its therapeutic applications. Front Pharmacol 13::871560

Papathomas TG, Gaal J, Corssmit EP, Oudijk L, Korpershoek E, Heimdal K, Bayley JP, Morreau H, van Dooren M, Papaspyrou K, Schreiner T, Hansen T, Andresen PA, Restuccia DF, van

Kessel I, van Leenders GJ, Kros JM, Looijenga LH, Hofland LJ, Mann W, van Nederveen FH, Mete O, Asa SL, de Krijger RR, Dinjens WN (2014) Non-pheochromocytoma (PCC)/paraganglioma (PGL) tumors in patients with succinate dehydrogenase-related PCC-PGL syndromes: a clinicopathological and molecular analysis. Eur J Endocrinol 170(1):1–12

Paresishvili T, Kakabadze Z (2023) Challenges and opportunities associated with drug delivery for the treatment of solid tumors. Oncol Rev 17:10577

Park J, Wang L, Ho P-C (2022) Metabolic guidance and stress in tumors modulate antigen-presenting cells. Oncogenesis 11(1):62

Pasini B, Stratakis CA (2009) SDH mutations in tumorigenesis and inherited endocrine tumours: lesson from the phaeochromocytoma-paraganglioma syndromes. J Intern Med 266(1):19–42

Patergnani S, Bouhamida E, Leo S, Pinton P, Rimessi A (2021) Mitochondrial oxidative stress and "Mito-inflammation": actors in the diseases. Biomedicines 9(2):2

Patra KC, Hay N (2014) The pentose phosphate pathway and cancer. Trends Biochem Sci 39(8):347–354

Payen VL, Mina E, Van Hée VF, Porporato PE, Sonveaux P (2020) Monocarboxylate transporters in cancer. Mol Metab 33:48–66

Pérez-González A, Bévant K, Blanpain C (2023) Cancer cell plasticity during tumor progression, metastasis and response to therapy. Nat Cancer 4(8):1063–1082. https://doi.org/10.1038/s43018-023-00595-y

Pérez-Tomás R, Pérez-Guillén I (2020) Lactate in the tumor microenvironment: AN essential molecule in cancer progression and treatment. Cancers 12(11):3244

Peruzzo R, Biasutto L, Szabò I, Leanza L (2016) Impact of intracellular ion channels on cancer development and progression. Eur Biophys J 45(7):685–707

Pollard PJ, Brière JJ, Alam NA, Barwell J, Barclay E, Wortham NC, Hunt T, Mitchell M, Olpin S, Moat SJ, Hargreaves IP, Heales SJ, Chung YL, Griffiths JR, Dalgleish A, McGrath JA, Gleeson MJ, Hodgson SV, Poulsom R, Rustin P, Tomlinson IP (2005) Accumulation of Krebs cycle intermediates and over-expression of HIF1α in tumors which result from germline FH and SDH mutations. Hum Mol Genet 14(15):2231–2239

Raez L, Papadopoulos K, Ricart A, Chiorean E, Dipaola R, Stein M, Lima C, Schlesselman J, Tolba K, Langmuir V, Kroll S, Jung D, Kurtoglu M, Rosenblatt J, Lampidis T (2012) A phase I dose-escalation trial of 2-Deoxy-d-glucose alone or combined with docetaxel in patients with advanced solid tumors. Cancer Chemother Pharmacol 71(2):523-530

Røsland GV, Dyrstad SE, Tusubira D, Helwa R, Tan TZ, Lotsberg ML, Pettersen IKN, Berg A, Kindt C, Hoel F, Jacobsen K, Arason AJ, Engelsen AST, Ditzel HJ, Lønning PE, Krakstad C, Thiery JP, Lorens JB, Knappskog S, Tronstad KJ (2019) Epithelial to mesenchymal transition (EMT) is associated with attenuation of succinate dehydrogenase (SDH) in breast cancer through reduced expression of SDHC. Cancer Metab 7:6

Rutter J, Winge DR, Schiffman JD (2010) Succinate dehydrogenase – assembly, regulation and role in human disease. Mitochondrion 10(4):393–401

Ryan DG, Murphy MP, Frezza C, Prag HA, Chouchani ET, O'Neill LA, Mills EL (2019) Coupling Krebs cycle metabolites to signalling in immunity and cancer. Nat Metab 1:16–33

Salehiomran MR, Nooreddini H, Baghdadi F (2013) Acute necrotizing encephalopathy of childhood; a case report. Iran J Child Neurol 7(2):51–54

Sanchez M, Hamel D, Bajon E, Duhamel F, Bhosle VK, Zhu T, Rivera JC, Dabouz R, Nadeau-Vallée M, Sitaras N, Tremblay DÉ, Omri S, Habelrih T, Rouget R, Hou X, Gobeil F, Joyal JS, Sapieha P, Mitchell G, Ribeiro-Da-Silva A, Mohammad Nezhady MA, Chemtob S (2022) The succinate receptor SUCNR1 resides at the endoplasmic reticulum and relocates to the plasma membrane in hypoxic conditions. Cells 11(14):2185

Saxena N, Maio N, Crooks DR, Ricketts CJ, Yang Y, Wei M-H, Fan TW-M, Lane AN, Sourbier C, Singh A, Killian JK, Meltzer PS, Vocke CD, Rouault TA, Linehan WM (2015) SDHB-deficient cancers: the role of mutations that impair iron Sulfur cluster delivery. JNCI J Natl Cancer Inst 108(1):djv287

Scagliola A, Mainini F, Cardaci S (2020) The tricarboxylic acid cycle at the crossroad between cancer and immunity. Antioxid Redox Signal 32(12):834–852

Schiliro C, Firestein BL (2021) Mechanisms of metabolic reprogramming in cancer cells supporting enhanced growth and proliferation. Cells 10(5):5
Schmidt DR, Patel R, Kirsch DG, Lewis CA, Vander Heiden MG, Locasale JW (2021) Metabolomics in cancer research and emerging applications in clinical oncology. CA Cancer J Clin 71(4):333–358
Sciacovelli M, Guzzo G, Morello V, Frezza C, Zheng L, Nannini N, Calabrese F, Laudiero G, Esposito F, Landriscina M, Defilippi P, Bernardi P, Rasola A (2013) The mitochondrial chaperone TRAP1 promotes neoplastic growth by inhibiting succinate dehydrogenase. Cell Metab 17(6):988–999
Selak MA, Armour SM, MacKenzie ED, Boulahbel H, Watson DG, Mansfield KD, Pan Y, Simon MC, Thompson CB, Gottlieb E (2005) Succinate links TCA cycle dysfunction to oncogenesis by inhibiting HIF-alpha prolyl hydroxylase. Cancer Cell 7(1):77–85
Senapati S, Mahanta AK, Kumar S, Maiti P (2018) Controlled drug delivery vehicles for cancer treatment and their performance. Signal Transduct Target Ther 3:7
Sever R, Brugge JS (2015) Signal transduction in cancer. Cold Spring Harb Perspect Med 5(4):a006098
Shen S, Clairambault J (2020) Cell plasticity in cancer cell populations. F1000Research 9:F1000 faculty Rev-635
Shi L, Pan H, Liu Z, Xie J, Han W (2017) Roles of PFKFB3 in cancer. Signal Transduct Target Ther 2:17044
Shin E, Koo JS (2021) Glucose metabolism and glucose transporters in breast cancer. Front Cell Dev Biol 9:728759
Siddiqui FA, Prakasam G, Chattopadhyay S, Rehman AU, Padder RA, Ansari MA, Irshad R, Mangalhara K, Bamezai RNK, Husain M, Ali SM, Iqbal MA (2018) Curcumin decreases Warburg effect in cancer cells by down-regulating pyruvate kinase M2 via mTOR-HIF1α inhibition. Sci Rep 8(1):1
Srivani G, Imran M, Merchant N, Mandala J, Nagaraju G (2022) Chapter 12 - Role of succinate dehydrogenase in hepatocellular carcinoma, In: Nagaraju GP, Peela S (eds) Theranostics and Precision Medicine for the Management of Hepatocellular Carcinoma, Academic Press, pp 167–180
Stincone A, Prigione A, Cramer T, Wamelink MMC, Campbell K, Cheung E, Olin-Sandoval V, Grüning N-M, Krüger A, Alam MT, Keller MA, Breitenbach M, Brindle KM, Rabinowitz JD, Ralser M (2015) The return of metabolism: biochemistry and physiology of the pentose phosphate pathway. Biol Rev Camb Philos Soc 90(3):927–963
Su L-J, Zhang J-H, Gomez H, Murugan R, Hong X, Xu D, Jiang F, Peng Z-Y (2019) Reactive oxygen species-induced lipid peroxidation in apoptosis, autophagy, and ferroptosis. Oxidative Med Cell Longev 2019:e5080843
Sullivan LB, Chandel NS (2014) Mitochondrial reactive oxygen species and cancer. Cancer Metab 2(1):17
Sun L, Zhang H, Gao P (2022) Metabolic reprogramming and epigenetic modifications on the path to cancer. Protein Cell 13(12):877–919
Takács-Vellai K, Farkas Z, Ősz F, Stewart GW (2021) Model systems in SDHx-related pheochromocytoma/paraganglioma. Cancer Metastasis Rev 40(4):1177–1201
Tefas LR, Barbălată C, Tefas C, Tomuță I (2021) Salinomycin-based drug delivery systems: overcoming the hurdles in cancer therapy. Pharmaceutics 13(8):1120
Teicher BA, Linehan WM, Helman LJ (2012) Targeting cancer metabolism. Clin Cancer Res 18(20):5537–5545
To KKW, Cho WCS (2020) Chapter 12 – drugs repurposed to potentiate immunotherapy for cancer treatment. In: To KKW, Cho WCS (eds) Drug repurposing in cancer therapy. Academic Press, pp 311–334
Torrens-Mas M, Oliver J, Roca P, Sastre-Serra J (2017) SIRT3: oncogene and tumor suppressor in cancer. Cancers 9(7):90
Tretter L, Patocs A, Chinopoulos C (2016) Succinate, an intermediate in metabolism, signal transduction, ROS, hypoxia, and tumorigenesis. Biochim Biophys Acta 1857(8):1086–1101

Tseng P-L, Wu W-H, Hu T-H, Chen C-W, Cheng H-C, Li C-F, Tsai W-H, Tsai H-J, Hsieh M-C, Chuang J-H, Chang W-T (2018) Decreased succinate dehydrogenase B in human hepatocellular carcinoma accelerates tumor malignancy by inducing the Warburg effect. Sci Rep 8:3081

Vander Heiden MG, Cantley LC, Thompson CB (2009) Understanding the Warburg effect: the metabolic requirements of cell proliferation. Science 324(5930):1029–1033

Wang G, Han JJ (2022) Connections between metabolism and epigenetic modifications in cancer. Med Rev 1(2):199–221

Wang Z-H, Peng W-B, Zhang P, Yang X-P, Zhou Q (2021) Lactate in the tumour microenvironment: from immune modulation to therapy. EBioMedicine 73:103627

Wang W, Rong Z, Wang G, Hou Y, Yang F, Qiu M (2023a) Cancer metabolites: promising biomarkers for cancer liquid biopsy. Biomark Res 11(1):66

Wang Y, Hu J, Wu S, Fleishman JS, Li Y, Xu Y, Zou W, Wang J, Feng Y, Chen J, Wang H (2023b) Targeting epigenetic and posttranslational modifications regulating ferroptosis for the treatment of diseases. Signal Transduct Target Ther 8(1):Article 1

Ward PS, Thompson CB (2012) Metabolic reprogramming: a cancer Hallmark even Warburg did not anticipate. Cancer Cell 21(3):297–308

Weber T, Lu M, Andera L, Lahm H, Gellert N, Fariss MW, Korinek V, Sattler W, Ucker DS, Terman A, Schröder A, Erl W, Brunk UT, Coffey RJ, Weber C, Neuzil J (2002) Vitamin E succinate is a potent novel antineoplastic agent with high selectivity and cooperativity with tumor necrosis factor-related apoptosis-inducing ligand (Apo2 ligand) *in vivo*. Clin Cancer Res 8(3):863–869

Wei Q, Qian Y, Yu J, Wong CC (2020) Metabolic rewiring in the promotion of cancer metastasis: mechanisms and therapeutic implications. Oncogene 39(39):39

Wei S, Yin D, Yu S, Lin X, Savani MR, Du K, Ku Y, Wu D, Li S, Liu H, Tian M, Chen Y, Bowie M, Hariharan S, Waitkus M, Keir ST, Sugarman ET, Deek RA, Labrie M, Khasraw M, Lu Y, Mills GB, Herlyn M, Wu K, Liu L, Wei Z, Flaherty KT, Abdullah K, Zhang G, Ashley DM (2022) Antitumor activity of a mitochondrial-targeted HSP90 inhibitor in gliomas. Clin Cancer Res 28(10):2180–2195

Wei Y, Ma X, Zhao J-C, Wang X-Q, Gao C-Q (2023) Succinate metabolism and its regulation of host-microbe interactions. Gut Microbes 15(1):2190300

Wigerup C, Påhlman S, Bexell D (2016) Therapeutic targeting of hypoxia and hypoxia-inducible factors in cancer. Pharmacol Ther 164:152–169

Wong CC, Qian Y, Yu J (2017) Interplay between epigenetics and metabolism in oncogenesis: mechanisms and therapeutic approaches. Oncogene 36(24):3359–3374

Wu KK (2023) Extracellular succinate: a physiological messenger and a pathological trigger. Int J Mol Sci 24(13):11165

Wu J-Y, Huang T-W, Hsieh Y-T, Wang Y-F, Yen C-C, Lee G-L, Yeh C-C, Peng Y-J, Kuo Y-Y, Wen H-T, Lin H-C, Hsiao C-W, Wu KK, Kung H-J, Hsu Y-J, Kuo C-C (2020) Cancer-derived succinate promotes macrophage polarization and cancer metastasis via succinate receptor. Mol Cell 77(2):213–227

Xiao M, Yang H, Xu W, Ma S, Lin H, Zhu H, Liu L, Liu Y, Yang C, Xu Y, Zhao S, Ye D, Xiong Y, Guan K-L (2012) Corrigendum: inhibition of -KG-dependent histone and DNA demethylases by fumarate and succinate that are accumulated in mutations of FH and SDH tumor suppressors. Genes Dev 26:1326–1338

Xiong L, Shen Y-Q, Jiang L-N, Zhu X-L, Yang W-C, Huang W, Yang G-F (2015) Succinate dehydrogenase: an ideal target for fungicide discovery. In: ACS symposium series, pp 175–194

Xu D, Jalal SI, Sledge GW, Meroueh SO (2016) Small-molecule binding sites to explore new targets in the cancer proteome. Mol BioSyst 12(10):3067–3087

Xu J, Ding L, Love PED (2017) Digital reproduction of historical building ornamental components: from 3D scanning to 3D printing. Autom Constr 76:85–96

Xu J, Zheng Y, Zhao Y, Zhang Y, Li H, Zhang A, Wang X, Wang W, Hou Y, Wang J (2022) Succinate/IL-1β Signaling Axis promotes the inflammatory progression of endothelial and exacerbates atherosclerosis. Front Immunol 13:817572

Yang Y, Sauve AA (2016) NAD^+ metabolism: bioenergetics, signaling and manipulation for therapy. Biochim Biophys Acta 1864(12):1787–1800

Yang M, Su H, Soga T, Kranc KR, Pollard PJ (2014) Prolyl hydroxylase domain enzymes: important regulators of cancer metabolism. Hypoxia 2:127–142

Yang J, Zhou Y, Li Y, Hu W, Yuan C, Chen S, Ye G, Chen Y, Wu Y, Liu J, Wang Y, Du J, Tong X (2022a) Functional deficiency of succinate dehydrogenase promotes tumorigenesis and development of clear cell renal cell carcinoma through weakening of ferroptosis. Bioengineered 13(4):11187–11207

Yang Y, Zhang M, Wang Y (2022b) The roles of histone modifications in tumorigenesis and associated inhibitors in cancer therapy. J Natl Cancer Cent 2(4):277–290

Ye D, Guan K-L, Xiong Y (2018) Metabolism, activity, and targeting of D- and L-2-hydroxyglutarates. Trends Cancer 4(2):151–165

You M, Xie Z, Zhang N, Zhang Y, Xiao D, Liu S, Zhuang W, Li L, Tao Y (2023) Signaling pathways in cancer metabolism: mechanisms and therapeutic targets. Signal Transduct Target Therap 8(1):Article 1

Zhang W, Lang R (2023) Succinate metabolism: a promising therapeutic target for inflammation, ischemia/reperfusion injury and cancer. Front Cell Dev Biol 11:1266973

Zhang J, Xiang H, Liu J, Chen Y, He R-R, Liu B (2020a) Mitochondrial Sirtuin 3: new emerging biological function and therapeutic target. Theranostics 10(18):8315–8342

Zhang J, Zhang Q, Yang Y, Wang Q (2020b) Association between succinate receptor SUCNR1 expression and immune infiltrates in ovarian cancer. Front Mol Biosci 7:150

Zhang Z, Huang Q, Zhao D, Lian F, Li X, Qi W (2023) The impact of oxidative stress-induced mitochondrial dysfunction on diabetic microvascular complications. Front Endocrinol 14:1112363

Zhao H, Li Y (2021) Cancer metabolism and intervention therapy. Mol Biomed 2(1):5

Zhao H, Wang A, Zhang Z (2020) LncRNA SDHAP1 confers paclitaxel resistance of ovarian cancer by regulating EIF4G2 expression via miR-4465. J Biochem 168

Zhong L, Li Y, Xiong L, Wang W, Wu M, Yuan T, Yang W, Tian C, Miao Z, Wang T, Yang S (2021) Small molecules in targeted cancer therapy: advances, challenges, and future perspectives. Signal Transduct Target Ther 6(1):1

Zimna A, Kurpisz M (2015) Hypoxia-inducible factor-1 in physiological and pathophysiological angiogenesis: applications and therapies. Biomed Res Int 2015:549412

RETRACTED CHAPTER: SDH and Its Impact on Neurodegenerative Diseases: Autophagy in Neuroinflammation Perspective

8

Sukesh Kumar Gupta, Archit Gupta, Dhruv Kansal, Devendra Kumar, and Vaishali Manikrao Patil

Abstract

Succinate dehydrogenase (SDH) occupies a central place in cellular energy production, linking the tricarboxylic cycle with the electron transport chain (ETC). Dysfunction of SDH can trigger mitochondrial impairment and disruptions in ATP generation, leading to the onset of neurodegeneration. The prevalence of neurodegenerative disorders such as Alzheimer's disease (AD), Parkinson's disease (PD), and Huntington's disease (HD) increases with age. Existing studies suggest that SDH in mitochondria play a key role in meeting high energy demands necessary for restoring brain tissue function. Therefore, SDH may act as a primary regulator in neuroprotection. This chapter presents recent findings on SDH activity and related pathways that may be critical for neuronal survival. Additionally, we explore the potential of SDH as a neuroprotective agent. SDH

The original version of this chapter was retracted: The retraction note to this chapter is available at https://doi.org/10.1007/978-981-96-6831-1_12

S. K. Gupta
School of Medicine, Wayne State University, Detroit, MI, USA

A. Gupta
SAV Pharmacy, Ghaziabad, Uttar Pradesh, India

D. Kansal
Central Drugs Standard Control Organization, New Delhi, India

D. Kumar
Department of Pharmaceutical Chemistry, School of Pharmacy and Technology Management, Shirpur Campus, Shirpur, Dhule, Maharashtra , India

V. M. Patil (✉)
Charak School of Pharmacy, Chaudhary Charan Singh University (CCSU), Meerut, Uttar Pradesh, India

V. M. Patil et al. (eds.), *Deciphering The Role of Succinate Dehydrogenase in Drug Discovery*, https://doi.org/10.1007/978-981-96-6831-1_8

contributes significantly on cellular and mitochondrial function and helps maintain homeostasis of the body. Increased SDH activity regulate lipid accumulation in the brain and excitotoxicity signaling dependent neurodegenerative disorders. Thus, SDH may represent a promising therapeutic candidate to ameliorate neurodegenerative diseases and aging.

Keywords

Succinate dehydrogenase (SDH) · Neurodegenerative disorders · Mitochondria · Neuroprotection · Mitophagy · Neuroinflammation

Abbreviations

AD	Alzheimer's disease
alpha-syn	Alpha-synuclein
AP	Amyloid peptide
ATP	Adenosine triphosphate
CMA	Chaperone-mediated autophagy
GIST	Gastrointestinal stromal tumor
HD	Huntington's disease
Htt	Huntingtin protein
LB	Lewy bodies
LV	Left ventricular
MCI	Moderate cognitive impairment
MPC	Mitochondrial pyruvate transporter
mtDNA	Mitochondrial DNA
NADH	Nicotinamide adenine dinucleotide
OXPHOS	Oxidative phosphorylation
PD	Parkinson's disease
PINK1	PTEN-initiated putative kinase 1
ROS	Reactive oxygen species
SDH	Succinate dehydrogenase
TCA	Tricarboxylic acid
UQ	Ubiquinone
8-OHdG	8-Hydroxy-deoxyguanosine

8.1 Introduction

Recent reports highlight the impact of neurodegenerative diseases throughout the world. In the United States, the most prevalent neurodegenerative diseases — Alzheimer's disease (AD) (affecting over 6.2 million people) and Parkinson's disease (PD) (over one million people)—highlight the urgent need to better understand the mechanisms underlying their development and to devise new treatment and prevention strategies (www.niehs.nih.gov 2022; www.fda.gov 2022). Various hypotheses have been proposed for the development of effective therapeutic interventions, including the correlation of mitophagy and neuroinflammation, various biochemical pathways (Charmpilas et al. 2022), mutations in associated proteins such as TAF8 (Wong et al. 2022), and CX3CR1 deficiency (Puntambekar et al. 2022). SDH plays an important role in the generation of oxidative stress, leading to neuroinflammation. Further understanding of SDH may help to explore the stipulated integral role of autophagy in neuroinflammation in the human brain (see Fig. 8.1). This chapter attempts to explore the impact of SDH on the pathogenesis of neurodegenerative disease. However, mitochondrial biogenesis and oxidative stress play a key role in neurodegenerative diseases.

8.2 Implications of Mitochondrial Function

Mitochondria are membrane-bound organelles found in nearly every eukaryotic cell. They are vital for sustaining life and act as gatekeepers of cell death, serving as regulators of cellular energy generation (Sica et al. 2016). They contain a remnant genome known as mitochondrial DNA (mtDNA), which encodes 13 proteins essential for the function of respiratory chain and is thought to have evolved from symbiotic predecessors. An inside and outside membrane separates and maintains the aqueous areas, the intermembrane space, and matrix in mitochondria. Mitochondria play a role in a variety of biological activities, such as calcium signaling and cellular fuctions like development, differentiation, regulation of cycle, and cell death. The maintenance of mitochondrial populations involves mechanisms such as fission and fusion, biogenesis, and autophagy, which regulates mitochondrial share and distributions. Mitochondrial dysfunction has been associated to metabolic and age-related diseases, neurological diseases, as well as heart and brain ischemia (Osellame et al. 2012; Nabi et al. 2022; Ruiz et al. 2013; Wu et al. 2019). Metabolic deficits cause mitochondrial dysfunction, oxidative stress, inflammation, leading to neuronal death (Yan et al. 2020).

Mitochondria are important controllers of cellular bioenergetics and are sensitive to alterations in environmental stimuli, namely, hormones, nutrition, and oxygen tension (Boutilier 2001; Civitarese and Ravussin 2008; Connor et al. 2005; Davidson and Duchen 2007; Melov 2000; Mookerjee et al. 2010; Semenza 2007). During this process, redox-sensitive cellular targets play an essential role, providing a sensitive and quick response to modification in metabolism and reactive oxygen species (ROS) fluxes. ROS were mostly investigated for their toxic effects on proteins, lipids, and DNA causing cell death and damage. However, studies support an increasing understanding of ROS' function in influencing cellular activities via redox-dependent signaling (Chen et al. 2004; Handy et al. 2009; Rasbach and Schnellmann 2007; Rigoulet et al. 2011; Ristow and Zarse 2010).

Mitochondrial disease is a category of hereditary energy metabolism disorders showing various symptoms, severity, and outcomes. They may be considered as one of the frequent classes of hereditary metabolic illness, with a minimum birth frequency (around 1/5000) (Mancuso et al. 2007; Neupert and Herrmann 2007; Calvo et al. 2006). Because almost all cells require oxidative phosphorylation (OXPHOS), mitochondrial disorders can affect any organ. These conditions may present to a

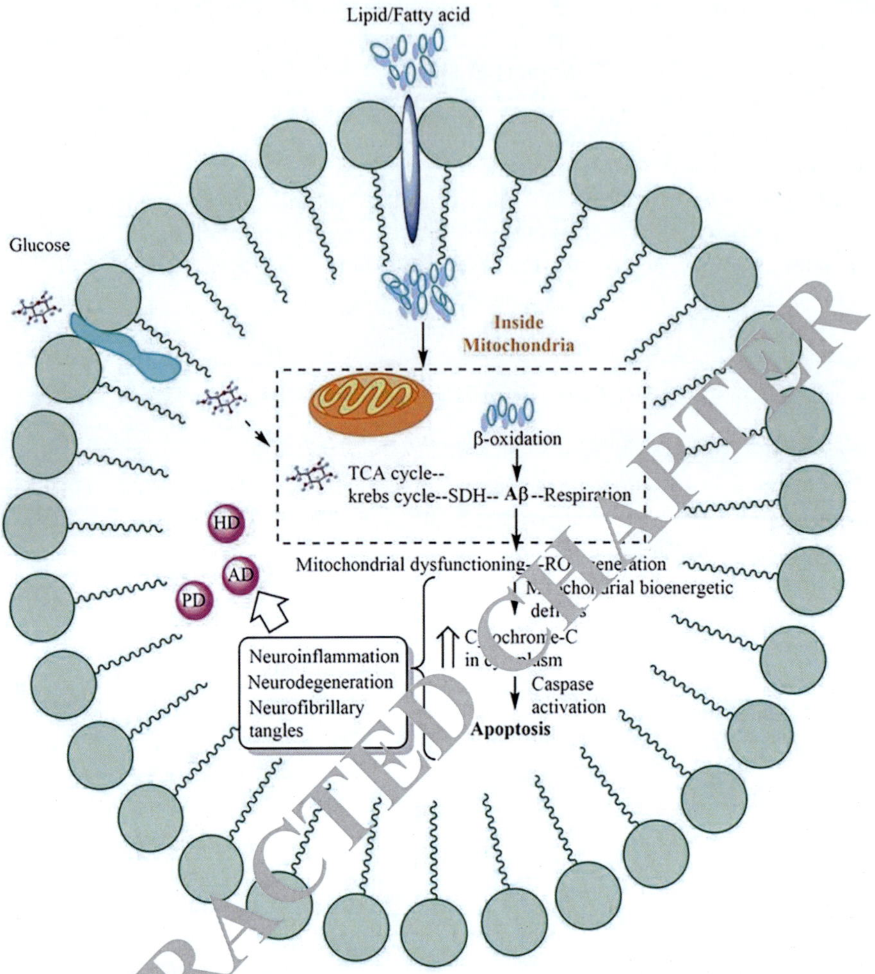

Fig. 8.1 Possible mechanism of SDH involved in neurodegenerative disorders. (*i*) Sequential transfer occurs in the plasma membranes of neurons, lipid/fatty acid, and glucose transfer. (*ii*) Oxidative stress and inflammation: mitochondria-associated oxidative stress and inflammatory responses play a key role in the development of neurodegenerative disease. Aβ directly disturbs the mitochondrial structure and function, promoting oxidative stress and inflammation, and contributes to the pathogenesis of neurodegenerative disorders and aging. (*iii*) Mitochondrial dysfunction: the TCA cycle and the electron transport chain are two major candidates involved in mitochondrial bioenergetics. Mitochondrial dysfunction leads to neurodegenerative disorders and aging. Two major biological functions that occur in mitochondria are the TCA cycle and respiration. *Aβ* beta amyloid, *AD* Alzheimer's disease, *PD* Parkinson's disease, *HD* Huntington's disease, *CNS* central nervous system, *SDH* Succinic dehydrogenase, *TCA* tricarboxylic acid

variety of pediatric subspecialists and are a part of the differential diagnosis in a wide range of clinical settings. Few afflicted patients have had a definitive molecular diagnosis (Debray et al. 2008).

Harman's free radical theory of aging remains one of the leading hypotheses in the field (Guarente 2008). According to Harman, oxygen-free radicals created during normal cellular respiration shows cumulative damage to molecules, resulting in organismal loss of functioning and, finally, death. Because free radicals or ROS are created in mitochondria during electron transport, mitochondria and aging have received a lot of attention (Lanza and Nair 2010). Mitochondrial activity and oxidative phosphorylation through the respiratory chain are two of life's most basic activities. Deficiencies in mitochondrial role and the activity of the respiratory chain's different protein complexes have been related to several clinical disorders (Schapira 2012). The complexes in mitochondria are known as nicotinamide adenine dinucleotide (NADH) that is, complex-I (ubiquinone oxidoreductase), complex-II (succinate dehydrogenase), complex-III (ubiquinol-cytochrome c oxidoreductase), complex-IV (cytochrome c oxidase), and complex-V (ATP synthase).

ComplexI is the first enzyme of electron transport chain (ETC). It oxidizes NADH, which is produced through the Krebs cycle in mitochondrial matrix and applies the two electrons to break ubiquinone to ubiquinol. It is the mitochondrial respiratory chain's biggest multimeric enzyme complex, responsible for electron transport and the creation of a proton gradient across inner membrane to fuel adenosine triphosphate (ATP) synthesis. Eukaryotic complex-I is made up of 14 conserved subunits that are similar to bacterial subunits, as well as more than 26 auxiliary subunits. Complex-I in mammals is made up of 45 subunits that must be assembled correctly to generate a fully functional mature complex. Eukaryotic complex-I is present in the inner membrane of mitochondria and protrudes into the matrix, forming an L-shaped structure. A hydrophilic peripheral arm and a hydrophobic membrane arm lie perpendicular to each other in this configuration. Multiple proteins have also been identified as complex-I biogenesis assembly factors, and numerous individuals with mutations in genes linked to complex-I deficiency and mitochondrial disorders have been uncovered (Swalwell et al. 2011; Rodenburg 2016; Wirth et al. 2016).

8.3 Complex-II

Complex-II, or SDH, present in the inner mitochondrial membrane, has a unique dual role in a cell's energy-producing activities (Van Vranken et al. 2015; Moosavi et al. 2019). It participates in the oxidation of succinate to fumarate (citric acid cycle), aerobic ETC, shortening of ubiquinone (coenzyme Q), and assists in the oxidative phosphorylation of ATP. It is a hetero-oligomer that is made up of four component proteins, that is, SDHA, SDHB, SDHC, and SDHD. The major catalytic domain supervising the oxidation of succinate to fumarate is formed by SDHA and SDHB. The electron transport process of ETC is governed by the membrane-anchoring subunits SDHC and SDHD. Furthermore, at least two additional proteins,

SDHAF1 and SDHAF2, aid it by flavination the SDHA subunit and boosting SDHB maturation, allowing the whole SDH complex to be assembled (Rasheed and Tarjan 2018).

Complex II deficiencies with a known genetic cause have been found in only a few diseases, including inborn paraganglioma–pheochromocytoma tumor syndrome, a subset of Leigh syndrome cases, gastrointestinal stromal tumor (GIST) primarily linked with Carney–Stratakis syndrome, and most recently, infantile leukoencephalopathy linked to SDHAF1 mutations (Hoekstra and Bayley 2013).

SDH/complex-II occupies a unique place among the electron transport system's other complexes as a key controller of ROS generation, either directly or indirectly through complex-I, downstream complex-III, or other mitochondrial ROS producers. Complex -II stands out as a target for preventing oxidative stress-mediated tissue impairment and cytoprotection against ischemic/reperfusion episodes, as well as a selective target for cancer treatment, due to its pivotal role as an operator of ROS synthesis, affecting both the NADH/NAD^+ ratios and the reduced/oxidized ubiquinone pool (UQH2/UQ ratio). The features of the SDH or complex-II system as a critical redox controller of ROS generation are discussed, as well as its involvement in post-ischemic oxidative tissue injury and its potential as a target for cancer cell-selective death by driving severe ROS overproduction (Ralph et al. 2011).

Malate and fumarate accumulate in the cell when succinate dehydrogenase is inhibited. Because cells require a modest concentration of fumarate to produce ATP, a high quantity of fumarate in the matrix can reduce ATP synthesis. Furthermore, the buildup of succinate in the cell is one of the primary causes of ROS production. SDH could control the amount of superoxide produced by the ETC's other complexes, I and III (Farshbaf and Kiani-Esfahani 2018). A recent review highlights its role in understanding disease pathology and development of therapeutics (Vanova et al. 2020).

8.4 Complex-III

Complex-III, also known as the cytochrome bc1 complex or Co Q-cytochrome c reductase, is a protein that transmits electrons from QH2 to cytochrome c (Chen et al. 2003; Acín-Pérez et al. 2004). With 11 subunits in each monomer, complex-III or ubiquinol is a symmetrical dimer. Cytochrome b (bL and bH), cytochrome c1, and a highly potent (2Fe–2S) cluster enveloped by an iron–sulfur protein are the catalytically active components. On both sides of cytochrome b, settled in the inner membrane of mitochondria, there are two coenzyme Q (CoQ) binding sites—one of which is the QH2 oxidation site (Qo) on the cytoplasmic side, which is associated with the less potent cytochrome bL, and the other is the Q-reduction site (Qi) on the matrix side, which is linked to cytochrome bH' high potential. The Q-cycle is responsible for complex-III's electron transfer mechanism. After delivering an electron to the (2Fe–2S) cluster, QH2 is oxidized to ubisemiquinone (QH−), and two protons from the matrix are continuously released into the mitochondrial intermembrane space (IMS). This electron is transported to cytochrome c1 by the (2Fe–2S) cluster, and then to cytochrome c, a mobile electron carrier. The second electron is then transferred to cytochrome bL, which transmits it to cytochrome bH at the Qi site, thanks to the highly reductive QH– produced at the Qo site. This electron is transferred by reduced cytochrome bH to CoQ of the Qi site, creating a new QH−. The second QH2 (dihydroxyquinone) molecule is oxidized at the Qo site, dislocating the other two protons to complete the Q-cycle. To create QH2, one electron is transmitted to the (2Fe–2S) cluster, another to cytochrome bH, and the final electron to QH− at the Qi site (Zhao et al. 2019).

8.5 Complex-IV

Complex-IV, also known as cytochrome c oxidase, converts cytochrome c to water by donating electrons from cytochrome c to the terminal electron acceptor, oxygen (Li et al. 2006). CuA, heme a (Fea), and a binuclear core comprising heme a3 (Fea3) and CuB make up mammalian complex-IV, which is made up of 13 distinct subunits with four redox-active metal centers. The core subunits are mitochondrial DNA-encoded subunits I, II, and III, whereas the auxiliary subunits are nuclear-coded subunits. Three of the four cofactors—heme and the binuclear center, which transmits electrons from heme a to oxygen—are found in subunit I. Subunit II and subunit III are on both ends of subunit I, with two CuA cofactors located on the intermembrane-facing side of subunit II. Subunit III is primarily engaged in proton pumping and helps to stabilize the other two core proteins. The allosteric ATP-mediated inhibition of complex-IV or cytochrome c oxidase, which depends on the ATP or ADP ratio, is used by nuclear-coded subunits to modulate physiological activity. Cytochrome c, like CoQ, is a mobile electron transporter that is electrostatically attached to outer surface of the inner mitochondrial membrane, allowing it to bind with complex-III's cytochrome c1 and accept electrons. Decreased cytochrome c passes over the membrane's surface and interacts with complex-IV's subunit II via electrostatic interactions, concurrently delivering electrons to the CuA site of subunit II, and subsequently passing electrons from heme a to the binuclear center of subunit I, where oxygen is broken to water. Four electrons from cytochrome c are transported nearly immediately to attach dioxygen; eight protons are drawn from the matrix—half are utilized to generate two H_2O molecules and other four are dropped over the membrane into the IMS (Zhao et al. 2019).

8.6 SDH Structure and Its Function

In the Kreb cycle, the oxidation of succinate to fumarate is catalyzed by SDH, which also produces FADH2. The SDH/complex-II crystal structure consists of a soluble enzymatic head group made up of an inner mitochondrial membrane matrix face bound heterodimer of SDHA and SDHB subunits. It catalyzes the electron

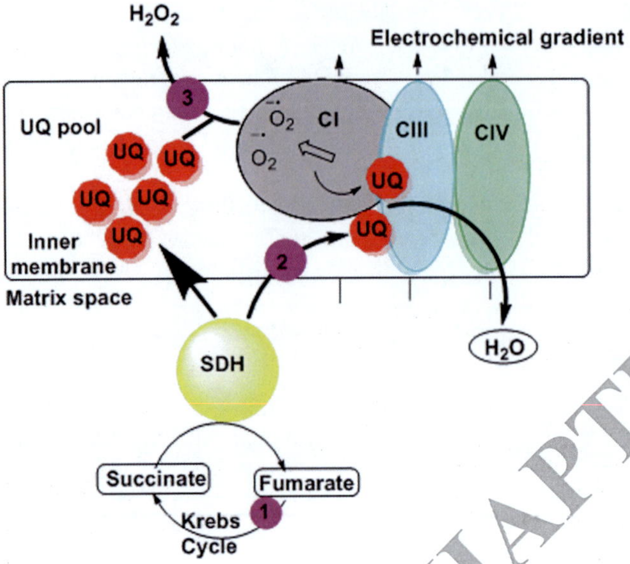

Fig. 8.2 Functions of succinate dehydrogenase (SDH) in the mitochondria. SDH plays a role in the conversion of succinate to fumarate in the Krebs cycle (Denoted by 1). Derived electrons in the respiratory chain complex-III lead to a decrease in oxygen levels and the formation of water (Denoted by 2). Assembling of electrochemical gradients across the mitochondrial inner membrane allows for the synthesis of ATP. On the other hand, electrons can move forward to the ubiquinone pool (UQ pool) and reduce the superoxide anions generated from the respiratory chain itself (Denoted by 3). Alteration in SDH activity would impede electron flow to both respiratory chain complex-III and the UQ pool resulting in the generation of reactive oxygen species that promote neuroinflammation in terms of neuronal loss and aging in the human brain. *UQ* ubiquinone; CI, CIII, CIV: the various complexes of the respiratory chain

transmits from succinate to fumarate reaction to break ubiquinone (UQ or Coenzyme Q) to ubiquinol (UQH2) as shown in Fig. 8.2. Subunit A is covalently attached to the FAD cofactor and subunit B contains three iron–sulfur (Fe–S) clusters in decreasing order of standard redox potential: [2Fe–2S], [4Fe–4S], and [3Fe–4S]. These clusters participate in the direct flow of electron transport reactions within the two soluble catalytic subunits (Zhao et al. 2019).

SDHC (larger cytochrome b binding subunit or CybL) and SDHD (smaller cytochrome b binding subunit or CybS) form the enzyme's foundation as an intramembranous complex within the inner mitochondrial membrane. The soluble SDH head group on the matrix face is connected via this integral membrane-anchoring domain. The membrane domain holds two UQ binding sites as lipophilic channels for phytyl the chain, one with a pocket for the quinoid head group closer to the matrix surface, known as the QP site, and the other extending down the reverse side of the SDHC or SDHD structure, known as the Qd site, with its quinoid pocket closer to the intramembranous space (Zhao et al. 2019).

8.7 Role of SDH in Mitochondrial Biogenesis

As mitochondria are not orchestrated from the beginning, they should multiply from existing ones to keep up with biogenesis. The amalgamation of external and internal mitochondrial layers, mitochondrial proteins, blend and import of atomic genome-encoded proteins, lipid import, oxidative phosphorylation, mtDNA replication, and mitochondrial fusion and fission appear to play a part in biogenesis. Typically, changes in physiological conditions, like expanded paces of ATP utilization, invigorate mitochondrial biogenesis for the cell's ability to create it to move toward its greatest capacity. Cell division and repair, undeveloped turn of events, physiological changes like thoughtful feeling, calorie limitation, work out, cool pressure, energy limit, chemicals (thyroid chemical, leptin, and erythropoietin), and mitochondrial sickness/harm like aggravation, hypoxia/ischemia, and oxidative/nitrosative pressure are among the primary triggers of mitochondrial biogenesis (Bezawork-Geleta et al. 2017). Mitochondrial harm manifests as decreased mitochondrial activity, mtRNA records, and protein union, as well as mtDNA disability.

8.8 Mitochondrial Neurodegeneration

For proper functioning of any cell or tissue, there is need of energy in the ATP form, which is generated by mitochondria—the powerhouse of cell. When there is any alteration in ATP generation, cellular homeostasis is affected. Neurons are sensitive to stress and degeneration if their mitochondrial activity and physiology are impaired. The ETC is linked to the Krebs cycle by SDH. As a result, SDH failure may affect mitochondrial functioning, ATP production, and energy hemostasis in the cell. SDH could modulate excitotoxicity in neurodegenerative illnesses through both direct and indirect mechanisms, in addition to lipid production (Martínez-Reyes and Chandel 2020). Furthermore, SDH mutation is linked to the beginning of neurodegenerative diseases like PD, AD, and Huntington's disease (Onyango et al. 2021; Huang et al. 2024).

8.9 Role of SDH in Mitochondrial Homeostasis

The enzyme complex succinate dehydrogenase is vital in the mitochondria because it not only participates in the Krebs cycle (where it degrades acetyl-CoA) but also functions as a member of the respiratory chain, which is required for ATP synthesis. In other words, it is an essential enzyme in intermediate metabolism and ATP synthesis during aerobic exercise. In the Krebs cycle, it helps convert succinate to fumarate. It provides electrons from the oxidation of succinate to complex-III of the ETC, enabling oxygen reduction and water formation. Electron transport produces an electrochemical gradient across the inner mitochondrial membrane, promoting ATP generation. Alternatively, electrons may be employed to reduce molecules in the ubiquinone pool, generating the reducing equivalents required to convert superoxide anions from either the same respiratory chain or exogenous sources. Thus, SDH plays an important role in mitochondria by maintaining its capability of generating adequate amount of ATP in the cell. In addition, it also helps maintain the reduced state of the mitochondrial ubiquinone (UQ) pool (Rustin et al. 2002).

8.10 Role of SDH in Free Radical Generation

The process involving complex-II in the Krebs cycle or tricarboxylic acid (TCA) cycle involves different reactions like hydration, dehydration, oxidation, reduction, and decarboxylation. Though these are electron-coupled reactions, their coupling with the ETC causes huge flux of free radicals. Additionally, antioxidants recycled by NADH via NADPH end up producing free radicals. Some studies have supported role of SDH in free radical generation (Vázquez-Memije et al. 2008).

The mitochondrial SDH complex catalyzes the Krebs cycle's oxidation of succinate to fumarate and supplies electrons to the ubiquinone pool in the respiratory chain. This process transmits electrons generated in the Krebs cycle to complex-III and creates an electrochemical gradient across the inner membrane of the mitochondria, enabling ATP production. Alternatively, electrons can be redirected to reduce the ubiquinone pool (UQ pool) and provide the reducing equivalents needed to reduce superoxide anions emanating from either an exogenous or endogenous source. When this succinate dehydrogenase enzyme is reduced, it hampers the electron flow in both complex-III and ubiquinone pool. Exceptionally responsive oxygen species (ROS) can harm NADH dehydrogenase, cytochrome c oxidase, and ATP synthase, making mitochondria energy creation shut down. Ca^{2+} transitions across the plasma layer and between intracellular compartments are properly controlled under typical physiological circumstances (Zorov et al. 2014).

8.11 Role of SDH in the Pathogenesis of Neurodegenerative Disorders

8.11.1 Alzheimer's Disease

Alzheimer's disease (AD) is the most prevalent form of dementia, with a growing morbidity rate in the aged population in recent decades. AD is a terrible neurological condition characterized by cognitive and motor decline. It progresses from moderate cognitive impairment (MCI) impacting memory, attention, and speech patterns to late-stage cognitive impairment affecting all areas of behavior as AD develops. Despite the rising number of AD cases, there is nothing in the way of therapy to halt or stop the disease's course. This is exacerbated by fact that Alzheimer's disease is essentially idiopathic, with multiple ideas linking the disease to a variety of physiological causes. Genetic heredity, diminished acetylcholine synthesis, the buildup of neurotoxic protein plaques of amyloid peptide (AP) and fibrillary tangles of hyperphosphorylated tau-protein, or irregular mitochondrial activity and dynamics are the primary explanations for the etiology of AD (the mitochondrial cascade hypothesis). The neuropathological results of late-stage tau tangle deposition and collection of APP have been a constant detection for AD pathology in all experimental attempts, while it is unclear whether these collections are a result or a causal agent for AD (Cadonic et al. 2016).

Quantitatively, PDHC and the Krebs cycle are the primary physiological sources of NADH, which is oxidized by most of the oxygen used by the body. The principal utilization of oxygen is the oxidation of NADH, and the key enzymes that create NADH from the Krebs cycle are involved in this process. The strong link between alterations in mitochondrial enzyme activity and CDR shows that these aberrations may be at the root of brain dysfunction. Although the relationship between PDHC activities and CDR is particularly strong ($r = 0.77$), each of the changes in the Krebs cycle enzyme activity or the total of dehydrogenases that are or are not decarboxylases correlated strongly with CDR. The fact that the increases or reductions are significantly associated with CDR shows that the changes in mitochondria in AD are regulated in a coordinated manner. Increases in dehydrogenase activity in the second half of the Krebs cycle may represent compensatory reactions to the Krebs cycle's reduced activity of decarboxylase dehydrogenases. Animal research backs up this theory. The KGDHC activity is reduced by chronic hypoxia, although SDH (complex-II) is increased by 66%. During hypoxia, an anaerobic mitochondrial process, including the conversion of oxaloacetate to succinate, the reverse reaction of malate dehydrogenase (MDH), and the electron transport chain via complex-I, can occur. Few research has looked at how multiple enzymes in the Krebs cycle work together to regulate each other. One bacterial study investigated RNA message levels following KGDHC suppression, but not enzyme activities (Bubber et al. 2005).

8.11.2 Parkinson's Disease

PD is a motor and non-motor symptom condition defined by dopaminergic neuron loss and the presence of Lewy bodies (LBs) (Iljina et al. 2016). The gradual degradation of dopaminergic neurons in the substantia nigra pars compacta of PD patients affects motor function. D1 and D2 DA receptors are both responsible for maintaining this activity. In dopaminergic treatment, restoring dopaminergic activity via D1 and D2 receptors helps improve motor symptoms in Parkinson's disease. In dopaminergic systems, several routes are implicated, including nigrostriatal, mesolimbic, mesocortical, tuberoinfundibular, and others, although DA blockage is best shown in the mesolimbic and mesocortical pathways. The existence of LBs in dopaminergic neurons of the substantia nigra pars compacta is another prominent symptom of Parkinson's disease (SNpc). LBs are mutations or aggregation of alpha-synuclein (alpha-syn) that develop insoluble fibrils, according to various research. Excessive synthesis of misfolded versions of ubiquitin proteins involved in protein recycling is also linked to LBs. Somatic mtDNA mutations have been found in a variety of tissues associated with aging and neurodegeneration, however the frequency of each mutation is generally quite low. Individual mutations can, however, clonally expand to large levels, resulting in respiratory chain–deficient cells that are identical to those found in heteroplasmic mtDNA illness. Although recent studies in the colon imply that a considerable fraction of cells may possess somatic mtDNA mutations, the potential impact of these somatic mtDNA mutations in aging or illness remains unknown. We hypothesized that because substantia nigra neurons experience age-related cell loss, have high oxidative capacity, and appear to be sensitive to mitochondrial dysfunction—as well as the presence of respiratory chain–deficient neurons—these cells may be especially vulnerable to the accumulation of somatic mtDNA mutations with age. We also wanted to see if there was clonal growth of somatic mtDNA mutations in substantia nigra neurons since Parkinson's disease is an age-related neurodegenerative illness with substantial cell loss in the substantia nigra (Bender et al. 2006).

8.11.3 Huntington's Disease

Huntington's disease (HD) is a neurodegenerative disorder characterized by involuntary irregular movements (chorea), cognitive impairments, and psychiatric symptoms. An aberrant amplification of CAG (cytosine, adenine, and guanine) repeats in an exon of the gene encoding the huntingtin protein (Htt), which provides a toxic function to the protein, causes the disorder. The selective decrease of medium spiny GABAergic neurons in striatum is the most dramatic neuropathological alteration in HD. Although the mechanisms driving striatal vulnerability in HD are unclear, there is persuasive evidence that mitochondrial abnormalities may play a key role. We will go through some recent research that backs up this theory. Studies of the harmful consequences of mutant Htt in animal models or cell culture indicate mitochondrial alterations such as reduced Ca^{2+} buffering capacity, membrane potential loss, and oxidative phosphorylation (OXPHOS) enzyme expression. Striatal neurons may be more susceptible to these problems. One theory posits that neurotransmission systems like dopamine and glutamate worsen striatal mitochondrial abnormalities. Mitochondrial dysfunction promotes Ca^{2+} homeostasis, which is connected to glutamate receptor-mediated excitotoxicity. Dopamine receptors also influence mutant Htt toxicity, at least in part by regulating mitochondrial complex-II expression. All these findings back up the theory that mitochondria, which operate as "sensors" of neurochemical environment, play an essential role in striatal degeneration in HD (Damiano et al. 2010).

The age at which Alzheimer's, Parkinson's, and Huntington's illnesses manifest as neurodegenerative disorders tends to rise. The major goals of research into novel treatment techniques are to alleviate clinical symptoms and preserve neurons from degeneration. Numerous studies show that mitochondria perform critical functions in high-energy-demanding tissues such as the brain. Mitochondrial dysfunction and physiology can render neurons more susceptible to stress and degeneration. The Krebs cycle is linked to the electron transport chain via SDH. As a result, SDH failure may decrease mitochondrial activity, ATP production, and energy homeostasis in the cell (Nicholls 2002; Hara and Snyder 2007; Garabadu et al. 2019). SDH might modulate excitotoxicity in neurodegenerative illnesses by direct and indirect mechanisms, in addition to lipid production. Furthermore, SDH mutations are linked to the development of neurodegenerative diseases. As a result, SDH may play an important role in neuroprotection (Garabadu et al. 2019; Furukawa ct al. 2006). This chapter discusses current data about SDH activity and associated pathways, which may play a role in neuronal survival. Furthermore, we also address all potential candidates for SDH as a neuroprotective agent. SDH inhibition causes striatal neuron loss and symptoms of HD. Another SDH inhibitor, malonate, can cause neuronal injury and degeneration. In the presence of malonate, neurodegeneration is caused by secondary excitotoxicity as well as mitochondrial membrane potential breakdown and cyt C release (Farshbaf and Kiani-Esfahani 2018; Fernandez-Gomez et al. 2005).

8.12 Involvement of SDH and Oxidative Stress in Alzheimer's Disease

Cell loss with expanded actuation of favorable to death qualities and flagging pathways, debilitated energy digestion, mitochondrial brokenness, and persistent oxidative pressure are among the biochemical, atomic, and cell irregularities that go before or with the neurodegeneration fountain of AD.

Expanded oxidative pressure is a notable and early element of Alzheimer's disease neurodegeneration. Oxidative pressure can cause an assortment of sub-atomic and metabolic changes in the cerebrum that are connected to Alzheimer's disease, as well as obstruct mental execution. Exploratory and clinical proof, then again, do not predominantly uphold oxidative pressure as the main part in the etiology of Alzheimer's disease. Transient hypoxic or ischemic harm, for instance, does not initiate Alzheimer's disease, and the subsequent loss of neurons is not restricted to the mind regions designated in the illness. Regardless, discoveries from an assortment of trial sources, as well as human exploration, have persuaded us that we really want to find out about how oxidative pressure adds to the etiology of Alzheimer's infection. Since past investigations have shown expanded NOS immunoreactivity in advertisement and other neurodegenerative illnesses, elevated degrees of NO can hinder complex-IV capacity, and NOX is an intense oxidant that can intervene DNA harm and potentiate the neurotoxic impacts of NO or Aβ and APP, we zeroed in our endeavors on examining NOS and NOX quality articulation (de la Monte and Wands 2006).

Amyloid plaques and neurofibrillary tangles were believed to be the significant reasons for oxidative pressure when oxidative harm was first found in AD minds. Following investigations by us and different gatherings showing oxidative harm in both minds and peripheral fluids and tissues of patients with moderate mental hindrance, which is considered the pre-obsessive condition going before promotion, have recently revised this idea. Tests in MCI patients, as well as creature and cell culture review, have laid out that oxidative pressure is perhaps the earliest recognizable cycle in promotion pathogenesis, happening before the development of amyloid plaques and the dementia stage. Thus, it is currently imagined that A beta peptides are first orchestrated as cell reinforcements to battle oxidative pressure and later transform into pro-oxidants solely after accumulation and testimony into amyloid plaques (Migliore and Coppedè 2009).

8.13 Role of SDH and Oxidative Stress in Parkinson's Disease

In eukaryotic cells, ETC in the mitochondria is an essential generator of ROS. Complex-I and III yield an unobtrusive measure of superoxide (O_2 •) as atomic oxygen is progressively decreased to water by the electron transport chain edifices. Superoxide can be changed to hydrogen peroxide (H_2O_2) by the protein manganese superoxide dismutase after it is produced inside the mitochondria (MnSOD). Different compounds found in the mitochondria (glutathione peroxidase 4, peroxiredoxin 3, peroxiredoxin 5) catalyze the further decrease of hydrogen peroxide to water and atomic oxygen. Under ordinary conditions, ROS partake in flagging occasions constrained by certain thiol deposits in proteins, which have the limit to govern enormous scope transcriptional changes, in addition to other things. In any case, conditions could make ROS age surpass a cell's cancer prevention agent ability. This present circumstance, known as oxidative pressure, harms cell macromolecules irreversibly and can prompt cell passing. Oxidative pressure markers, including raised measures of oxidatively harmed lipids, proteins, and DNA, have all been found in the SNpc of PD patients samples (Hauser and Hastings 2013).

Loss-of-work changes in DJ-1 reason genetic beginning stage PD and influence the oxidative pressure reaction to breakdown. DJ-1 safeguards dopaminergic neurons against oxidative pressure and the deadly impacts of freak human-synuclein by upregulating glutathione creation and expanding heat shock protein articulation. DJ-1 overexpression in neuroblastoma cells supported cell flexibility to rotenone or 6-hydroxydopamine poisonousness and diminished intracellular ROS, as per a new report. DJ-1 levels were brought down by RNA hushing, which made the contrary difference. Moreover, when neuroblastoma cells were presented to rotenone or 6-hydroxydopamine, DJ-1 mRNA and protein levels expanded. Pretreatment with a cancer prevention agent switched these modifications, proposing that oxidative pressure was the reason for DJ-1 overexpression. Neurotoxins openness brought about overexpression of DJ-1, yet additionally cell rearrangement of DJ-1, as proven by DJ-1 movement into mitochondria. The association of DJ-1 in brain protection against oxidative attacks is upheld by these discoveries. Moreover, a few proteins connected to familial PD, for example, PTEN-initiated putative kinase 1 (PINK1), DJ-1, alpha-synuclein, leucine-rich recurrent kinase 2, and potentially parkin, are mitochondrial proteins or are related with mitochondria, and all collaborate with oxidative pressure and free extreme harm pathways, featuring the significance of oxidative pressure and mitochondrial brokenness in the infection pathogenesis (Browne et al. 1999).

8.14 Implications of SDH and Oxidative Stress in Huntington's Disease

DNA strand breaks can be brought about by free extremist instigated oxidative harm to DNA. Various examinations have observed higher DNA strand breaks in HD striatal neurons, proposing that both apoptotic and necrotic cell passing cycles are involved. Moreover, the level of DNA fracture in the HD striatum is decidedly connected with the span of the polyglutamine development in huntingtin, as indicated by a new report. Studies report huge expansions in DNA fracture in HD striatal and cortical neurons when contrasted with age-matched controls utilizing in situ end marking strategies. Moreover, in light of the fact that practically zero ISEL naming was found inside cell cores, it shows that mitochondrial DNA (mtDNA) is more inclined to fracture than atomic DNA (nDNA).

Unreasonable oxidation of DNA bases, like deoxyguanosine, can likewise be brought about by oxidative harm (to create 8-hydroxy-deoxyguanosine; 8-OHdG). Involving high-performance liquid chromatography (HPLC) identification procedures in after death tissue, we found significant expansions in degrees of 8-OHdG in nDNA in the caudate of grade 4 HD patients. In similar subject gathering, no progressions in nDNA 8-OHdG content were found in mind regions affected less significantly by HD, like the front facing and parietal cortices and cerebellum (Browne et al. 1999).

Mitochondria play a part in both the creation of receptive oxygen species (ROS) and their guideline inside cells. Thus, mHtt impacts on mitochondria ought to be anticipated to influence oxidative harm in cells, and proof of expanded ROS creation and oxidative harm markers have been found in HD posthumous mind tissue, creature models, and HD cell lines. This topic will not be discussed here since it has recently been completely investigated somewhere else, including in these Proceedings (Browne 2008).

8.15 Role of SDH in Neuroinflammation

Complex-II digestion controls mononuclear phagocyte work, and the metabolites they make play significant flagging jobs in intrinsic and versatile insusceptible frameworks, as well as aggravation. Persistent oxidative pressure has been connected to neuroinflammation as both a cause and an outcome. Receptive oxygen and nitrogen species, supportive of provocative cytokines, and glutamate are all neurotoxic, and microglial cells are the significant wellspring of them. Moreover, astrocyte function is generally centered around the conservation of neuronal settings, for example, the extracellular milieu and synapse and particle buffering. These activities need energy; thus, the presence of it is crucial for work of SDH in the astrocyte. Accordingly, changes in astrocyte SDH capacity might impact the astrocyte's neuroprotective abilities. Therefore, obviously SDH lack and neuroinflammation are connected. Therefore, SDH digestion may be a promising objective for decreasing neuroinflammation in an assortment of neurological ailments (Garabadu et al. 2019).

The pathogenesis of different NDs, including AD and PD, is impacted by changes in complex-II (SDH)—pyruvate digestion. A higher pyruvate level in the cerebrospinal liquid has been recognized as an indicative for Alzheimer's infection, while a comparative peculiarity has been identified in the blood serum of Parkinson's sickness patients. Mitochondrial pyruvate digestion is constrained by mitochondrial pyruvate transporters (MPCs), which control pyruvate carbon stream generally speaking. The MPC is an inward layer carrier that helps pyruvate get into mitochondria from the cytoplasm. It is a vital controller of mitochondrial substrate use, and limitations in mitochondrial pyruvate take-up can potentiate the utilization of unsaturated fats and an assortment of amino acids to fuel cell energetics and biosynthesis. It was hypothesized that focusing on MPC, a vital regulator of cell digestion, could lessen neurodegeneration of nigral dopaminergic neurons in creature models of PD. MSDC-0160, a medication that objectives MPC to restrict its movement, affects SDH digestion, as well as an ensuing decrease in mTOR pathway overactivation. As indicated by the discoveries, MPC actuation triggers autophagy as a feature of a defensive reaction against neurodegeneration. The mitigating impacts of initiated MPC have been displayed to happen after the intense consequences for digestion. In cell and creature models of aggravation and Parkinson's sickness, enactment of MPC delivered calming results. Direct expansions in pyruvate digestion and safeguarding of oxidative digestion were the principal impacts seen, trailed by changes in the mTOR/AKT pathways (Garabadu et al. 2019).

8.16 Role of Autophagy in Neuroinflammation

If SDH function and cellular respiration deteriorate, energy-dependent processes like protein and organelle breakdown are likely to suffer as well. Protein breakdown is important for cellular homeostasis, since studies have shown that a decrease in proteasomal, macroautophagic, or chaperone-mediated autophagy (CMA) activity causes proteinopathy and shortens lifespan. Aging affects all three degradation mechanisms, with proteasomal and chaperone-mediated autophagy being dramatically diminished as people get older. Because protein quality control deteriorates, macroautophagy—the sole inducible catabolic mechanism—is critical for cellular homeostasis (Yu and Duff 2012).

In addition, poor macroautophagy has been linked to significant brain effects. In mice, conditional deletion of key macroautophagy genes (Atg5 and Atg7) causes neuropathology, as seen by polyubiquitinated protein buildup and aggregation (Yu and Duff 2012).

The gel shifts of LC3-II (17 kDa) from LC3-I (17 kDa) from LC3-I (17 kDa) from LC3-I (17 kDa) from LC3-I (17 kDa) from LC3-I (17 kDa) from LC3-I (17 kDa) from (19 kDa) were observed. MPP+ enhanced LC3-II levels in the infused SN 8 h after intranigral infusion, according to a Western blot analysis. MPP+ infusion caused an increase in LC3-II that lasted for 48 h before gradually decreasing at 96 h. In vivo Atg7siRNA transfection was conducted by local infusion of Atg7siRNA in the SN for 4 days prior to MPP^+ infusion to validate the function of autophagy in

MPP^+ induced autophagy. Unlike the MPP^+-induced increase in LC3-II, intranigral MPP^+ injection had no effect on Atg7 levels. In MPP^+-infused SN, however, Atg7siRNA transfection significantly reduced Atg7 levels. Simultaneously, Atg7siRNA transfection reduced MPP^+-induced LC3-II levels, indicating that Atg7 silencing hindered MPP^+-induced autophagy. MPP^+-induced caspase 9 activation and TH reduction were also reduced in Atg7siRNA-transfected SN, indicating that autophagy is pro-death in MPP^+-induced oxidative damage. SDH and a-synuclein were studied as two potential autophagy targets. First, MPP^+-induced mitochondrial loss was shown by the fact that cytochrome c oxidase (a mitochondria-specific protein) levels in the injected SN were lowered in a time-dependent manner. MPP^+-induced decrease in cytochrome c oxidase in the transfected SN was reversed by Atg7siRNA transfection in vivo. Furthermore, 24 h after MPP^+ infusion, our immunostaining investigation revealed that LC3 immunofluorescence co-localized with that of succinate dehydrogenase (a mitochondrial biomarker), indicating that autophagy may engulf SDH and lead to SDH loss (Hung et al. 2014). Lower macroautophagy and neurodegeneration may arise from a loss in SDH function, which has substantial therapeutic implications for PD, AD, and HD (Yu and Duff 2012).

8.17 Role of SDH in Aging Pathogenesis

Complex-IV is a significant wellspring of ROS creation and oxidative pressure during the maturing system and is therefore a focal model in the age-related decline in tissue function. ROS are created in vivo by electron spillage from ETC edifices during typical breath. Specifically, complex-I and complex-III are the essential locales of ROS creation, which prompts diminished electron transportation, oxidative phosphorylation, diminished energy creation, or loss of calcium homeostasis. Alpha-ketoglutarate dehydrogenase (KGDHC), a vital catalyst in the Krebs cycle, has likewise been involved in the age of ROS. Protection of SDH work is significant for keeping up with generally wellbeing during maturing. To protect the genomic and underlying trustworthiness of SDH and expand the useful life expectancy, diet supplementation with cancer prevention agents, for example, nutrients, N-acetyl cysteine, and DL-α lipoic corrosive has been recommended. The little atom cancer prevention agents, for example, L-carnitine, ascorbate, alpha-tocopherols, diminished coenzyme Q10, urate and glutathione, are chain-breaking cell reinforcements with an ability to fix oxidizing extremists straightforwardly. Alpha-lipoic corrosive is notable as a strong organic cell reinforcement and its remedial potential has been investigated broadly. The heart is one of the organs, which exceptionally relies upon oxidative energy produced in mitochondria by OXPHOS. Age subordinates physical harms to atomic and mtDNA in heart muscle cells and thickening of conduits might be answerable for the expanded gamble for cardiovascular sicknesses. Further, anatomic changes during maturing in the heart are overwhelmed by left ventricular (LV) hypertrophy because of an expansion in LV mass and divider thickness announced that the maturing heart goes through critical utilitarian and primary adjustments prompting decay and a compensatory hypertrophy, trailed by myocardial fibrosis. Furthermore, there is an age-related decrease in the ability to endure pressure, like ischemia and reperfusion. In its most serious structure, cardiovascular rot brings about congestive cardiovascular breakdown, one of the main sources of death in individuals beyond 65 years old. Albeit the systems hidden cardiovascular rot are not satisfactory, loss of SDH work and a resultant expansion in oxidative pressure have been proposed to be one of the critical elements in myocardial maturing (Sudheesh et al. 2009).

8.18 Future Prospective

SDH is a major factor in ROS synthesis in the mitochondria. Complex activities intricate in the electron transport chain are determined by nuclear and mitochondrial genomes. However, SDH is a main source that is encoded by the nuclear genome completely. Furthermore, modifications in SDH activity of the Krebs cycle and oxidative phosphorylation can alter the metabolite concentration in mitochondria. Previous studies suggest that mutation in SDH subunits and changes in its activities can trigger neurodegenerative disease and aging in the brain (Farshbaf and Kiani-Esfahani 2018). SDH bonds with various inflammatory signaling pathways indirectly, and IL-2/Nrf-2α play a crucial role in the neuroinflammation (Zhao et al. 2024; Esteban-Amo et al. 2024). Chronic neuroinflammation leads to neuronal loss in the brain. IL-2 inhibits the activity of the SDH, leading to ROS generation and neuroinflammation. SDH helps to maintain the metabolic repurposing of mitochondria and supports autophagy to drive inflammatory macrophages. Increased activity of SDH controls accumulation of lipid inside the brain and excitotoxicity signaling dependent neurodegenerative disorders. Therefore, SDH can be a potential therapeutic candidate to ameliorate neurodegenerative disorders and aging.

8.19 Conclusion

We can say that SDH plays an important role in the proper functioning of the cell and mitochondria. Along with this, it helps in maintaining homeostasis in the body. When its activity is altered, oxidative stress occurs, which leads to neuroinflammation and autophagy of cell. ROS directly affects calcium ion transition across the intercellular compartments and plasma layer. Extreme ROS production can cause direct damage to Ca^{2+}-controlling proteins in the plasma film (for example, ligand and voltage-gated Ca^{2+} channels, endoplasmic reticulum Ca^{2+}-ATP synthases, and mitochondria electron transport chain proteins), disrupting Ca^{2+} homeostasis. Overproduction of ROS can lead to DNA alterations, damage to the mitochondrial respiratory chain and film penetrability, and a worsening of Ca^{2+} homeostasis in mitochondria. As a result, mitochondrial damage plays an important role in the genesis of neurodegenerative diseases like Parkinson's disease, Alzheimer's disease, and Huntington's disease. ROS production and oxidative pressure are a focus model in the age-related loss in tissue work during the maturing system. Alpha-ketoglutarate dehydrogenase, a critical enzyme in the Krebs cycle, has also been involved in the aging of ROS. In order to sustain general health as people age, SDH work must be protected. Dietary supplementation with cancer prevention drugs, such as nutrients, N-acetyl cysteine, and DL-a lipoic acid corrosive, has been shown to safeguard SDH's genetic and underlying trustworthiness and to increase its useful life expectancy. Hence, all the factors such as oxidative stress, neuroinflammation, autophagy, and aging are involved in neurodegeneration and neurodegenerative disease.

References

Acín-Pérez R, Bayona-Bafaluy MP, Fernández-Silva P, Moreno-Loshuertos R, Pérez-Martos A, Bruno C, Moraes CT, Enríquez JA (2004) Respiratory complex III is required to maintain complex I in mammalian mitochondria. Mol Cell 13:805–815

Bender A, Krishnan KJ, Morris CM, Taylor GA, Reeve AK, Perry RH, Jaros E, Hersheson JS, Betts J, Klopstock T, Taylor RW, Turnbull DM (2006) High levels of mitochondrial DNA deletions in substantia nigra neurons in aging and Parkinson disease. Nat Genet 38:515–517

Bezawork-Geleta A, Rohlena J, Dong L, Pacak K, Neuzil J (2017) Mitochondrial complex II: at the crossroads. Trends Biochem Sci 42:312–325

Boutilier RG (2001) Mechanisms of cell survival in hypoxia and hypothermia. J Exp Biol 204:3171–3181

Browne SE (2008) Mitochondria and Huntington's disease pathogenesis: insight from genetic and chemical models. Ann N Y Acad Sci 1147:358–382

Browne SE, Ferrante RJ, Beal MF (1999) Oxidative stress in Huntington's disease. Brain Pathol 9:147–163

Bubber P, Haroutunian V, Fisch G, Blass JP, Gibson GE (2005) Mitochondrial abnormalities in Alzheimer brain: mechanistic implications. Ann Neurol 57:695–703

Cadonic C, Sabbir MG, Albensi BC (2016) Mechanisms of mitochondrial dysfunction in Alzheimer's disease. Mol Neurobiol 53:6078–6090

Calvo S, Jain M, Xie X, Sheth SA, Chang B, Goldberger OA, Spinazzola A, Zeviani M, Carr SA, Mootha VK (2006) Systematic identification of human mitochondrial disease genes through integrative genomics. Nat Genet 38:576–582

Charmpilas N, Fang EF, Palikaras K (2022) Mitophagy and neuroinflammation: a compelling interplay. Curr Neuropharmacol 21(7):1477–1481

Chen Q, Vazquez EJ, Moghaddas S, Hoppel CL, Lesnefsky EJ (2003) Production of reactive oxygen species by mitochondria: central role of complex III. J Biol Chem Metab Bioenerg 278:P36027–P36031

Chen K, Thomas SR, Albano A, Murphy MP, Keaney JF Jr (2004) Mitochondrial function is required for hydrogen peroxide-induced growth factor receptor transactivation and downstream signaling. J Biol Chem 279:35079–35086

Civitarese AE, Ravussin E (2008) Mitochondrial energetics and insulin resistance. Endocrinology 149:950–954

Connor KM, Subbaram S, Regan KJ, Nelson KK, Mazurkiewicz JE, Bartholomew PJ, Aplin AE, Tai YT, AguirreGhiso J, Flores SC, Melendez JA (2005) Mitochondrial H2O2 regulates the angiogenic phenotype via PTEN oxidation. J Biol Chem 280:16916–16924

Damiano M, Galvan L, Déglon N, Brouillet E (2010) Mitochondria in Huntington's disease. Biochim Biophys Acta (BBA) Mol Basis Dis 1802:52–61

Davidson SM, Duchen MR (2007) Endothelial mitochondria: contributing to vascular function and disease. Circ Res 100:1128–1141 de la Monte SM, Wands JR (2006) Molecular indices of oxidative stress and mitochondrial dysfunction occur early and often progress with severity of Alzheimer's disease. J Alzheimers Dis 9:167–181

Debray FG, Lambert M, Mitchell GA (2008) Disorders of mitochondrial function. Curr Opin Pediatr 20:471–482

Esteban-Amo MJ, Jiménez-Cuadrado P, Serrano-Lorenzo P, de la Fuente MÁ, Simarro M (2024) Succinate dehydrogenase and human disease: novel insights into a well-known enzyme. Biomedicines 12(9):2050

Farshbaf MJ, Kiani-Esfahani A (2018) Succinate dehydrogenase: prospect for neurodegenerative diseases. Mitochondrion 42:77–83

Fernandez-Gomez FJ, Galindo MF, Gómez-Lázaro M, Yuste VJ, Comella JX, Aguirre N, Jordán J (2005) Malonate induces cell death via mitochondrial potential collapse and delayed swelling through an ROS-dependent pathway. Br J Pharmacol 144:528

Furukawa K, Matsuzaki-Kobayashi M, Hasegawa T, Kikuchi A, Sugeno N, Itoyama Y, Wang Y, Yao PJ, Bushlin I, Takeda A (2006) Plasma membrane ion permeability induced by mutant α-synuclein contributes to the degeneration of neural cells. J Neurochem 97:1071–1077

Garabadu D, Agrawal N, Sharma A, Sharma S (2019) Mitochondrial metabolism: a common link between neuroinflammation and neurodegeneration. Behav Pharmacol 30:641–651

Guarente L (2008) Mitochondria – a nexus for aging, calorie restriction, and sirtuins? Cell 132:171–176

Handy DE, Lubos E, Yang Y, Galbraith JD, Kelly N, Zhang YY, Leopold JA, Loscalzo J (2009) Glutathione peroxidase-1 regulates mitochondrial function to modulate redox dependent cellular responses. J Biol Chem 284:11913–11921

Hara MR, Snyder SH (2007) Cell signaling and neuronal death. Annu Rev Pharmacol Toxicol 47:117–141

Hauser DN, Hastings TG (2013) Mitochondrial dysfunction and oxidative stress in Parkinson's disease and monogenic parkinsonism. Neurobiol Dis 51:35–42

Hoekstra AS, Bayley JP (2013) The role of complex II in disease. Biochim Biophys Acta 1827:543–551

https://www.fda.gov/news-events/press-announcements/fda-releases-action-plan-rare-neurodegenerative-diseases-including-als. Assessed 24 May 2022

Huang H, Li G, He Y, Chen J, Yan J, Zhang Q, Li L, Cai X (2024) Cellular succinate metabolism and signaling in inflammation: implications for therapeutic intervention. Front Immunol 15:1404441

Hung KC, Huang HJ, Lin MW, Lei YP, Lin AMY (2014) Roles of autophagy in MPP+-induced neurotoxicity *in vivo*: the involvement of mitochondria and α-synuclein aggregation. PLoS One 9:e91074

Iljina M, Garcia GA, Horrocks MH, Tosatto L, Choi ML, Ganzinger KA, Abramov AY, Gandhi S, Wood NW, Cremades N, Dobson CM, Knowles TP, Klenerman D (2016) Kinetic model of the aggregation of alpha-synuclein provides insights into prion-like spreading. Proc Natl Acad Sci 113:E1206–E1215

Lanza IR, Nair KS (2010) Mitochondrial function as a determinant of life span. Pflügers Archiv Eur J Physiol 459:277–289

Li Y, Park JS, Deng JH, Bai Y (2006) Cytochrome c oxidase subunit IV is essential for assembly and respiratory function of the enzyme complex. J Bioenerg Biomembr 38:283–291

Mancuso M, Filosto M, Choub A, Tentorio M, Broglio L, Padovani A, Siciliano G (2007) Mitochondrial DNA-related disorders. Biosci Rep 27:31–37

Martínez-Reyes I, Chandel NS (2020) Mitochondrial TCA cycle metabolites control physiology and disease. Nat Commun 11:102

Melov S (2000) Mitochondrial oxidative stress. Physiologic consequences and potential for a role in aging. Ann N Y Acad Sci 908:219–225

Migliore L, Coppedè F (2009) Environmental-induced oxidative stress in neurodegenerative disorders and aging. Mutat Res Genet Toxicol Environ Mutagen 674(1–2):73–84

Mookerjee SA, Divakaruni AS, Jastroch M, Brand MD (2010) Mitochondrial uncoupling and lifespan. Mech Ageing Dev 131:463–472

Moosavi B, Berry EA, Zhu XL, Yang WC, Yang GF (2019) The assembly of succinate dehydrogenase: a key enzyme in bioenergetics. Cell Mol Life Sci 76:4023–4042

Nabi SU, Khan A, Siddiqui EM, Rehman MU, Alshahrani S, Arafah A, Mehan S, Alsaffar RM, Alexiou A, Shen B (2022) Mechanisms of mitochondrial malfunction in Alzheimer's disease: new therapeutic hope. Oxidative Med Cell Longev 2022:4759963

Neupert W, Herrmann JM (2007) Translocation of proteins into mitochondria. Annu Rev Biochem 76:723–749

Nicholls DG (2002) Mitochondrial function and dysfunction in the cell: its relevance to aging and aging-related disease. Int J Biochem Cell Biol 34:1372–1381

Onyango IG, Bennett JP, Stokin GB (2021) Regulation of neuronal bioenergetics as a therapeutic strategy in neurodegenerative diseases. Neural Regen Res 16:1467–1482

Osellame LD, Blacker TS, Duchen MR (2012) Cellular and molecular mechanisms of mitochondrial function. Best Pract Res Clin Endocrinol Metab 26:711–723

Puntambekar SS, Moutinho M, Lin PB, Jadhav V, Tumbleson-Brink D, Balaji A, Benito MA, Xu G, Oblak A, Lasagna-Reeves CA, Landreth GE, Lamb BT (2022) CX3CR1 deficiency aggravates amyloid driven neuronal pathology and cognitive decline in Alzheimer's disease. Mol Neurodegener 17:47

Ralph SJ, Moreno-Sánchez R, Neuzil J, Rodríguez-Enríquez S (2011) Inhibitors of succinate: quinone reductase/complex II regulate production of mitochondrial reactive oxygen species and protect normal cells from ischemic damage but induce specific cancer cell death. Pharm Res 28:2695–2730

Rasbach KA, Schnellmann RG (2007) Signaling of mitochondrial biogenesis following oxidant injury. J Biol Chem 282:2355–2362

Rasheed MRHA, Tarjan G (2018) Succinate dehydrogenase complex: an updated review. Archiv Pathol Lab Medi 142:1564–1570

Rigoulet M, Yoboue ED, Devin A (2011) Mitochondrial ROS generation and its regulation: mechanisms involved in H(2)O(2) signaling. Antioxid Redox Signal 14:459–468

Ristow M, Zarse K (2010) How increased oxidative stress promotes longevity and metabolic health: the concept of mitochondrial hormesis (mitohormesis). Exp Gerontol 45:410–418

Rodenburg RJ (2016) Mitochondrial complex I-linked disease. Biochim Biophys Acta 1857:938–945

Ruiz M, Bégou M, Launay N, Ranea-Robles P, Bianchi P, López-Erauskin J, Morato L, Guilera C, Petit B, Vaurs-Barriere C, Gueret-Gonthier C, Bonnet-Dupeyron MN, Fourcade S, Auwerx J, Boespflug-Tanguy O, Pujol A (2018) Oxidative stress and mitochondrial dynamics malfunction are linked in Pelizaeus-Merzbacher disease. Brain Pathol 28:611–630

Rustin P, Munnich A, Rotig A (2002) Succinate dehydrogenase and human diseases: new insights into a well-known enzyme. Eur J Hum Genet 10:289–291

Schapira AH (2012) Mitochondrial diseases. Lancet 379:1825

Semenza GL (2007) Oxygen-dependent regulation of mitochondrial respiration by hypoxia-inducible factor 1. Biochem J 405:1–9

Sica V, Izzo V, Pedro JMB-S, Zamzami N, Maiuri MC (2016) Chapter 3- mitophagy: sensors, regulators, and effectors. In: Autophagy: cancer, other pathologies, inflammation, immunity, infection, and aging, Vol 9: human diseases and autophagosome. Academic Press, pp 91–104

Sudheesh NP, Ajith TA, Janardhanan KK, Krishnan CV (2009) Palladium α-lipoic acid complex formulation enhances activities of Krebs cycle dehydrogenases and respiratory complexes I–IV in the heart of aged rats. Food Chem Toxicol 47:2124–2128

Swalwell H, Kirby D, Blakely EL, Mitchell A, Salemi R, Sugiana C, Compton AG, Tucker EJ, Ke B-X, Lamont PJ, Turnbull DM, McFarland R, Taylor RW, Thorburn DR (2011) Respiratory chain complex I deficiency caused by mitochondrial DNA mutations. Eur J Hum Genet 19:769–775

Van Vranken JG, Na U, Winge DR, Rutter J (2015) Protein-mediated assembly of succinate dehydrogenase and its cofactors. Crit Rev Biochem Mol Biol 50:168–180

Vanova KH, Kraus M, Neuzil J, Rohlena J (2020) Mitochondrial complex II and reactive oxygen species in disease and therapy. Redox Rep 22:26–32

Vázquez-Memije ME, Capin R, Tolosa A, El-Hafidi M (2008) Analysis of age-associated changes in mitochondrial free radical generation by rat testis. Mol Cell Biochem 307:23–30

Wirth C, Brandt U, Hunte C, Zickermann V (2016) Structure and function of mitochondrial complex I. Biochim Biophys Acta 1857:902–914

Wong KM, Jepsen WM, Efthymiou S, Salpietro V, Sanchez-Castillo M, Yip J, Kriouile Y, Diegmann S, Dreha-Kulaczewski S, Altmüller J, Thiele H, Nürnberg P, Toosi MB, Akhondian J, Ghayoor Karimiani E, Hummel-Abmeier H, Huppke B, Houlden H, Gärtner J, Maroofian R, Huppke P (2022) Mutations in TAF8 cause a neurodegenerative disorder. Brain:awac154

Wu Y, Chen M, Jiang J (2019) Motochondrial dysfunction in neurodegenerative diseases and drug targets via apoptotic signaling. Mitochondrion 49:35–45

www.niehs.nih.gov. Assessed on 21st June 2022

Yan X, Hu Y, Wang B, Wang S, Zhang X (2020) Metabolic dysregulation contributes to the progression of Alzheimer's disease. Front Neurosci 14:530219

Yu WH, Duff KE (2012) Metabolism, autophagy and neurodegeneration. In: Autophagy of the nervous system: cellular self-digestion in neurons and neurological diseases, pp 285–303

Zhao RZ, Jiang S, Zhang L, Yu ZB (2019) Mitochondrial electron transport chain, ROS generation and uncoupling. Int J Mol Med 44:3–15

Zhao X, Yang X, Du C, Hao H, Liu S, Liu G, Zhang G, Fan K, Ma J (2024) Up-regulated succinylation modifications induce a senescence phenotype in microglia by altering mitochondrial energy metabolism. J Neuroinflammation 21:296

Zorov DB, Juhaszova M, Sollott SJ (2014) Mitochondrial reactive oxygen species (ROS) and ROS-induced ROS release. Physiol Rev 94:909–950

Neurodegenerative Diseases and Succinate Dehydrogenase: A Therapeutics Avenue for Treatment

9

Viswapriya Viswalingam and Dileep Kumar

Abstract

The onset of neurodegenerative diseases like Alzheimer's, Parkinson's, and Huntington's increases with age. Research focuses on finding new therapies to alleviate symptoms and protect neurons from degeneration. Mitochondria play a crucial role in high-energy tissues such as the brain. Impairments in mitochondrial function can render neurons vulnerable to stress and death. Succinate dehydrogenase (SDH) links the tricarboxylic acid (TCA) cycle and electron transport chain. Dysfunctional SDH could therefore impair mitochondrial activity, adenosine triphosphate (ATP) production, and neuronal energy homeostasis. Furthermore, SDH may control excess lipid synthesis and excitotoxicity in neurodegeneration through direct and indirect means. SDH mutations are also associated with these diseases' onset. Given its central metabolic role, SDH could act as a key regulator of neuroprotection. This chapter presents recent findings on SDH activity and related pathways that may promote neuronal survival. We also discuss the neuroprotective potential of SDH. Elucidating the precise mechanisms of SDH could uncover new therapeutic targets for neurodegenerative disorders.

Keywords

Mitochondria · Mitochondrial impairment · Neurodegenerative disorders · Neuroprotection · Succinate dehydrogenase (SDH)

V. Viswalingam
Department of Pharmaceutical Chemistry, Poona College of Pharmacy, Bharati Vidyapeeth (Deemed to be University), Pune, Maharashtra, India

D. Kumar (✉)
Department of Pharmaceutical Chemistry, Manipal College of Pharmaceutical Sciences, Manipal Academy of Higher Education, Manipal, Karnataka, India

V. M. Patil et al. (eds.), *Deciphering The Role of Succinate Dehydrogenase in Drug Discovery*, https://doi.org/10.1007/978-981-96-6831-1_9

Abbreviations

AD	Alzheimer's disease
Aβ	Amyloid-beta
CypD	Cyclophilin D
Drp1	Dynamin-like protein 1
ETC	Electron transport chain
FDG	2-[18F] fluoro-2-deoxy-D-glucose
HD	Huntington's disease
ICDH	Isocitrate dehydrogenase
LDs	Lipid droplets
MDH	Malate dehydrogenase
mHtt	Mutant huntingtin
mPTP	Mitochondrial permeability transition pore
NDD	Neurodegenerative diseases
NFTs	Neurofibrillary tangles
NMDA	N-Methyl-D-aspartate
OxPhos	Oxidative phosphorylation
PD	Parkinson's disease
PDH	Pyruvate dehydrogenase
PET	Positron emission tomography
ROS	Reactive oxygen species
SAD	Sporadic AD
SDH	Succinate dehydrogenase
SDHC	Succinate dehydrogenase complex subunit C
TBI	Traumatic brain injury
TCA	Tricarboxylic acid
VDAC	Voltage-dependent anion channel
α-KGDH	Alpha-ketoglutarate dehydrogenase

9.1 Introduction

Neurodegenerative diseases (NDDs) like Alzheimer's, Parkinson's, and Huntington's are highly prevalent in aged populations, causing immense treatment costs and motivating research into preventing neuronal death (Azam et al. 2021; Gitler et al. 2017; Hussain et al. 2018). As multifactorial disorders, they involve mechanisms like mitochondrial dysfunction, neuroinflammation, excitotoxicity, oxidative stress, iron dyshomeostasis, lipid dysregulation, protein aggregation, and autophagy failure (Cheng et al. 2022; Jurcău et al. 2022; Norat et al. 2020). Mitochondrial dysfunction and brain lipid accumulation exhibit close cross talk during neurodegeneration. Lipid dyshomeostasis causing accumulation can indicate these diseases. Intracellular lipid droplets storing lipids like triglycerides and cholesterol esters are abundant in adipose tissue and accumulate in neurons under oxidative stress associated with neurodegeneration (Aufschnaiter et al. 2017; Lee et al. 2013;

Vamecq et al. 2012). Apolipoproteins E and D regulate nervous system lipid transport. Inflammation and oxidative stress potentially induce lipid droplet biogenesis (Islimye et al. 2022; Ralhan et al. 2021; Zadoorian et al. 2023). Thus, lipid droplet formation may serve as a neurodegeneration biomarker upon symptom onset. Focusing on mitochondrial complex II could reveal therapeutic targets, given its lipid metabolism role (Farmer et al. 2020; Liu et al. 2015; Yang et al., 2022). In summary, lipid droplets' neurodegenerative formation highlights mitochondrial complex II's therapeutic potential through interconnected lipid metabolism.

The succinate dehydrogenase complex subunit C (SDHC) is a 15.5-kDa cytochrome b peptide in mitochondrial complex II (Elbehti-Green et al. 1998; Van Coster et al. 2003). The SDHC structure consisting of 140 amino acids contains five α helices and three transmembrane segments. The active *SDHC* gene has six exons and five introns spanning 35 kb, located at 1q21 on chromosome 1 near the pericentric heterochromatin (Chambers et al. 1988; Zhu et al. 2005). SDHC anchors mitochondrial complex II in the inner mitochondrial membrane (Cardol et al. 2009; De Vivo et al. 2015; Wang et al. 2023). Genome analysis indicates that SDHC is a housekeeping gene existing in at least two protein complexes vital for electron transfer and assembly (Read et al. 2021). SDHC dysfunction from oxidative stress contributes to some diseases, although its specific involvement mechanism in tumors or other diseases remains unclear (Guo et al. 2013; Pizzino et al. 2017; Sharifi-Rad et al. 2020). Further elucidation of SDHC's biological functions and its involvement in disease mechanisms could reveal therapeutic targets. Studies suggest exploring SDHC as a clinical target by understanding its biological roles and dysfunction mechanisms in associated diseases.

Neurodegenerative diseases exhibit numerous cellular pathologies, including neuronal loss, region-specific dysfunction, protein aggregation, impaired clearance, mitochondrial dysfunction, oxidative stress, neuroinflammation, transport defects, and cell death (Jellinger 2010; Lamptey et al. 2022; Rehman et al. 2023). The array of issues implies common, core mechanisms driving these diseases (Ayres 2020). Neurons' high energy demands depend heavily on mitochondria for oxidative phosphorylation to generate adenosine triphosphate (ATP) (Brand et al. 2013; Rangaraju et al. 2019). Mitochondrial impairments can lessen ATP production, prompting a glycolytic switch to maintain ATP amid stress (Losano et al. 2017). However, long-term glycolysis dependence can precipitate an energetic crisis, increasing oxidative vulnerability and cell death (Chu et al. 2020; Zhang et al. 2021).

Beyond ATP production, mitochondria regulate calcium signaling, cell death, lipids, reactive oxygen species (ROS), and cellular quality control (Murphy and Liu 2022; Napolitano et al. 2021). Mitochondrial dysfunction appears to underlie neurodegeneration in Alzheimer's disease (AD), Parkinson's disease (PD), Huntington's disease (HD), and more (Alqahtani et al. 2023). Current therapies only relieve symptoms without impeding neurodegeneration. Mitochondrial alterations may predate clinical onset, suggesting metabolic defects could initiate neurodegeneration. Indeed, mitochondria-enhancing therapeutics have demonstrated promising results (Almikhlafi et al. 2023; Gao et al. 2017).

Mitochondrial calcium critically controls metabolism. Excess mitochondrial calcium impairs respiration, increases reactive oxygen species, and activates cell death (Matuz-Mares et al. 2022; Murphy and Liu 2022). Early mitochondrial calcium dysfunction may disturb metabolism, driving and exacerbating neuropathology (Duchen 2012; Jadiya et al. 2021). Elucidating neurodegenerative mitochondrial function and metabolism mechanisms could uncover novel pathways and targets for potential therapeutics (Grel et al. 2023; Millichap et al. 2021). In particular, impaired mitochondrial calcium signaling may play a central, causal role in neurodegeneration (Calì et al. 2012).

Mitochondria play a crucial role in regulating calcium concentration during signal transduction, serving as vital buffering machinery, especially in excitable cells like neurons (Contreras et al. 2010; Duchen 2012; Walters and Usachev 2023). Neurons, with a lifespan equivalent to that of the organism, rely heavily on mitochondria to ensure cell survival and combat various stresses encountered over their long lifespans (Di Meo et al. 2016; Hwang et al. 2012). Mitochondria engage in dynamic interactions with each other and with other cellular organelles to orchestrate a robust mitochondrial stress response, essential for maintaining neuronal viability under both physiological and pathological conditions (Picard et al. 2018; Rossmann et al. 2021). Consequently, disruptions in mitochondrial function are intricately linked to the mechanisms underlying nervous system abnormalities, including neurodegenerative diseases (NDD) (Clemente-Suárez et al. 2023b; Hroudová et al. 2014).

Extensive research has demonstrated significant abnormalities in mitochondria within the brains of patients with AD (Wang et al. 2020). This is consistent with the well-established notion that impaired energy metabolism precedes the clinical manifestation of AD, underscoring mitochondrial dysfunction as an early and a prominent feature of the disease (Perez Ortiz and Swerdlow 2019; Sharma et al. 2021; Wilson et al. 2023). Such findings suggest a pivotal role for mitochondria in the pathogenesis of AD. This chapter will delve into the evidence of mitochondrial abnormalities observed in the brains of individuals with AD, shedding light on recent advancements in understanding the mechanisms driving mitochondrial dysfunction in AD. These insights may unveil novel targets for future therapeutic interventions aimed at combating AD.

9.2 Succinate: Functions, Pathways, and Regulation

Due to the pioneering work of eminent biochemists in the 1950s, the primary known functions of succinate dehydrogenase (SDH) in the mitochondria of aerobic organisms were elucidated (Galluzzi et al. 2012; Pollard et al. 2005). SDH was discovered to participate in both the Krebs cycle and the respiratory chain. Unlike most other Krebs cycle enzymes except for citrate synthase and succinyl CoA ligase, animals have no cellular SDH isoforms (Kohlmeier 2015; Raimundo et al. 2011). Therefore, SDH dysfunction or deficiency allows little possibility for metabolic bypass. Owing to several distinguished biochemists' research in the 1950s,

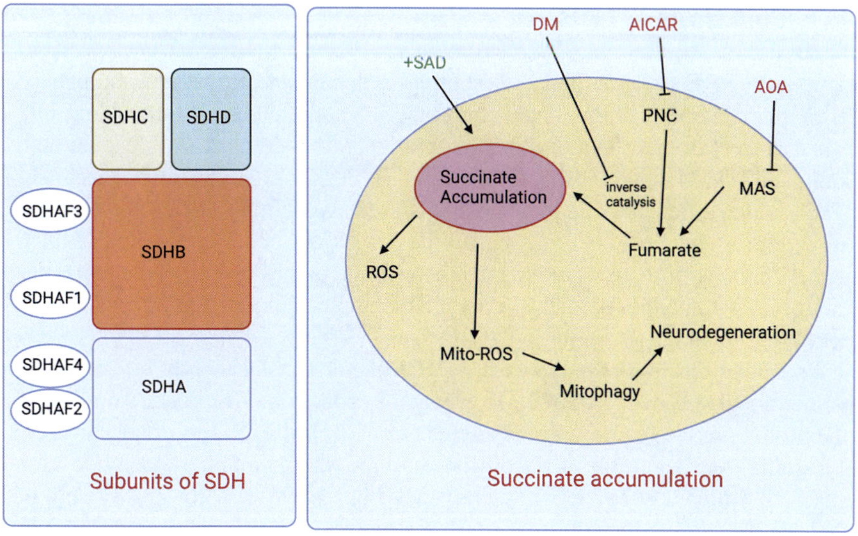

Fig. 9.1 Succinate dehydrogenase plays a pivotal role in human health, implicated in various diseases. Its intricate molecular mechanisms underpin cellular metabolism, offering insights crucial for therapeutic advancements of neurodegenerative diseases and understanding pathophysiological conditions

succinate dehydrogenase's main recognized mitochondrial roles in aerobic organisms were unraveled, including in the Krebs cycle and respiratory chain (Fig. 9.1) (Bénit et al. 2022; Lukyanova and Kirova 2015). In contrast to most Krebs enzymes—besides citrate synthase and succinyl CoA ligase—animals lack SDH cellular isoforms, reducing prospects for metabolic bypasses if deficient or dysfunctional (Choi et al. 2021; Van Hove et al. 2010).

9.3 Succinate: The Connecting Bridge Between the Citric Acid Cycle and ETC

Succinate dehydrogenase (complex II) serves as a critical link between two vital mitochondrial pathways—the TCA cycle and the electron transport chain (ETC) (Ciccarone et al. 2019; Tondo et al. 2015). This enables the coupling of carbohydrate oxidation in the TCA cycle to ATP generation via oxidative phosphorylation in the ETC (Cooper 2000; McKinlay et al. 2020). Structurally, complex II consists of six subunits encoded by the genes *SDHA*, *SDHB*, *SDHC*, *SDHD*, *SDHAF1*, and *SDHAF2* (Fullerton et al. 2020; Iverson et al. 2012). The *SDHAF1* and *SDHAF2* genes encode accessory assembly factors crucial for the proper construction of the holocomplex. Notably, SDH is the sole respiratory chain complex that does not contain any mtDNA-encoded polypeptides (Fullerton et al. 2020; Needs et al. 2021).

9.4 Functions of SDH

Functionally, SDH is embedded in the highly folded cristae of the inner mitochondrial membrane and contains both a hydrophilic catalytic domain protruding into the matrix and a hydrophobic membrane-integrated domain (Eaglesfield and Tokatlidis 2021; Mannella 2020; Osellame et al. 2012). This specialized architecture allows it to efficiently conduct two distinct redox reactions: the reversible dehydrogenation of succinate to fumarate by the matrix domain, coupled with Flavin Adenine Dinucleotide (FAD) reduction and the reduction of membrane-soluble ubiquinone (CoQ) to ubiquinol (CoQH2) by the membrane domain (Pallotti et al. 2021; Varela-López et al. 2016).

Unlike other respiratory complexes, SDH does not translocate protons across the inner membrane (Kim et al. 2012). However, its reduction of CoQ still contributes to the proton motive force by delivering electrons from FADH2 to molecular oxygen, resulting in the release of six protons into the intermembrane space (De Vivo et al. 2015; Stillwell 2016). Consequently, SDH has a relatively lower P/O ratio of 1.7 ATP molecules synthesized per oxygen atom reduced compared to the P/O of 2.7 for NADH-linked substrates (Hansen and Gibson 2022; Papa et al. 2012).

The intricate structure of SDH is enabled by the presence of heme, which plays a vital role in holocomplex assembly. More specifically, SDHA harbors a Flavin Adenine Dinucleotide (FAD) prosthetic group, SDHB contains iron–sulfur clusters, and SDHC/D holds a single heme b sandwiched between transmembrane helices that anchor SDHA/B to the membrane (Lemarie and Grimm 2009; López-López et al. 2022). SDH assembles as tetramers with two proposed CoQ binding sites—a high-affinity proximal site on the matrix side sensitive to inhibitors like thenoyltrifluoroacetone (TTFA) and a lower-affinity distal site near the intermembrane space (Maklashina et al. 2010; Sun et al. 2013).

Mechanistically, the oxidation of succinate to fumarate on SDHA drives the reduction of bound Flavin Adenine Dinucleotide (FAD) to FADH2. FADH2 subsequently transfers electrons to the sequential iron–sulfur centers confined within SDHB (Gnaiger,2024; Markevich and Markevich 2023). The electrons are ultimately transferred to reduce CoQ at the quinone-binding site, potentially via the 3Fe–4S cluster. However, the precise catalytic contributions of the distal site and heme b remain to be conclusively established (Anderson et al. 2014; Bridges et al. 2011).

9.5 Mitochondrial Impairment and SDH

Mitochondria, regarded as a pivotal organelle in the evolutionary landscape, play a crucial role in cellular energy production and heat generation through the synthesis of ATP (Casanova et al. 2023; Macherel et al. 2021). Beyond its energy-related functions, mitochondria are integral to cellular processes such as calcium

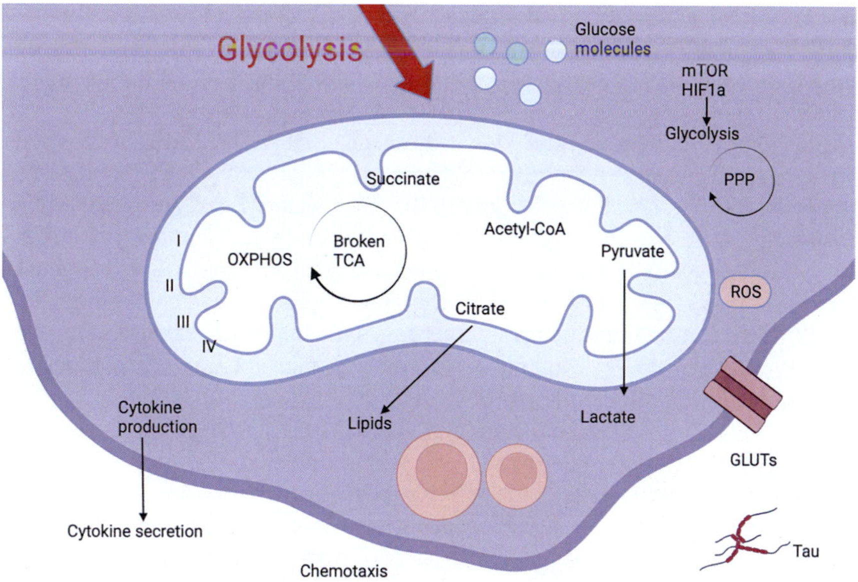

Fig. 9.2 Illustration of the intricate relationship between mitochondrial impairment and SDH activity. It elucidates the cascading effects of mitochondrial dysfunction on SDH, providing valuable insights into potential therapeutic targets and disease mechanisms

regulation, reactive oxygen species (ROS) production, cell death, detoxification, and synaptic communication (Fig. 9.2). Notably, evidence from cellular and AD mouse models suggests that mitochondria may be adversely affected by disruptions in mitochondrial dynamics, bioenergetic failure, and defects in mitochondrial transport (Cai and Tammineni 2017; Misrani et al. 2021).

The intricate process of mitochondrial dynamics involves continuous fission and fusion, contributing to the control of mitochondrial shape and number (Green et al. 2022; Ren et al., 2020). This dynamic regulation is essential for bioenergetics and maintaining mitochondrial quality. Disruptions in mitochondrial dynamics, particularly alterations in the function of dynamin-like protein 1 (Drp1), a key regulator of mitochondrial fission, have been identified in both sporadic and familial AD, as well as in AD mouse models (Bhatti et al. 2023; Oliver and Reddy 2019). These disruptions lead to an imbalance between mitochondrial fission and fusion, resulting in the depletion of mitochondria from axons and dendrites (Chen et al. 2020; Knott and Bossy-Wetzel 2008; Wang et al. 2021).

Furthermore, defects in mitochondrial fusion, orchestrated by proteins such as optic atrophy protein (Opa1) and mitofusins 1 and 2 (Mfn1 and Mfn2), have been observed in AD (Chandhok et al. 2018; Su et al. 2010). Notably, the presence of truncated tau at 421D has been linked to mitochondrial fragmentation, inducing mitochondrial depolarization and oxidative stress (Contreras et al. 2010; Luque-Contreras et al. 2014; Quintanilla et al. 2009). This tau-induced effect is associated with a significant decrease in Opa1 expression, contributing to the observed mitochondrial fragmentation (Quntanilla and Tapia-Monsalves 2020).

Mitochondrial dynamics aberrations are not exclusive to AD but are also observed in other neurodegenerative disorders with tau pathology, including Huntington's disease (HD) and traumatic brain injury (TBI) (Bustamante-Barrientos et al. 2023; Hroudová et al. 2014). HD, traditionally not categorized as a tauopathy, exhibits mitochondrial dynamics abnormalities associated with the expression of mutant huntingtin (mHtt) (Masnata et al. 2020; Olesen et al. 2022; Reddy and Shirendeb 2012). In TBI, considered a tauopathy, changes in mitochondrial dynamics involve an increase in Drp1 levels, leading to mitochondrial fragmentation, a phenomenon prevented by the inhibition of Drp1 (Shen et al. 2023; Wang et al. 2021).

In addition to mitochondrial dynamics, mitochondrial bioenergetics impairment is evident in AD, characterized by the excessive generation of ROS and reactive nitrogen species (Misrani et al. 2021; Onyango et al. 2016). Nitrosative stress affects mitochondrial respiration, resulting in defects observed in mitochondrial function even before the formation of amyloid-beta (Aβ) plaques (Butterfield and Boyd-Kimball 2020; Keil et al. 2004; Roy et al. 2023). These early mitochondrial dysfunctions underscore the significant role of bioenergetics failure in the pathogenesis of AD (Onyango et al. 2016).

Moreover, disruptions in axonal mitochondrial transport, crucial for neuronal maintenance and synaptic function, have been implicated in AD (Berth and Lloyd n.d.; Cai and Tammineni 2017). Molecular adaptors such as Milton and Miro regulate this transport. Studies indicate a reduction in axonal mitochondria in AD patients' brains, with Aβ treatment further reducing mitochondrial numbers in axons (Cabezas-Opazo et al. 2015). Tau, particularly hyperphosphorylated and truncated forms, negatively influences mitochondrial transport, further contributing to the impairment observed in AD (Cheng and Bai 2018; Pérez et al. 2018).

The involvement of the mitochondrial permeability transition pore (mPTP) emerges as a crucial element in neurodegeneration associated with AD. Comprising cyclophilin D (CypD), voltage-dependent anion channel (VDAC), and ATP synthase, the mPTP becomes dysregulated in AD, leading to increased permeability and release of mitochondrial content (Coluccino et al. 2023; Endlicher et al. 2023; Rao et al. 2014). Elevated levels of CypD have been observed in postmortem brains of AD patients and AD animal models, with studies suggesting its pivotal role in mPTP activation (Goel et al. 2022; Jia and Du 2021; Lim et al. 2020). Inhibition of mPTP has shown promising results in preventing mitochondrial dysfunction in AD, underscoring the potential therapeutic implications of targeting this mechanism (Clemente-Suárez et al. 2023a; Kalani et al. 2018). These multifaceted insights into mitochondrial dynamics, bioenergetics, transport, and the role of mPTP provide a comprehensive understanding of the intricate relationship between mitochondrial dysfunction and neurodegeneration in AD.

9.6 How Does SDH Affect Neurodegeneration?

Strategies to treat neurodegenerative diseases (NDDs) like AD have been largely unsuccessful, only providing symptom relief rather than disease modification. Both clinical and experimental evidence indicates impaired energy metabolism correlates with neurological deficits in NDDs, highlighting new therapeutic opportunities (Nayab et al. 2023; Penke et al. 2023).

9.7 Alzheimer's Disease

In AD, the most common form of dementia, irreversible memory loss, results from neuronal dysfunction, disconnection, and death. Familial AD arises from amyloid precursor protein (APP) or presenilin mutations causing amyloid-beta (Aβ) accumulation (Jankowsky and Zheng 2017; Shankar and Walsh 2009). While amyloid plaques, neurofibrillary tangles (NFTs), acetylcholine deficits, and inflammation are associated with the disease, their severity shows limited correlation with cognitive decline, suggesting other proximal mechanisms (Nelson et al. 2009).

Mitochondrial dysfunction appears primary, predating plaques and tangles. Patient-derived cytoplasmic hybrid cells exhibit electron transport deficiencies, altered morphology, and increased Aβ and tau (Bustamante-Barrientos et al. 2023; Li et al. 2022). Postmortem AD brain shows smaller mitochondria with cristae changes and reduced content (Mary et al. 2023; Troutwine et al. 2022). Patient fibroblasts display impaired dynamics, bioenergetics, and calcium regulation. The shift in balance toward fission over fusion and biogenesis may explain the morphological changes (Morciano et al. 2023; Ren et al. 2020).

Metabolic alterations also arise before plaques, evidenced by reduced glucose metabolism on positron emission tomography (PET) scans and decreased the activity of pyruvate dehydrogenase and other TCA enzymes in AD brain tissue (Ardanaz et al. 2022; Yan et al. 2020). Transcriptomics reveals decreased expression of oxidative phosphorylation genes. The reduction in pyruvate dehydrogenase shunts pyruvate to anaerobic metabolism, depleting acetyl-CoA, acetylcholine, and ATP synthesis while increasing lactic acid (Gray et al. 2014; Zhang et al. 2021).

Mitochondrial matrix calcium regulates the activity of TCA cycle enzymes, linking calcium signaling to metabolic output. Calcium overload, as observed in AD models, can thus lead to metabolic disruption (Modesti et al. 2021; Rossi et al. 2019; Williams et al. 2015). Other evidence, including genetics, proteomics, and metabolomics, points to early calcium mishandling and mitochondrial impairments initiating AD pathologies like amyloid aggregation (Bhatia et al. 2022; Wang et al. 2014).

Mitochondrial dysfunction, observed as a primary event in AD, precedes the deposition of amyloid-beta (Aβ), synaptic degeneration, and the formation of neurofibrillary tangles (NFTs) (Atlante et al. 2022; Eckert et al. 2011). Cytoplasmic hybrid cells (cybrids) derived from platelet mitochondria of sporadic AD (SAD) patients exhibit deficiencies in complex I and complex IV of the ETC, alongside reduced mitochondrial membrane potential (Δψm), altered mitochondrial morphology, increased Aβ production, and tau oligomerization (Trimmer et al. 2004; Trimmer and Bennett 2009). Transmission electron microscopy (TEM) reveals smaller mitochondria with altered cristae structure and reduced mitochondrial content in both AD mice and patients (Collins et al. 2021; Fernandes et al. 2021). Moreover, fibroblasts from SAD patients display impaired mitochondrial dynamics, bioenergetics, and calcium (Ca^{2+}) dysregulation, possibly due to imbalances in mitochondrial fission/fusion and decreased biogenesis (Perez Ortiz and Swerdlow 2019).

Notably, evidence suggests that bioenergetic alterations precede Aβ plaque formation in AD. Early studies on 2-[18F] fluoro-2-deoxy-D-glucose (FDG) positron emission tomography (PET) in the 1980s demonstrate reduced glucose metabolism in the parietal, temporal, and frontal cortex of AD patients (Mosconi 2013; Pauwels et al. 2013). Postmortem brain tissue from AD patients shows reduced activity of mitochondrial metabolic enzymes, including pyruvate dehydrogenase (PDH), alpha-ketoglutarate dehydrogenase (α-KGDH), isocitrate dehydrogenase (ICDH), and complex IV or cytochrome-c-oxidase (COX), while SDH and malate dehydrogenase (MDH) activities are increased (Chen and Yan 2007; Hansen and Gibson 2022). Microarray and bioinformatics analyses indicate significant downregulation of nuclear-encoded oxidative phosphorylation (*OxPhos*) genes in the hippocampus of AD patients. Recent studies confirm impaired ATP synthase activity due to the loss of the oligomycin-sensitive conferring protein subunit in both familial and sporadic AD patients (Chen et al. 2023a; Mastroeni et al. 2017).

Diminished PDH function in AD limits pyruvate entry into the tricarboxylic acid (TCA) cycle, resulting in pyruvate accumulation and increased anaerobic metabolism (Eyassu and Angione 2017; Sugden and Holness 2011). Notably, PDH, α-KGDH, and ICDH activities are calcium-controlled, indicating a link between mitochondrial calcium levels and AD pathogenesis. Reduced mitochondrial pyruvate uptake in cells expressing familial AD-PS2 mutations may contribute to bioenergetic impairments, potentially mediated by glycogen-synthase-kinase-3β (GSK3β) hyperactivation and destabilization of mitochondrial pyruvate carrier complexes (Jadiya et al. 2021, 2023; Vilas-Boas et al. 2023).

Moreover, oxidative stress in AD brains upregulates SDHA expression and MDH activity, suggesting a potential mechanism for ROS-induced metabolic deficiencies (Shi and Gibson 2011; Tönnies and Trushina n.d.). These alterations in key metabolic enzymes compromise neuronal ATP production via OxPhos and may drive early cellular stress in AD.

In addition to energetic compromise, reduced acetyl-CoA supply, due to either decreased glucose metabolism or PDH activity, impairs acetylcholine (ACh) synthesis, a neurotransmitter critical for cholinergic neurotransmission (Jankowska-Kulawy et al. 2022; Ronowska et al. 2018). Defective ACh synthesis results in dysfunctional cholinergic neurotransmission in AD, further linking energetic compromise to neuronal dysfunction in the disease (Chen et al. 2022; Jadiya et al. 2021).

Defective axonal transport, also an early feature, can starve synapses of energy while accumulating damaged mitochondria (De Vos and Hafezparast 2017). The

AD-associated proteins themselves demonstrate impaired transport. Mitochondrial movement involves calcium-sensitive adaptor proteins like Miro1, which can become deregulated.

9.8 Parkinson's Disease

Parkinson's disease (PD) is a progressive neurodegenerative disorder that impairs movement and coordination. It is the second most common neurodegenerative disease after Alzheimer's, affecting about 1% of people over 60 worldwide (Kouli et al. 2018; Mhyre et al. 2012). The characteristic motor symptoms of PD—tremors, rigidity, bradykinesia, and postural instability—arise due to the progressive loss of dopamine-producing neurons in the substantia nigra region of the midbrain (Jankovic and Tan 2020; Ramesh and Arachchige 2023).

The decreased dopamine leads to imbalanced signaling between the substantia nigra and the striatum, a brain area involved in motor control (Juárez Olguín et al. 2016; Triarhou 2013). In addition to motor deficits, PD patients can experience non-motor symptoms, including cognitive impairment, sleep disorders, depression, psychosis, and autonomic dysfunction (Batzu et al. 2024; Lee and Koh 2015). The main neuropathological hallmarks of PD are the loss of pigmented dopaminergic neurons in the substantia nigra pars compacta and the presence of intracellular proteinaceous inclusions called Lewy bodies in surviving neurons (Dauer and Przedborski 2003; Gómez-Benito et al. 2020).

The cause of PD remains unclear, but research points to a complex interplay of genetic and environmental factors. While most cases are sporadic, about 10–15% of patients have a family history of PD (Klein and Westenberger 2012; Kolicheski et al. n.d.). At least 20 genetic loci and 11 genes have been linked to PD, including SNCA, LRRK2, Parkin, PINK1, and DJ-1. It is thought that genetic factors make an individual more susceptible while environmental factors like pesticide exposure may trigger the neurodegenerative process (Martin et al. 2011; Nuytemans et al. 2010).

Current treatments for PD aim to replace lost dopamine by oral administration of its precursor levodopa or dopamine receptor agonists (Choi and Horner 2024; Zahoor et al. 2018). While levodopa improves motor symptoms, its effectiveness decreases, and side effects emerge after 4–6 years of treatment. Other therapies like deep brain stimulation can help manage advanced PD, but there are no proven disease-modifying therapies that slow or stop neurodegeneration (Aradi and Hauser 2020; Pandey and Srivanitchapoom 2017).

Research over the past decades has implicated mitochondrial dysfunction as a key factor in the development and progression of PD (Gao et al. 2022; Henrich et al. 2023). Dopaminergic neurons have unusually high energy requirements and are therefore particularly vulnerable to defects in mitochondrial metabolism and dynamics; multiple lines of evidence point to impaired activity of mitochondrial complex I in PD patients. PD-linked proteins like alpha-synuclein, Parkin, PINK1, and DJ-1 are involved in pathways that impact mitochondrial health and quality control (Golpich et al. 2016; Haddad and Nakamura 2015).

The brain exhibits high activity in lipid synthesis, with excess lipids being stored in the form of lipid droplets (LDs), which serve as cytoplasmic lipid storage organelles. LD densities are typically low under normal conditions, but during neurological disorders such as PD, LDs become more abundant. Therefore, the accumulation of LDs in the brain may serve as an early hallmark of PD onset. In muscle tissue, SDH activity reflects the cell's capacity for the TCA cycle. The analysis of specimens from patients with hereditary myopathy has revealed high LD accumulation in their myofibrils, which correlates with decreased SDH activity. This reduction in SDH activity may lead to lower ATP levels, elevated malate levels, and reduced citrate levels.

Malate is transported into the cytoplasm via the malate–aspartate shuttle, located in the mitochondrial membrane, where it appears as aspartate and α-ketoglutarate. Increased levels of aspartate in the cytoplasm activate the mammalian target of rapamycin (mTOR), converting it into its active form. mTOR functions by activating various targets within the cell. Mitochondrial dysfunction likely contributes to PD pathogenesis through several mechanisms: reduced ATP production, impaired calcium buffering, increased oxidative stress, activation of cell death pathways, and deficient mitochondrial trafficking and turnover (Chen et al. 2019; Nicoletti et al. 2021; Park et al. 2018). Restoring mitochondrial function by enhancing biogenesis, transport, calcium handling, and antioxidant defenses represents a promising neuroprotective approach. Clinical trials are underway to test advanced mitochondrial therapeutics like SS-31, which concentrates on the inner mitochondrial membrane and improves energetics (Kuang et al. 2023; Simmons et al. 2020).

Elucidating the role of mitochondria in PD may uncover new drug targets and biomarkers for the early diagnosis and tracking of disease progression. However, more research is needed to unravel the precise molecular events linking mitochondrial dysfunction to the death of dopaminergic neurons (Li et al. 2021; Moon and Paek 2015). Advancing our understanding of the mitochondrial mechanisms underpinning PD pathogenesis could pave the way for novel neuroprotective therapies to slow or stop this debilitating disease.

9.9 Huntington's Disease

Huntington's disease (HD) is an inherited neurodegenerative disorder caused by an expanded CAG trinucleotide repeat in the huntingtin gene (*HTT*) (Ajitkumar and De Jesus 2024). This results in a mutant huntingtin protein (mHtt) that misfolds and aggregates in neurons. HD is clinically characterized by progressive motor, cognitive, and psychiatric dysfunction mainly due to the loss of GABAergic medium spiny neurons in the striatum (Irfan et al. 2022; Schulte and Littleton 2011).

The "energy impairment" hypothesis of HD pathogenesis was first proposed in the 1980s based on the observations of reduced brain glucose utilization and weight loss in HD patients (Dubinsky 2017; Jiang et al. 2023). PET imaging studies have consistently shown decreased glucose uptake in HD brains, even in presymptomatic children, suggesting metabolic defects precede overt symptoms (Feigin et al. 2001; Han et al. 2021; Pagano et al. 2016).

Mitochondrial dysfunction is a hallmark feature of HD, a neurodegenerative disorder characterized by the progressive loss of motor control, cognitive decline, and psychiatric symptoms (Franco-Iborra et al. 2018; Perez Ortiz and Swerdlow 2019). Mitochondria, often referred to as the powerhouse of the cell, are essential organelles responsible for generating energy in the form of adenosine triphosphate (ATP) through oxidative phosphorylation. In HD, mitochondrial morphology is commonly observed to be abnormal, indicating a disruption in mitochondrial structure and function (Bhatti et al. 2023).

The huntingtin protein (htt), whose mutated form is the underlying cause of HD, plays a crucial role in regulating mitochondrial dynamics. Htt controls both the movement of mitochondria within the cell, known as anterograde and retrograde transport, and the processes of mitochondrial fission and fusion (Costa and Scorrano 2012; Reddy and Shirendeb 2012). Normal htt interacts with various proteins involved in these processes to maintain the balance between mitochondrial division and fusion, ensuring the proper function and distribution of mitochondria within the cell (Adebayo et al. 2021; Elgass et al. 2013; Scott and Youle 2010). However, mutations in the *htt* gene disrupt these interactions, leading to aberrant mitochondrial dynamics and impaired mitochondrial function (Jurcău et al. 2022).

Furthermore, htt is localized in the outer membrane of mitochondria, where it plays a role in regulating mitochondrial membrane potential, a critical factor in maintaining mitochondrial function (Schenkel and Bakovic 2014; Zorova et al. 2018). Mutations in htt can compromise mitochondrial membrane potential, further contributing to mitochondrial dysfunction in HD. In addition to structural and functional abnormalities in mitochondria, HD is also associated with energy metabolic deficits. Studies using imaging techniques such as positron emission tomography (PET) have revealed alterations in glucose metabolism in specific brain regions of HD patients, particularly the caudate putamen and cortex (Blázquez et al. 2022). These metabolic changes occur even in presymptomatic stages of the disease and are correlated with disease progression.

Various metabolic pathways, including glycolysis and the tricarboxylic acid (TCA) cycle, appear to be affected in HD (Kozlowski and Pallardy 1997; Veech 2013). Alterations in the expression of glucose transporter proteins, such as GLUT3, have been observed in HD mouse models, suggesting the dysregulation of glucose metabolism (Besson et al. 2015; McDonald et al. 2023). Moreover, abnormalities in enzyme complexes involved in oxidative phosphorylation, such as mitochondrial complex II, have been detected in postmortem studies of HD brains (Aidt et al. 2013; Campesan et al. 2023).

The identification of metabolic abnormalities in HD underscores the importance of targeting energy metabolism as a potential therapeutic strategy (Duan 2013; Mochel and Haller 2011). Modulating glucose metabolism pathways to compensate for mitochondrial dysfunction may hold promise for slowing disease progression and alleviating symptoms in HD patients (Chen et al. 2019; Chen et al., 2023b).

Furthermore, exploring interventions that improve neuronal energy metabolism could provide valuable insights into the pathogenesis of neurodegenerative diseases beyond HD.

Multiple studies indicate that key TCA cycle and oxidative phosphorylation (OxPhos) enzymes are downregulated in HD patients, including PDH, SDH, and complexes II, III, and IV (Dubinsky 2017; Lazzarino et al. 2019). HD patients show increased lactate production presymptomatically, reflecting a shift from OxPhos to glycolysis. Chronic inhibition of SDH replicates HD-like pathology in animal models, indicating that TCA cycle deficits can drive neurodegeneration (Dai et al. 2023; Martin et al. 2007). Treating HD mouse models with energy-boosting compounds like coenzyme Q and creatine increases longevity and improves motor function, highlighting the potential of targeting metabolism therapeutically (Yang et al. 2009).

Various other mitochondrial changes have been described in HD. The analysis of HD patient-derived stem cells revealed altered mitochondrial morphology, decreased respiration, reduced ATP and complex III activity, apoptosis activation, and glycolysis upregulation (Harmuth et al. 2018; Lopes et al. 2022). Proteomic profiling in HD stem cells found ETC protein decreases before huntingtin changes. Together, these findings indicate early mitochondrial impairment in HD (Agrawal and Fox 2019; Machiela and Southwell n.d.). Mitochondrial dysfunction also contributes to glutamate-mediated excitotoxicity in HD through defective calcium homeostasis (Verma et al. 2022; Zhao et al. 2019). Mitochondria of an HD patient show increased susceptibility to permeability transition pore opening, calcium overload, swelling, and oxidative stress. Impaired axonal transport reported in HD may arise from insufficient ATP production due to mitochondrial deficits (Dai et al. 2023; Verma et al. 2022).

In summary, studies demonstrate mitochondrial dysfunction and metabolic defects arise before HD symptoms and appear to play a central role in driving progression. Alterations in intracellular calcium signaling lead to remodeling of mitochondrial calcium exchange machinery as an initial compensatory response. However, this ultimately promotes events culminating in neurodegeneration (Britti et al. 2018; Garbincius and Elrod 2022). Strategies to enhance mitochondrial function, energy metabolism, calcium handling, and transport could have therapeutic potential to slow HD progression (Camara et al. 2010). Elucidating the precise sequence of mitochondrial pathobiology in HD could uncover new biomarkers and drug targets for this devastating disease.

9.10 Prion's Disease

Loss of synapses is a common feature in neurodegenerative diseases like prion disorders; however, the mechanisms involved are not fully understood. This study has explored whether mitochondrial abnormalities might contribute to the pathology in a mouse model of prion disease (Lepeta et al. 2016; Soto and Satani 2011). The regions of synaptic loss containing neuronal mitochondria with structural defects in the inner membrane and impaired function of respiratory complex IV, however, there being no reduction in mitochondrial density compared to healthy mice, even at disease stages when synaptic loss having begun (Eckert et al. 2011; Ge et al. 2020). This suggests that the mitochondrial changes precede the outright loss of mitochondria.

As the disease progresses, increasing numbers of neuronal mitochondria exhibit morphological abnormalities and complex IV deficits, indicating impaired respiratory capacity (Bell et al. 2021; Grünewald et al. 2019). The defects possibly are due to nitric oxide–mediated damage, as nitric oxide production is elevated in this model and can inhibit complex IV (Förstermann and Sessa 2012; Iwakiri and Kim 2015). While the precise mechanisms require further research, these findings implicate mitochondrial dysfunction in the early pathogenesis of prion disease, which could trigger and worsen synaptic loss. Therapies targeting mitochondrial function therefore possibly slow disease progression.

9.11 SDH and Its Therapeutic Targets

The brain, distinguished by its heightened sensitivity to oxidative stress compared to other bodily organs, is characterized by a naturally elevated oxidative metabolism, a scarcity of antioxidant enzymes, and a greater membranous composition (Garbarino et al. 2015; Gemma et al. 2007). The imbalance between the body's oxidative and antioxidant systems, resulting in oxidative stress, is the primary mechanism driving injury during brain ischemia (Jelinek et al. 2021). Reactive oxygen species (ROS) predominantly originate from mitochondrial SDH and NADPH oxidase (Hernansanz-Agustín and Enríquez 2021). While ROS are normally produced during aerobic metabolism under physiological conditions, ischemia-reperfusion injuries lead to alterations in the expression and activity of enzymes both promoting (e.g., nitric oxide synthase [NOS], COX, xanthine dehydrogenase/xanthine oxidase, coenzyme II oxidase, and monoamine oxidase [MAO]) and inhibiting (e.g., superoxide dismutase [SOD], catalase [CAT], peroxidase, and glutathione peroxidase [GSH-Px]) ROS production (Granger and Kvietys 2015; Xiang et al. 2021). This disruption in ROS equilibrium results in their accumulation within the body (Schieber and Chandel 2014; Sharma et al. 2012). Importantly, brain ischemia prompts persistent high expression of inducible nitric oxide synthase (iNOS) during inflammatory responses and neuronal activity, leading to excessive NO generation (Iova et al. 2023; Wang et al. 2022b). Excessive NO can combine with superoxide anions, forming highly reactive peroxynitrite, which may further impair SOD function, exacerbating brain ischemic damage (Fujii et al. 2022; Mittal et al. 2014).

Associated with significant defects in mitochondrial respiratory enzymes, Huntington's disease (HD) involves impairment of mitochondrial succinate dehydrogenase (SDH, complex II) and aconitase, accompanied by protein aggregates hindering mitochondrial function and axonal trafficking, leading to mitochondrial fragmentation and fusion inhibition (Damiano et al. 2010; Su et al. 2010). SDH inhibitors, such as 3-nitropropionic acid and malonate, induce medium spiny neuronal loss and evoke clinical and pathological features resembling HD in both rodents and nonhuman primates (Túnez et al. 2010; Túnez and Santamaría 2009).

Studies on HD transgenic mice reveal that high doses of CoQ10 notably prolong survival, enhance motor performance and grip strength, and reduce brain atrophy in R6/2 HD mice in a dose-dependent manner (Smith et al. 2006). Combining CoQ10 with minocycline in the R6/2 mouse model of HD demonstrates significantly improved behavioral measures, reduced neuropathological deficits, prolonged survival, and mitigated striatal neuron atrophy compared to either treatment alone. Likewise, combining CoQ10 with remacemide (an N-methyl-D-aspartate [NMDA] antagonist) shows significantly improved motor performance and increased survival in both R6/2 and N-171–82Q transgenic mouse models of HD. These compounds were investigated separately and in combination in 340 HD patients, where CoQ10 administration resulted in a 14% decrease in disease progression, while remacemide showed no efficacy (Smith et al. 2006).

In a study conducted by Dharshini et al., potential druggable targets for AD and PD were explored using BA9 RNA-seq data. Variants were identified and compared with genome-wide association studies (GWAS), revealing 167 variants common between AD and PD, absent in control samples. The effect of these GWAS variants was evaluated using various in silico tools, with 128 variants found in noncoding regions affecting regulatory elements in gene expression. The SZRD1 (rs138678090(C/T)) and YWHAB (rs188983062(G/C)) variants were found to affect hsa-miR-301a-3p and hsa-miR-212-3p miRNA binding, both downregulated in both the diseases. Identified 3′ untranslated regions (UTR) variants disrupted regulatory binding motifs of 47 transcription factors (TFs) involved in AD and PD pathogenesis, such as DNMT1 and HDAC1, regulating DNA methylation and histone acetylation affecting downstream signaling genes (Dharshini et al. 2021).

BA9-specific coexpression and functional interaction networks were constructed, revealing functional associations between variants, differentially expressed genes (DEGs), and TFs. For instance, the CREBBP TF regulatory motif was disrupted by the STARD10 (rs148235301) gene variant, activating genes *FLT1* and *FGFR1*, upregulated in disease conditions involved in vascular smooth muscle and endothelial pathways, crucial for maintaining vascular integrity and blood–brain barrier (BBB) restoration (M. Yang et al. 2023).

Impairment in vascular–glial communication due to reduced regional cerebral blood flow in the motor cortex was observed in AD and PD, leading to an energy crisis in neurons unable to meet their energy requirements for functional homeostasis (Monti et al. 2022; Muddapu et al. 2020). MAPK1, FGFR1, and FLT1 were identified as promising therapeutic targets for AD and PD, with potential to reduce neuroinflammation and extend neuronal survival (Q. Wang et al. 2015).

Modulating the FLT1 pathway may provide therapeutic vascular endothelial growth factor (VEGF) concentration, activating the VEGFR2 pathway to reduce BBB permeability and leaky blood vessels (Croll et al. 2004; Shibuya 2011). The proposed targets for treating neurodegeneration belong to the kinase family, albeit designing selective inhibitors is challenging due to domain functionality and cross talk between various kinases (Bhujbal and Hah 2023; Mahato and Sidorova 2020). Understanding the kinome's role and designing inhibitors could open new avenues for treating central nervous system (CNS) degeneration, reducing pathological neuroinflammation and increasing vasculature integrity to protect neurons from chronic insult.

9.12 Future Prospects

A more comprehensive comprehension of the cellular and molecular mechanisms underlying neuronal mitochondrial metabolism alterations in neurodegenerative diseases (NDDs) is imperative. Numerous challenging inquiries persist, including the central role of mitochondrial defects across diverse NDDs with varying etiologies and pathologies, the contribution of mitochondrial dysfunction to protein aggregation, and the interplay between metabolic defects and neurodegeneration (Gao et al. 2022; Jadiya et al. 2023). Additionally, elucidating the initiating events of mitochondrial dysfunction, neuronal sensing of bioenergetic crises, and the impact of cell-specific metabolic profiles on disease progression remain crucial (Melser et al. 2015; Murphy et al. 2016). Understanding upstream and downstream signaling pathways during bioenergetic crises in different NDDs and identifying optimal models for deciphering these events for translation into humans are also pivotal areas of investigation (Catozzi et al. 2016; Roux and Topisirovic 2018).

Neurodegeneration involves compromised cellular events that heavily rely on high ATP levels, such as postsynaptic signaling, axonal transport, protein clearance mechanisms, and neurotransmission (Wareham et al. 2022; Wilson et al. 2023). Evidence suggests that mitochondrial calcium (mCa^{2+}) and metabolic impairments are primary cellular defects in NDD pathogenesis, with mitochondrial dysfunction potentially serving as a central mechanism in disease progression (Jung et al. 2020; Zhang et al. 2022). However, understanding how mitochondrial dysfunction directly influences protein aggregation and brain region- and cell type–specific dysfunction in NDDs and discerning whether mitochondrial dysfunction is causal or consequential to the underlying pathology remain elusive (Swerdlow 2020). Proposed feedback loops between mitochondrial defects and disease pathology offer insights into NDD mechanisms, including potential effects on protein aggregation through ATP-dependent proteostasis machinery and mitochondrial proteostasis failure contributing to pathology-specific protein aggregation (Lu and Guo 2020; Quiles and Gustafsson 2020).

Mitochondria, essential organelles for energy generation, play critical roles in Parkinson's disease. The electron transport chain (ETC) complex within mitochondria comprises four complexes, alterations in which can lead to the generation of ROS and depletion of adenosine triphosphate (ATP) (Keane et al. 2011; Winklhofer and Haass 2010). While most complexes are encoded by mitochondrial DNA (mtDNA), complex II is encoded by the nuclear genome. Defects in complex II have been associated with neuronal injuries, and abnormalities in succinate dehydrogenase (SDH) activity, a component of complex II, have been reported in various neurodegenerative disorders, including PD (Carelli and Chan 2014; Goetzman et al. 2023).

The main hallmarks of neurodegenerative disorders like PD include lipid accumulation and excitotoxicity (Bloomingdale et al. 2022; Jellinger 2010). Mitochondrial complex II plays a dual role in regulating both lipid metabolism and excitotoxicity, directly and indirectly. Activation of the mammalian target of rapamycin (mTOR), a key signaling protein in PD, can inhibit SDH activity, which serves as a link between oxidative phosphorylation (OxPhos) and the Krebs cycle (de la Cruz López et al. 2019; Gu et al. 2022). Additionally, fatty acid oxidation is implicated in the Krebs cycle through the SDH complex. Thus, inhibition of SDH by active mTOR decreases fatty acid oxidation, leading to lipid accumulation (Duarte Hospital et al. 2023; Lee et al. 2022). Conversely, mTOR can induce the activity of sterol regulatory element-binding protein (SREBP), responsible for lipid biosynthesis. Evidence suggests that lipid accumulation is an early hallmark of neurodegeneration, highlighting the potential of SDH as an effective therapeutic target in neurological diseases like PD and aging (Laplante and Sabatini 2009; Porstmann et al. 2008).

Manipulating SDH activity, through either overexpression or inhibition, could potentially control lipid overload in neurodegenerative disorders (Johri and Beal 2012; Wang et al. 2016). Additionally, increasing the protein level of SDH may help regulate excitotoxicity and N-methyl-D-aspartate (NMDA)-dependent signaling. The identification and characterization of SDH protein offer valuable insights into its protective mechanisms against lipid overload and ROS generation in neurons, positioning it as a promising therapeutic candidate for mitigating PD progression (Benske et al. 2022; Wang et al. 2022a).

Selecting appropriate experimental models that faithfully replicate human disease pathology is another significant challenge in NDD research (Lama et al. 2021; Zhao and Bhattacharyya 2018). While animal models have historically been invaluable, their limitations in translating findings to humans underscore the need for alternative models such as postmortem human brains, human induced pluripotent stem cells (iPSCs), and organoids (Kim and Chang 2023; Labau et al. 2022). These models offer unique advantages and complement animal models by providing insights into human pathophysiology and enabling in vitro studies. However, each model has its limitations, necessitating a multifaceted approach to model NDD pathogenesis accurately (Fischell and Fishman 2021; Uddin et al. 2021). Generating robust animal models that mimic familial or nonfamilial forms of NDDs, coupled with precise causal experimentation, will enhance our understanding of mitochondrial metabolism's impact on disease pathogenesis (Lang et al. 2022; Trancikova et al. 2012). In summary, advancing our understanding of metabolic regulation, identifying mitochondrial targets, and elucidating the temporal sequence of pathological events are crucial for developing novel therapeutic strategies against NDDs.

9.13 Conclusion

Abnormalities in energy metabolism and mitochondrial function are implicated in neurodegenerative diseases. Dysfunctional mitochondria significantly reduce neuronal energy supply, potentially exacerbating neurodegeneration. Consequently, targeting energy metabolism defects emerges as a rational therapeutic approach. Manipulating energy metabolism could prove highly effective in treating neurodegenerative diseases. Hypoxia, utilized as a treatment strategy, triggers an evolutionarily conserved adaptive response in mammals, reducing reliance on mitochondrial oxidative metabolism. This metabolic shift may offer therapeutic benefits for neurodegenerative diseases.

There is substantial evidence that SDH is a major source of ROS generation in mitochondria. SDH is unique among electron transport chain complexes in that it is entirely encoded by the nuclear genome, rather than by the mitochondrial genome. By linking the TCA cycle and oxidative phosphorylation, changes in SDH activity or assembly can influence TCA metabolite concentrations. Studies show mutations in SDH subunits and inhibition of its activity trigger neurodegeneration in the brain. SDH interacts with various signaling pathways indirectly, including the mTOR/SREBP pathway that plays a key role in lipid synthesis. SDH regulates mTOR/SREBP both directly and indirectly. For instance, mTOR inhibits SDH activity, leading to reduced fatty acid oxidation and lipid accumulation. Previous research indicates that SDH could be an effective therapeutic target for neurological diseases and aging. Increasing SDH activity may help control lipid overloading and excitotoxicity, as well as NMDA-dependent signaling, in neurodegenerative disorders. Identifying and characterizing the SDH protein offer important new insights into the mechanisms by which SDH protects neurons from lipid overloading and ROS production. Overall, SDH represents a promising candidate for ameliorating neurodegeneration.

In summary, existing evidence points to the importance of SDH in mitochondrial ROS generation, lipid metabolism, and neuroprotection. Further research into its mechanisms of action could uncover novel therapeutic approaches to treat neurological diseases by targeting SDH. However, further preclinical studies are necessary to evaluate the safety and efficacy of hypoxic exposure as a potential treatment for human diseases associated with mitochondrial dysfunction.

References

Adebayo M, Singh S, Singh AP, Dasgupta S (2021) Mitochondrial fusion and fission: the fine-tune balance for cellular homeostasis. FASEB J 35(6):e21620. https://doi.org/10.1096/fj.202100067R

Agrawal S, Fox JH (2019) Novel proteomic changes in brain mitochondria provide insights into mitochondrial dysfunction in mouse models of Huntington's disease. Mitochondrion 47:318–329. https://doi.org/10.1016/j.mito.2019.03.004

Aidt FH, Nielsen SMB, Kanters J, Pesta D, Nielsen TT, Nørremølle A, Hasholt L, Christiansen M, Hagen CM (2013) Dysfunctional mitochondrial respiration in the striatum of the Huntington's disease transgenic R6/2 mouse model. PLoS Curr 5. https://doi.org/10.1371/currents.hd.d8917b4862929772c5a2f2a34ef1c201

Ajitkumar A, De Jesus O (2024) Huntington disease. In: StatPearls. StatPearls. http://www.ncbi.nlm.nih.gov/books/NBK559166/

Almikhlafi MA, Karami MM, Jana A, Alqurashi TM, Majrashi M, Alghamdi BS, Ashraf GM (2023) Mitochondrial medicine: a promising therapeutic option against various neurodegenerative disorders. Curr Neuropharmacol 21(5):1165–1183. https://doi.org/10.2174/1570159X20666220830112408

Alqahtani T, Deore SL, Kide AA, Shende BA, Sharma R, Dadarao Chakole R, Nemade LS, Kishor Kale N, Borah S, Shrikant Deokar S, Behera A, Dhawal Bhandari D, Gaikwad N, Kalam Azad A, Ghosh A (2023) Mitochondrial dysfunction and oxidative stress in Alzheimer's disease, and Parkinson's disease, Huntington's disease and amyotrophic lateral sclerosis – an updated review. Mitochondrion 71:83–92. https://doi.org/10.1016/j.mito.2023.05.007

Anderson RF, Shinde SS, Hille R, Rothery RA, Weiner JH, Rajagukguk S, Maklashina E, Cecchini G (2014) Electron-transfer pathways in the Heme and Quinone-binding domain of complex ii (succinate dehydrogenase). Biochemistry 53(10):1637–1646. https://doi.org/10.1021/bi401630m

Aradi SD, Hauser RA (2020) Medical management and prevention of motor complications in Parkinson's disease. Neurotherapeutics 17(4):1339–1365. https://doi.org/10.1007/s13311-020-00889-4

Ardanaz CG, Ramírez MJ, Solas M (2022) Brain metabolic alterations in Alzheimer's disease. Int J Mol Sci 23(7):3785. https://doi.org/10.3390/ijms23073785

Atlante A, Valenti D, Latina V, Amadoro G (2022) Dysfunction of mitochondria in Alzheimer's disease: ANT and VDAC interact with toxic proteins and aid to determine the fate of brain cells. Int J Mol Sci 23(14):7722. https://doi.org/10.3390/ijms23147722

Aufschnaiter A, Kohler V, Diessl J, Peselj C, Carmona-Gutierrez D, Keller W, Büttner S (2017) Mitochondrial lipids in neurodegeneration. Cell Tissue Res 367(1):125–140. https://doi.org/10.1007/s00441-016-2463-1

Ayres JS (2020) The biology of physiological health. Cell 181(2):250–269. https://doi.org/10.1016/j.cell.2020.03.036

Azam S, Haque ME, Balakrishnan R, Kim I-S, Choi D-K (2021) The ageing brain: molecular and cellular basis of neurodegeneration. Front Cell Dev Biol 9. https://www.frontiersin.org/articles/10.3389/fcell.2021.683459

Batzu L, Podlewska A, Gibson L, Chaudhuri KR, Aarsland D (2024) Chapter Two – a general clinical overview of the non-motor symptoms in Parkinson's disease: neuropsychiatric symptoms. In: Duty S, Iravani M (eds) International review of neurobiology, vol 174. Academic Press, pp 59–97. https://doi.org/10.1016/bs.irn.2023.11.001

Bell SM, Barnes K, De Marco M, Shaw PJ, Ferraiuolo L, Blackburn DJ, Venneri A, Mortiboys H (2021) Mitochondrial dysfunction in Alzheimer's disease: a biomarker of the future? Biomedicines 9(1):63. https://doi.org/10.3390/biomedicines9010063

Bénit P, Goncalves J, El Khoury R, Rak M, Favier J, Gimenez-Roqueplo A-P, Rustin P (2022) Succinate dehydrogenase, succinate, and superoxides: a genetic, epigenetic, metabolic, environmental explosive crossroad. Biomedicines 10(8):1788. https://doi.org/10.3390/biomedicines10081788

Benske TM, Mu T-W, Wang Y-J (2022) Protein quality control of N-methyl-D-aspartate receptors. Front Cell Neurosci 16:907560. https://doi.org/10.3389/fncel.2022.907560

Berth SH, Lloyd TE (n.d.) Disruption of axonal transport in neurodegeneration. J Clin Invest 133(11):e168554. https://doi.org/10.1172/JCI168554

Besson MT, Alegría K, Garrido-Gerter P, Barros LF, Liévens J-C (2015) Enhanced neuronal glucose transporter expression reveals metabolic choice in a HD Drosophila model. PLoS ONE 10(3):e0118765. https://doi.org/10.1371/journal.pone.0118765

Bhatia S, Rawal R, Sharma P, Singh T, Singh M, Singh V (2022) Mitochondrial dysfunction in Alzheimer's disease: opportunities for drug development. Curr Neuropharmacol 20(4):675–692. https://doi.org/10.2174/1570159X19666210517114016

Bhatti JS, Kaur S, Mishra J, Dibbanti H, Singh A, Reddy AP, Bhatti GK, Reddy PH (2023) Targeting dynamin-related protein-1 as a potential therapeutic approach for mitochondrial dysfunction in Alzheimer's disease. Biochim Biophys Acta (BBA) Mol Basis Dis 1869(7):166798. https://doi.org/10.1016/j.bbadis.2023.166798

Bhujbal SP, Hah J-M (2023) An innovative approach to address neurodegenerative diseases through kinase-targeted therapies: potential for designing covalent inhibitors. Pharmaceuticals 16(9):9. https://doi.org/10.3390/ph16091295

Blázquez E, Hurtado-Carneiro V, LeBaut-Ayuso Y, Velázquez E, García-García L, Gómez-Oliver F, Ruiz-Albusac JM, Ávila J, Pozo MÁ (2022) Significance of brain glucose hypometabolism, altered insulin signal transduction, and insulin resistance in several neurological diseases. Front Endocrinol 13:873301. https://doi.org/10.3389/fendo.2022.873301

Bloomingdale P, Karelina T, Ramakrishnan V, Bakshi S, Véronneau-Veilleux F, Moye M, Sekiguchi K, Meno-Tetang G, Mohan A, Maithreye R, Thomas VA, Gibbons F, Cabal A, Bouteiller J, Geerts H (2022) Hallmarks of neurodegenerative disease: a systems pharmacology perspective. CPT Pharmacometrics Syst Pharmacol 11(11):1399–1429. https://doi.org/10.1002/psp4.12852

Brand MD, Orr AL, Perevoshchikova IV, Quinlan CL (2013) The role of mitochondrial function and cellular bioenergetics in ageing and disease. Br J Dermatol 169(0 2):1–8. https://doi.org/10.1111/bjd.12208

Bridges H, Bill E, Hirst J (2011) Mossbauer spectroscopy on respiratory complex I: the iron-sulfur cluster ensemble in the NADH-reduced enzyme is partially oxidized. Biochemistry 51:149–158. https://doi.org/10.1021/bi201644x

Britti E, Delaspre F, Tamarit J, Ros J (2018) Mitochondrial calcium signalling and neurodegenerative diseases. Neuronal Signal 2(4):NS20180061. https://doi.org/10.1042/NS20180061

Bustamante-Barrientos FA, Luque-Campos N, Araya MJ, Lara-Barba E, de Solminihac J, Pradenas C, Molina L, Herrera-Luna Y, Utreras-Mendoza Y, Elizondo-Vega R, Vega-Letter AM, Luz-Crawford P (2023) Mitochondrial dysfunction in neurodegenerative disorders: potential therapeutic application of mitochondrial transfer to central nervous system-residing cells. J Transl Med 21(1):613. https://doi.org/10.1186/s12967-023-04493-w

Butterfield DA, Boyd-Kimball D (2020) Mitochondrial oxidative and nitrosative stress and Alzheimer disease. Antioxidants 9(9):818. https://doi.org/10.3390/antiox9090818

Cabezas-Opazo FA, Vergara-Pulgar K, Pérez MJ, Jara C, Osorio-Fuentealba C, Quintanilla RA (2015) Mitochondrial dysfunction contributes to the pathogenesis of Alzheimer's disease. Oxidative Med Cell Longev 2015:509654. https://doi.org/10.1155/2015/509654

Cai Q, Tammineni P (2017) Mitochondrial aspects of synaptic dysfunction in Alzheimer's disease. J Alzheimer's Dis 57(4):1087–1103. https://doi.org/10.3233/JAD-160726

Calì T, Ottolini D, Brini M (2012) Mitochondrial Ca2+ and neurodegeneration. Cell Calcium 52(1):73–85. https://doi.org/10.1016/j.ceca.2012.04.015

Camara AKS, Lesnefsky EJ, Stowe DF (2010) Potential therapeutic benefits of strategies directed to mitochondria. Antioxid Redox Signal 13(3):279–347. https://doi.org/10.1089/ars.2009.2788

Campesan S, del Popolo I, Marcou K, Straatman-Iwanowska A, Repici M, Boytcheva KV, Cotton VE, Allcock N, Rosato E, Kyriacou CP, Giorgini F (2023) Bypassing mitochondrial defects rescues Huntington's phenotypes in Drosophila. Neurobiol Dis 185:106236. https://doi.org/10.1016/j.nbd.2023.106236

Cardol P, Figueroa F, Remacle C, Franzén L-G, González-Halphen D (2009) Chapter 13 – oxidative phosphorylation: building blocks and related components. In: Harris EH, Stern DB, Witman GB (eds) The Chlamydomonas sourcebook, 2nd edn. Academic Press, pp 469–502. https://doi.org/10.1016/B978-0-12-370873-1.00021-6

Carelli V, Chan DC (2014) Mitochondrial DNA: impacting central and peripheral nervous systems. Neuron 84(6):1126–1142. https://doi.org/10.1016/j.neuron.2014.11.022

Casanova A, Wevers A, Navarro-Ledesma S, Pruimboom L (2023) Mitochondria: it is all about energy. Front Physiol 14:1114231. https://doi.org/10.3389/fphys.2023.1114231

Catozzi S, Di-Bella JP, Ventura AC, Sepulchre J-A (2016) Signaling cascades transmit information downstream and upstream but unlikely simultaneously. BMC Syst Biol 10(1):84. https://doi.org/10.1186/s12918-016-0303-2

Chambers JC, Kenan D, Martin BJ, Keene JD (1988) Genomic structure and amino acid sequence domains of the human La autoantigen. J Biol Chem 263(34):18043–18051. https://doi.org/10.1016/S0021-9258(19)81321-2

Chandhok G, Lazarou M, Neumann B (2018) Structure, function, and regulation of mitofusin-2 in health and disease. Biol Rev Camb Philos Soc 93(2):933–949. https://doi.org/10.1111/brv.12378

Chen JX, Yan SD (2007) Amyloid-β-induced mitochondrial dysfunction. J Alzheimer's Dis 12(2):177. https://doi.org/10.3233/jad-2007-12208

Chen C, Turnbull DM, Reeve AK (2019) Mitochondrial dysfunction in Parkinson's disease – cause or consequence? Biology 8(2):38. https://doi.org/10.3390/biology8020038

Chen Y, Guo S, Tang Y, Mou C, Hu X, Shao F, Yan W, Wu Q (2020) Mitochondrial fusion and fission in neuronal death induced by cerebral ischemia-reperfusion and its clinical application: a mini-review. Med Sci Monit 26:e928651-1–e928651-12. https://doi.org/10.12659/MSM.928651

Chen Z-R, Huang J-B, Yang S-L, Hong F-F (2022) Role of cholinergic signaling in Alzheimer's disease. Molecules 27(6):1816. https://doi.org/10.3390/molecules27061816

Chen L, Zhou M, Li H, Liu D, Liao P, Zong Y, Zhang C, Zou W, Gao J (2023a) Mitochondrial heterogeneity in diseases. Signal Transduct Target Ther 8(1):1. https://doi.org/10.1038/s41392-023-01546-w

Chen X, Ji Y, Liu R, Zhu X, Wang K, Yang X, Liu B, Gao Z, Huang Y, Shen Y, Liu H, Sun H (2023b) Mitochondrial dysfunction: roles in skeletal muscle atrophy. J Transl Med 21(1):503. https://doi.org/10.1186/s12967-023-04369-z

Cheng Y, Bai F (2018) The association of tau with mitochondrial dysfunction in Alzheimer's disease. Front Neurosci 12:163. https://doi.org/10.3389/fnins.2018.00163

Cheng R, Dhorajia V, Kim J, Kim Y (2022) Mitochondrial iron metabolism and neurodegenerative diseases. Neurotoxicology 88:88–101. https://doi.org/10.1016/j.neuro.2021.11.003

Choi J, Horner KA (2024) Dopamine agonists. In: StatPearls. StatPearls. http://www.ncbi.nlm.nih.gov/books/NBK551686/

Choi I, Son H, Baek J-H (2021) Tricarboxylic Acid (TCA) cycle intermediates: regulators of immune responses. Life (Basel) 11(1):1. https://doi.org/10.3390/life11010069

Chu Y, Chang Y, Lu W, Sheng X, Wang S, Xu H, Ma J (2020) Regulation of autophagy by glycolysis in cancer. Cancer Manag Res 12:13259–13271. https://doi.org/10.2147/CMAR.S279672

Ciccarone F, Di Leo L, Ciriolo MR (2019) TCA cycle aberrations and cancer. In: Boffetta P, Hainaut P (eds) Encyclopedia of cancer, 3rd edn. Academic Press, pp 429–436. https://doi.org/10.1016/B978-0-12-801238-3.65066-3

Clemente-Suárez VJ, Martín-Rodríguez A, Yáñez-Sepúlveda R, Tornero-Aguilera JF (2023a) Mitochondrial transfer as a novel therapeutic approach in disease diagnosis and treatment. Int J Mol Sci 24(10):8848. https://doi.org/10.3390/ijms24108848

Clemente-Suárez VJ, Redondo-Flórez L, Beltrán-Velasco AI, Ramos-Campo DJ, Belinchón-deMiguel P, Martinez-Guardado I, Dalamitros AA, Yáñez-Sepúlveda R, Martín-Rodríguez A, Tornero-Aguilera JF (2023b) Mitochondria and brain disease: a comprehensive review of pathological mechanisms and therapeutic opportunities. Biomedicines 11(9):2488. https://doi.org/10.3390/biomedicines11092488

Collins HE, Kane MS, Litovsky SH, Darley-Usmar VM, Young ME, Chatham JC, Zhang J (2021) Mitochondrial morphology and mitophagy in heart diseases: qualitative and quantitative analyses using transmission electron microscopy. Front Aging 2:670267. https://doi.org/10.3389/fragi.2021.670267

Coluccino G, Muraca VP, Corazza A, Lippe G (2023) Cyclophilin D in mitochondrial dysfunction: a key player in neurodegeneration? Biomolecules 13(8):1265. https://doi.org/10.3390/biom13081265

Contreras L, Drago I, Zampese E, Pozzan T (2010) Mitochondria: The calcium connection. Biochimica et Biophysica Acta 1797(6):607–618. https://doi.org/10.1016/j.bbabio.2010.05.005

Cooper GM (2000) The mechanism of oxidative phosphorylation. In: The cell: a molecular Approach, 2nd edn. Sinauer Associates. https://www.ncbi.nlm.nih.gov/books/NBK9885/

Costa V, Scorrano L (2012) Shaping the role of mitochondria in the pathogenesis of Huntington's disease. EMBO J 31(8):1853–1864. https://doi.org/10.1038/emboj.2012.65

Croll SD, Goodman JH, Scharfman HE (2004) Vascular Endothelial Growth Factor (VEGF) in seizures: a double-edged sword. Adv Exp Med Biol 548:57–68

Dai Y, Wang H, Lian A, Li J, Zhao G, Hu S, Li B (2023) A comprehensive perspective of Huntington's disease and mitochondrial dysfunction. Mitochondrion 70:8–19. https://doi.org/10.1016/j.mito.2023.03.001

Damiano M, Galvan L, Déglon N, Brouillet E (2010) Mitochondria in Huntington's disease. Biochim Biophys Acta (BBA) Mol Basis Dis 1802(1):52–61. https://doi.org/10.1016/j.bbadis.2009.07.012

Dauer W, Przedborski S (2003) Parkinson's disease: mechanisms and models. Neuron 39(6):889–909. https://doi.org/10.1016/S0896-6273(03)00568-3

de la Cruz López KG, Toledo Guzmán ME, Sánchez EO, García Carrancá A (2019) mTORC1 as a regulator of mitochondrial functions and a therapeutic target in cancer. Front Oncol 9:1373. https://doi.org/10.3389/fonc.2019.01373

De Vivo DC, Paradas C, DiMauro S (2015) Chapter 41 – mitochondrial encephalomyopathies. In: Darras BT, Jones HR, Ryan MM, De Vivo DC (eds) Neuromuscular disorders of infancy, childhood, and adolescence, 2nd edn. Academic Press, pp 796–833. https://doi.org/10.1016/B978-0-12-417044-5.00041-X

De Vos KJ, Hafezparast M (2017) Neurobiology of axonal transport defects in motor neuron diseases: Opportunities for translational research? Neurobiol Dis 105:283–299. https://doi.org/10.1016/j.nbd.2017.02.004

Dharshini SAP, Jemimah S, Taguchi YH, Gromiha MM (2021) Exploring common therapeutic targets for neurodegenerative disorders using transcriptome study. Front Genet 12:639160. https://doi.org/10.3389/fgene.2021.639160

Di Meo S, Reed TT, Venditti P, Victor VM (2016) Role of ROS and RNS sources in physiological and pathological conditions. Oxidative Med Cell Longev 2016:1245049. https://doi.org/10.1155/2016/1245049

Duan W (2013) Targeting Sirtuin-1 in Huntington's disease: rationale and current status. CNS Drugs 27(5):345–352. https://doi.org/10.1007/s40263-013-0055-0

Duarte Hospital C, Tête A, Debizet K, Imler J, Tomkiewicz-Raulet C, Blanc EB, Barouki R, Coumoul X, Bortoli S (2023) SDHi fungicides: an example of mitotoxic pesticides targeting the succinate dehydrogenase complex. Environ Int 180:108219

Dubinsky JM (2017) Towards an understanding of energy impairment in Huntington's disease brain. J Huntington's Dis 6(4):267–302

Duchen MR (2012) Mitochondria, calcium-dependent neuronal death and neurodegenerative disease. Pflugers Arch 464(1):111–121

Eaglesfield R, Tokatlidis K (2021) Targeting and insertion of membrane proteins in mitochondria. Front Cell Dev Biol 99:803205

Eckert A, Schmitt K, Götz J (2011) Mitochondrial dysfunction – the beginning of the end in Alzheimer's disease? Separate and synergistic modes of tau and amyloid-β toxicity. Alzheimer's Res Ther 3(3):15

Elbehti-Green A, Au HC, Mascarello JT, Ream-Robinson D, Scheffler IE (1998) Characterization of the human SDHC gene encoding one of the integral membrane proteins of succinate–quinone oxidoreductase in mitochondria1The first two authors contributed equally to this paper.1. Gene 213(1):133–140. https://doi.org/10.1016/S0378-1119(98)00186-3

Elgass K, Pakay J, Ryan MT, Palmer CS (2013) Recent advances into the understanding of mitochondrial fission. Biochim Biophys Acta 1833(1):150–161

Endlicher R, Drahota Z, Štefková K, Červinková Z, Kučera O (2023) The mitochondrial permeability transition pore – current knowledge of its structure, function, and regulation, and optimized methods for evaluating its functional state. Cells 12(9):9

Eyassu F, Angione C (2017) Modelling pyruvate dehydrogenase under hypoxia and its role in cancer metabolism. R Soc Open Sci 4(10):170360. https://doi.org/10.1098/rsos.170360

Farmer BC, Walsh AE, Kluemper JC, Johnson LA (2020) Lipid droplets in neurodegenerative disorders. Front Neurosci 14:742. https://doi.org/10.3389/fnins.2020.00742

Feigin A, Leenders KL, Moeller JR, Missimer J, Kuenig G, Spetsieris P, Antonini A, Eidelberg D (2001) Metabolic network abnormalities in early Huntington's disease: an [18F]FDG PET study. J Nucl Med 42(11):1591–1595

Fernandes T, Resende R, Silva DF, Marques AP, Santos AE, Cardoso SM, Domingues MR, Moreira PI, Pereira CF (2021) Structural and functional alterations in Mitochondria-Associated Membranes (MAMs) and in mitochondria activate stress response mechanisms in an *in vitro* model of Alzheimer's disease. Biomedicines 9(8):881. https://doi.org/10.3390/biomedicines9080881

Fischell JM, Fishman PS (2021) A multifaceted approach to optimizing AAV delivery to the brain for the treatment of neurodegenerative diseases. Front Neurosci 15:747726. https://doi.org/10.3389/fnins.2021.747726

Förstermann U, Sessa WC (2012) Nitric oxide synthases: regulation and function. Eur Heart J 33(7):829–837. https://doi.org/10.1093/eurheartj/ehr304

Franco-Iborra S, Vila M, Perier C (2018) Mitochondrial quality control in neurodegenerative diseases: focus on Parkinson's disease and Huntington's disease. Front Neurosci 12. https://www.frontiersin.org/journals/neuroscience/articles/10.3389/fnins.2018.00342

Fujii J, Homma T, Osaki T (2022) Superoxide radicals in the execution of cell death. Antioxidants 11(3):501. https://doi.org/10.3390/antiox11030501

Fullerton M, McFarland R, Taylor RW, Alston CL (2020) The genetic basis of isolated mitochondrial complex II deficiency. Mol Genet Metab 131(1):53–65. https://doi.org/10.1016/j.ymgme.2020.09.009

Galluzzi L, Kepp O, Trojel-Hansen C, Kroemer G (2012) Mitochondrial control of cellular life, stress, and death. Circ Res 111(9):1198–1207. https://doi.org/10.1161/CIRCRESAHA.112.268946

Gao J, Wang L, Liu J, Xie F, Su B, Wang X (2017) Abnormalities of mitochondrial dynamics in neurodegenerative diseases. Antioxidants 6(2):2. https://doi.org/10.3390/antiox6020025

Gao X-Y, Yang T, Gu Y, Sun X-H (2022) Mitochondrial dysfunction in Parkinson's disease: from mechanistic insights to therapy. Front Aging Neurosci 14:885500. https://doi.org/10.3389/fnagi.2022.885500

Garbarino VR, Orr ME, Rodriguez KA, Buffenstein R (2015) Mechanisms of oxidative stress resistance in the brain: lessons learned from hypoxia tolerant extremophilic vertebrates. Arch Biochem Biophys 576:8–16. https://doi.org/10.1016/j.abb.2015.01.029

Garbincius JF, Elrod JW (2022) Mitochondrial calcium exchange in physiology and disease. Physiol Rev 102(2):893–992. https://doi.org/10.1152/physrev.00041.2020

Ge P, Dawson VL, Dawson TM (2020) PINK1 and Parkin mitochondrial quality control: a source of regional vulnerability in Parkinson's disease. Mol Neurodegener 15(1):20. https://doi.org/10.1186/s13024-020-00367-7

Gemma C, Vila J, Bachstetter A, Bickford PC (2007) Oxidative stress and the aging brain: from theory to prevention. In: Riddle DR (ed) Brain aging: models, methods, and mechanisms. CRC Press/Taylor and Francis. http://www.ncbi.nlm.nih.gov/books/NBK3869/

Gitler AD, Dhillon P, Shorter J (2017) Neurodegenerative disease: models, mechanisms, and a new hope. Dis Model Mech 10(5):499–502. https://doi.org/10.1242/dmm.030205

Gnaiger E (2024) Complex II ambiguities – FADH2 in the electron transfer system. J Biol Chem 300(1). https://doi.org/10.1016/j.jbc.2023.105470

Goel P, Chakrabarti S, Goel K, Bhutani K, Chopra T, Bali S (2022) Neuronal cell death mechanisms in Alzheimer's disease: an insight. Front Mol Neurosci 15. https://www.frontiersin.org/articles/10.3389/fnmol.2022.937133

Goetzman E, Gong Z, Zhang B, Muzumdar R (2023) Complex II biology in aging, health, and disease. Antioxidants 12(7):7. https://doi.org/10.3390/antiox12071477

Golpich M, Amini E, Mohamed Z, Azman Ali R, Mohamed Ibrahim N, Ahmadiani A (2016) Mitochondrial dysfunction and biogenesis in neurodegenerative diseases: pathogenesis and treatment. CNS Neurosci Ther 23(1):5–22. https://doi.org/10.1111/cns.12655

Gómez-Benito M, Granado N, García-Sanz P, Michel A, Dumoulin M, Moratalla R (2020) Modeling Parkinson's disease with the Alpha-synuclein protein. Front Pharmacol 11. https://www.frontiersin.org/journals/pharmacology/articles/10.3389/fphar.2020.00356

Granger DN, Kvietys PR (2015) Reperfusion injury and reactive oxygen species: the evolution of a concept. Redox Biol 6:524–551. https://doi.org/10.1016/j.redox.2015.08.020

Gray LR, Tompkins SC, Taylor EB (2014) Regulation of pyruvate metabolism and human disease. Cell Mol Life Sci 71(14):2577–2604. https://doi.org/10.1007/s00018-013-1539-2

Green A, Hossain T, Eckmann DM (2022) Mitochondrial dynamics involves molecular and mechanical events in motility, fusion and fission. Front Cell Dev Biol 10:1010232. https://doi.org/10.3389/fcell.2022.1010232

Grel H, Woznica D, Ratajczak K, Kalwarczyk E, Anchimowicz J, Switlik W, Olejnik P, Zielonka P, Stobiecka M, Jakiela S (2023) Mitochondrial dynamics in neurodegenerative diseases: unraveling the role of fusion and fission processes. Int J Mol Sci 24(17):13033. https://doi.org/10.3390/ijms241713033

Grünewald A, Kumar KR, Sue CM (2019) New insights into the complex role of mitochondria in Parkinson's disease. Prog Neurobiol 177:73–93. https://doi.org/10.1016/j.pneurobio.2018.09.003

Gu X, Wang W, Yang Y, Lei Y, Liu D, Wang X, Wu T (2022) The effect of metabolites on mitochondrial functions in the pathogenesis of skeletal muscle aging. Clin Interv Aging 17:1275–1295. https://doi.org/10.2147/CIA.S376668

Guo C, Sun L, Chen X, Zhang D (2013) Oxidative stress, mitochondrial damage and neurodegenerative diseases. Neural Regen Res 8(21):2003–2014. https://doi.org/10.3969/j.issn.1673-5374.2013.21.009

Haddad D, Nakamura K (2015) Understanding the susceptibility of dopamine neurons to mitochondrial stressors in Parkinson's disease. FEBS Lett 589(24, Part A):3702–3713. https://doi.org/10.1016/j.febslet.2015.10.021

Han R, Liang J, Zhou B (2021) Glucose metabolic dysfunction in neurodegenerative diseases – new mechanistic insights and the potential of hypoxia as a prospective therapy targeting metabolic reprogramming. Int J Mol Sci 22(11):5887. https://doi.org/10.3390/ijms22115887

Hansen GE, Gibson GE (2022) The α-Ketoglutarate dehydrogenase complex as a hub of plasticity in neurodegeneration and regeneration. Int J Mol Sci 23(20):12403. https://doi.org/10.3390/ijms232012403

Harmuth T, Prell-Schicker C, Weber JJ, Gellerich F, Funke C, Drießen S, Magg JCD, Krebiehl G, Wolburg H, Hayer SN, Hauser S, Krüger R, Schöls L, Riess O, Hübener-Schmid J (2018) Mitochondrial morphology, function and homeostasis are impaired by expression of an N-terminal Calpain cleavage fragment of Ataxin-3. Front Mol Neurosci 11:368. https://doi.org/10.3389/fnmol.2018.00368

Henrich MT, Oertel WH, Surmeier DJ, Geibl FF (2023) Mitochondrial dysfunction in Parkinson's disease – a key disease hallmark with therapeutic potential. Mol Neurodegener 18(1):83. https://doi.org/10.1186/s13024-023-00676-7

Hernansanz-Agustín P, Enríquez JA (2021) Generation of reactive oxygen species by mitochondria. Antioxidants 10(3):415. https://doi.org/10.3390/antiox10030415

Hroudová J, Singh N, Fišar Z (2014) Mitochondrial dysfunctions in neurodegenerative diseases: relevance to Alzheimer's disease. Biomed Res Int 2014:175062. https://doi.org/10.1155/2014/175062

Hussain R, Zubair H, Pursell S, Shahab M (2018) Neurodegenerative diseases: regenerative mechanisms and novel therapeutic approaches. Brain Sci 8(9):177. https://doi.org/10.3390/brainsci8090177

Hwang AB, Jeong D-E, Lee S-J (2012) Mitochondria and organismal longevity. Curr Genomics 13(7):519–532. https://doi.org/10.2174/138920212803251427

Iova O-M, Marin G-E, Lazar I, Stanescu I, Dogaru G, Nicula CA, Bulboacă AE (2023) Nitric oxide/nitric oxide synthase system in the pathogenesis of neurodegenerative disorders – an overview. Antioxidants 12(3):753. https://doi.org/10.3390/antiox12030753

Irfan Z, Khanam S, Karmakar V, Firdous SM, El Khier BSIA, Khan I, Rehman MU, Khan A (2022) Pathogenesis of Huntington's disease: an emphasis on molecular pathways and prevention by natural remedies. Brain Sci 12(10):1389. https://doi.org/10.3390/brainsci12101389
Islimye E, Girard V, Gould AP (2022) Functions of stress-induced lipid droplets in the nervous system. Front Cell Dev Biol 10:863907. https://doi.org/10.3389/fcell.2022.863907
Iverson TM, Maklashina E, Cecchini G (2012) Structural basis for malfunction in complex II. J Biol Chem 287(42):35430–35438. https://doi.org/10.1074/jbc.R112.408419
Iwakiri Y, Kim MY (2015) Nitric oxide in liver diseases. Trends Pharmacol Sci 36(8):524–536. https://doi.org/10.1016/j.tips.2015.05.001
Jadiya P, Garbincius JF, Elrod JW (2021) Reappraisal of metabolic dysfunction in neurodegeneration: focus on mitochondrial function and calcium signaling. Acta Neuropathol Commun 9:124. https://doi.org/10.1186/s40478-021-01224-4
Jadiya P, Cohen HM, Kolmetzky DW, Kadam AA, Tomar D, Elrod JW (2023) Neuronal loss of NCLX-dependent mitochondrial calcium efflux mediates age-associated cognitive decline. iScience 26(3):106296. https://doi.org/10.1016/j.isci.2023.106296
Jankovic J, Tan EK (2020) Parkinson's disease: Etiopathogenesis and treatment. J Neurol Neurosurg Psychiatry 91(8):795–808. https://doi.org/10.1136/jnnp-2019-322338
Jankowska-Kulawy A, Klimaszewska-Łata J, Gul-Hinc S, Ronowska A, Szutowicz A (2022) Metabolic and cellular compartments of Acetyl-CoA in the healthy and diseased brain. Int J Mol Sci 23(17):17. https://doi.org/10.3390/ijms231710073
Jankowsky JL, Zheng H (2017) Practical considerations for choosing a mouse model of Alzheimer's disease. Mol Neurodegener 12(1):89. https://doi.org/10.1186/s13024-017-0231-7
Jelinek M, Jurajda M, Duris K (2021) Oxidative stress in the brain: basic concepts and treatment strategies in stroke. Antioxidants 10(12):1886. https://doi.org/10.3390/antiox10121886
Jellinger KA (2010) Basic mechanisms of neurodegeneration: a critical update. J Cell Mol Med 14(3):457–487. https://doi.org/10.1111/j.1582-4934.2010.01010.x
Jia K, Du H (2021) Mitochondrial permeability transition: a pore intertwines brain aging and Alzheimer's disease. Cells 10(3):3. https://doi.org/10.3390/cells10030649
Jiang A, Handley RR, Lehnert K, Snell RG (2023) From pathogenesis to therapeutics: a review of 150 Years of Huntington's disease research. Int J Mol Sci 24(16):13021. https://doi.org/10.3390/ijms241613021
Johri A, Beal MF (2012) Mitochondrial dysfunction in neurodegenerative diseases. J Pharmacol Exp Ther 342(3):619–630. https://doi.org/10.1124/jpet.112.192138
Juárez Olguín H, Calderón Guzmán D, Hernández García E, Barragán Mejía G (2016) The role of dopamine and its dysfunction as a consequence of oxidative stress. Oxidative Med Cell Longev 2016:9730467. https://doi.org/10.1155/2016/9730467
Jung H, Kim SY, Canbakis Cecen FS, Cho Y, Kwon S-K (2020) Dysfunction of mitochondrial Ca2+ regulatory machineries in brain aging and neurodegenerative diseases. Front Cell Dev Biol 8. https://www.frontiersin.org/articles/10.3389/fcell.2020.599792
Jurcău MC, Andronie-Cioara FL, Jurcău A, Marcu F, Ţiț DM, Pașcalău N, Nistor-Cseppentö DC (2022) The link between oxidative stress, mitochondrial dysfunction and neuroinflammation in the pathophysiology of Alzheimer's disease: therapeutic implications and future perspectives. Antioxidants 11(11):2167. https://doi.org/10.3390/antiox11112167
Kalani K, Yan SF, Yan SS (2018) Mitochondrial permeability transition pore: a potential drug target for neurodegeneration. Drug Discov Today 23(12):1983–1989. https://doi.org/10.1016/j.drudis.2018.08.001
Keane PC, Kurzawa M, Blain PG, Morris CM (2011) Mitochondrial dysfunction in Parkinson's disease. Parkinson's Dis 2011:716871. https://doi.org/10.4061/2011/716871
Keil U, Bonert A, Marques CA, Scherping I, Weyermann J, Strosznajder JB, Müller-Spahn F, Haass C, Czech C, Pradier L, Müller WE, Eckert A (2004) Amyloid β-induced changes in nitric oxide production and mitochondrial activity lead to apoptosis*. J Biol Chem 279(48):50310–50320. https://doi.org/10.1074/jbc.M405600200

Kim S, Chang M-Y (2023) Application of human brain organoids – opportunities and challenges in modeling human brain development and neurodevelopmental diseases. Int J Mol Sci 24(15):12528. https://doi.org/10.3390/ijms241512528

Kim HJ, Khalimonchuk O, Smith PM, Winge DR (2012) Structure, function, and assembly of heme centers in mitochondrial respiratory complexes. Biochim Biophys Acta 1823(9):1604–1616. https://doi.org/10.1016/j.bbamcr.2012.04.008

Klein C, Westenberger A (2012) Genetics of Parkinson's disease. Cold Spring Harb Perspect Med 2(1):a008888. https://doi.org/10.1101/cshperspect.a008888

Knott AB, Bossy-Wetzel E (2008) Impairing the mitochondrial fission and fusion balance: a new mechanism of neurodegeneration. Ann N Y Acad Sci 1147:283–292. https://doi.org/10.1196/annals.1427.030

Kohlmeier M (2015) Chapter 8 – amino acids and nitrogen compounds. In: Kohlmeier M (ed) Nutrient metabolism, 2nd edn. Academic Press, pp 265–477. https://doi.org/10.1016/B978-0-12-387784-0.00008-0

Kolicheski A, Turcano P, Tamvaka N, McLean PJ, Springer W, Savica R, Ross OA (n.d.) Early-onset Parkinson's disease: creating the right environment for a genetic disorder. J Parkinsons Dis 12(8):2353–2367. https://doi.org/10.3233/JPD-223380

Kouli A, Torsney KM, Kuan W-L (2018) Parkinson's disease: etiology, neuropathology, and pathogenesis. In: Stoker TB, Greenland JC (eds) Parkinson's disease: pathogenesis and clinical aspects. Codon. http://www.ncbi.nlm.nih.gov/books/NBK536722/

Kozlowski TT, Pallardy SG (1997) Chapter 6 – enzymes, energetics, and respiration. In: Kozlowski TT, Pallardy SG (eds) Physiology of woody plants, 2nd edn. Academic Press, pp 134–158. https://doi.org/10.1016/B978-012424162-6/50023-5

Kuang G, Halimitabrizi M, Edziah A-A, Salowe R, O'Brien JM (2023) The potential for mitochondrial therapeutics in the treatment of primary open-angle glaucoma: a review. Front Physiol 14:1184060. https://doi.org/10.3389/fphys.2023.1184060

Labau JIR, Andelic M, Faber CG, Waxman SG, Lauria G, Dib-Hajj SD (2022) Recent advances for using human induced-pluripotent stem cells as pain-in-a-dish models of neuropathic pain. Exp Neurol 358:114223. https://doi.org/10.1016/j.expneurol.2022.114223

Lama J, Buhidma Y, Fletcher EJR, Duty S (2021) Animal models of Parkinson's disease: a guide to selecting the optimal model for your research. Neuronal Signal 5(4):NS20210026. https://doi.org/10.1042/NS20210026

Lamptey RNL, Chaulagain B, Trivedi R, Gothwal A, Layek B, Singh J (2022) A review of the common neurodegenerative disorders: current therapeutic approaches and the potential role of nanotherapeutics. Int J Mol Sci 23(3):3. https://doi.org/10.3390/ijms23031851

Lang M, Grünewald A, Pramstaller PP, Hicks AA, Pichler I (2022) A genome on shaky ground: exploring the impact of mitochondrial DNA integrity on Parkinson's disease by highlighting the use of cybrid models. Cell Mol Life Sci 79(5):283. https://doi.org/10.1007/s00018-022-04304-3

Laplante M, Sabatini DM (2009) An emerging role of mTOR in lipid biosynthesis. Curr Biol 19(22):R1046–R1052. https://doi.org/10.1016/j.cub.2009.09.058

Lazzarino G, Amorini AM, Signoretti S, Musumeci G, Lazzarino G, Caruso G, Pastore FS, Di Pietro V, Tavazzi B, Belli A (2019) Pyruvate dehydrogenase and tricarboxylic acid cycle enzymes are sensitive targets of traumatic brain injury induced metabolic derangement. Int J Mol Sci 20(22):22. https://doi.org/10.3390/ijms20225774

Lee HM, Koh S-B (2015) Many faces of Parkinson's disease: non-motor symptoms of Parkinson's disease. J Mov Disord 8(2):92–97. https://doi.org/10.14802/jmd.15003

Lee S-J, Zhang J, Choi AMK, Kim HP (2013) Mitochondrial dysfunction induces formation of lipid droplets as a generalized response to stress. Oxidative Med Cell Longev 2013:e327167. https://doi.org/10.1155/2013/327167

Lee S, Xu H, Van Vleck A, Mawla AM, Li AM, Ye J, Huising MO, Annes JP (2022) β-Cell succinate dehydrogenase deficiency triggers metabolic dysfunction and insulinopenic diabetes. Diabetes 71(7):1439–1453. https://doi.org/10.2337/db21-0834

Lemarie A, Grimm S (2009) Mutations in the heme b-binding residue of SDHC inhibit assembly of respiratory chain complex II in mammalian cells. Mitochondrion 9:254–260. https://doi.org/10.1016/j.mito.2009.03.004

Lepeta K, Lourenco MV, Schweitzer BC, Martino Adami PV, Banerjee P, Catuara-Solarz S, de La Fuente Revenga M, Guillem AM, Haidar M, Ijomone OM, Nadorp B, Qi L, Perera ND, Refsgaard LK, Reid KM, Sabbar M, Sahoo A, Schaefer N, Sheean RK et al (2016) Synaptopathies: synaptic dysfunction in neurological disorders – a review from students to students. J Neurochem 138(6):785–805. https://doi.org/10.1111/jnc.13713

Li J-L, Lin T-Y, Chen P-L, Guo T-N, Huang S-Y, Chen C-H, Lin C-H, Chan C-C (2021) Mitochondrial function and Parkinson's disease: from the perspective of the electron transport chain. Front Mol Neurosci 14:797833. https://doi.org/10.3389/fnmol.2021.797833

Li Y, Xia X, Wang Y, Zheng JC (2022) Mitochondrial dysfunction in microglia: a novel perspective for pathogenesis of Alzheimer's disease. J Neuroinflammation 19(1):248. https://doi.org/10.1186/s12974-022-02613-9

Lim JW, Lee J, Pae AN (2020) Mitochondrial dysfunction and Alzheimer's disease: prospects for therapeutic intervention. BMB Rep 53(1):47–55. https://doi.org/10.5483/BMBRep.2020.53.1.279

Liu L, Zhang K, Sandoval H, Yamamoto S, Jaiswal M, Sanz E, Li Z, Hui J, Graham BH, Quintana A, Bellen HJ (2015) Glial lipid droplets and ROS induced by mitochondrial defects promote neurodegeneration. Cell 160(1):177–190. https://doi.org/10.1016/j.cell.2014.12.019

Lopes C, Ferreira IL, Maranga C, Beatriz M, Mota SI, Sereno J, Castelhano J, Abrunhosa A, Oliveira F, De Rosa M, Hayden M, Laço MN, Januário C, Castelo Branco M, Rego AC (2022) Mitochondrial and redox modifications in early stages of Huntington's disease. Redox Biol 56:102424. https://doi.org/10.1016/j.redox.2022.102424

López-López A, Keech O, Rouhier N (2022) Maturation and assembly of iron-sulfur cluster-containing subunits in the mitochondrial complex I from plants. Front Plant Sci 13. https://www.frontiersin.org/journals/plant-science/articles/10.3389/fpls.2022.916948

Losano JDA, Padín JF, Méndez-López I, Angrimani DSR, García AG, Barnabe VH, Nichi M (2017) The stimulated glycolytic pathway is able to maintain ATP levels and kinetic patterns of bovine epididymal sperm subjected to mitochondrial uncoupling. Oxidative Med Cell Longev 2017:1682393. https://doi.org/10.1155/2017/1682393

Lu B, Guo S (2020) Mechanisms linking mitochondria dysfunction and proteostasis failure. Trends Cell Biol 30(4):317–328. https://doi.org/10.1016/j.tcb.2020.01.008

Lukyanova LD, Kirova YI (2015) Mitochondria-controlled signaling mechanisms of brain protection in hypoxia. Front Neurosci 9. https://www.frontiersin.org/journals/neuroscience/articles/10.3389/fnins.2015.00320

Luque-Contreras D, Carvajal K, Toral-Rios D, Franco-Bocanegra D, Campos-Peña V (2014) Oxidative stress and metabolic syndrome: cause or consequence of Alzheimer's disease? Oxidative Med Cell Longev 2014:497802. https://doi.org/10.1155/2014/497802

Macherel D, Haraux F, Guillou H, Bourgeois O (2021) The conundrum of hot mitochondria. Biochimica et Biophysica Acta Bioenerg 1862(2):148348. https://doi.org/10.1016/j.bbabio.2020.148348

Machiela E, Southwell AL (n.d.) Biological aging and the cellular pathogenesis of Huntington's disease. J Huntington's Dis 9(2):115–128. https://doi.org/10.3233/JHD-200395

Mahato AK, Sidorova YA (2020) RET receptor tyrosine kinase: role in neurodegeneration, obesity, and cancer. Int J Mol Sci 21(19):7108. https://doi.org/10.3390/ijms21197108

Maklashina E, Rajagukguk S, McIntire W, Cecchini G (2010) Mutation of the heme axial ligand of escherichia coli succinate-quinone reductase: implications for heme ligation in mitochondrial complex II from yeast. Biochim Biophys Acta 1797:747–754. https://doi.org/10.1016/j.bbabio.2010.01.019

Mannella CA (2020) Consequences of folding the mitochondrial inner membrane. Front Physiol 11. https://www.frontiersin.org/journals/physiology/articles/10.3389/fphys.2020.00536

Markevich NI, Markevich LN (2023) Computational modeling analysis of kinetics of fumarate reductase activity and ROS production during reverse electron transfer in mitochondrial respiratory complex II. Int J Mol Sci 24(9):9. https://doi.org/10.3390/ijms24098291

Martin W, Wieler M, Hanstock C (2007) Is brain lactate increased in Huntington's disease? J Neurol Sci 263:70–74. https://doi.org/10.1016/j.jns.2007.05.035

Martin I, Dawson VL, Dawson TM (2011) Recent advances in the genetics of Parkinson's disease. Annu Rev Genomics Hum Genet 12:301–325. https://doi.org/10.1146/annurev-genom-082410-101440

Mary A, Eysert F, Checler F, Chami M (2023) Mitophagy in Alzheimer's disease: molecular defects and therapeutic approaches. Mol Psychiatry 28(1):1. https://doi.org/10.1038/s41380-022-01631-6

Masnata M, Salem S, de Rus Jacquet A, Anwer M, Cicchetti F (2020) Targeting tau to treat clinical features of Huntington's disease. Front Neurol 11:580732. https://doi.org/10.3389/fneur.2020.580732

Mastroeni D, Khdour OM, Delvaux E, Nolz J, Olsen G, Berchtold N, Cotman C, Hecht SM, Coleman PD (2017) Nuclear but not mitochondrial-encoded oxidative phosphorylation genes are altered in aging, mild cognitive impairment, and Alzheimer's disease. Alzheimer's Dement 13(5):510–519. https://doi.org/10.1016/j.jalz.2016.09.003

Matuz-Mares D, González-Andrade M, Araiza-Villanueva MG, Vilchis-Landeros MM, Vázquez-Meza H (2022) Mitochondrial calcium: effects of its imbalance in disease. Antioxidants 11(5):801. https://doi.org/10.3390/antiox11050801

McDonald TS, Lerskiatiphanich T, Woodruff TM, McCombe PA, Lee JD (2023) Potential mechanisms to modify impaired glucose metabolism in neurodegenerative disorders. J Cereb Blood Flow Metab 43(1):26–43. https://doi.org/10.1177/0271678X221135061

McKinlay JB, Cook GM, Hards K (2020) Chapter Four – microbial energy management – a product of three broad tradeoffs. In: Poole RK (ed) Advances in microbial physiology, vol 77. Academic Press, pp 139–185. https://doi.org/10.1016/bs.ampbs.2020.09.001

Melser S, Lavie J, Bénard G (2015) Mitochondrial degradation and energy metabolism. Biochim Biophys Acta 1853(10 Part B):2812–2821. https://doi.org/10.1016/j.bbamcr.2015.05.010

Mhyre TR, Boyd JT, Hamill RW, Maguire-Zeiss KA (2012) Parkinson's disease. Subcell Biochem 65:389–455. https://doi.org/10.1007/978-94-007-5416-4_16

Millichap LE, Damiani E, Tiano L, Hargreaves IP (2021) Targetable pathways for alleviating mitochondrial dysfunction in neurodegeneration of metabolic and non-metabolic diseases. Int J Mol Sci 22(21):11444. https://doi.org/10.3390/ijms222111444

Misrani A, Tabassum S, Yang L (2021) Mitochondrial dysfunction and oxidative stress in Alzheimer's disease. Front Aging Neurosci 13. https://www.frontiersin.org/articles/10.3389/fnagi.2021.617588

Mittal M, Siddiqui MR, Tran K, Reddy SP, Malik AB (2014) Reactive oxygen species in inflammation and tissue injury. Antioxid Redox Signal 20(7):1126–1167. https://doi.org/10.1089/ars.2012.5149

Mochel F, Haller RG (2011) Energy deficit in Huntington disease: why it matters. J Clin Invest 121(2):493–499. https://doi.org/10.1172/JCI45691

Modesti L, Danese A, Angela Maria Vitto V, Ramaccini D, Aguiari G, Gafà R, Lanza G, Giorgi C, Pinton P (2021) Mitochondrial Ca2+ signaling in health, disease and therapy. Cells 10(6):1317. https://doi.org/10.3390/cells10061317

Monti G, Gomes Moreira D, Richner M, Mutsaers HAM, Ferreira N, Jan A (2022) GLP-1 receptor agonists in neurodegeneration: neurovascular unit in the spotlight. Cells 11(13):2023. https://doi.org/10.3390/cells11132023

Moon HE, Paek SH (2015) Mitochondrial dysfunction in Parkinson's disease. Exp Neurobiol 24(2):103–116. https://doi.org/10.5607/en.2015.24.2.103

Morciano G, Boncompagni C, Ramaccini D, Pedriali G, Bouhamida E, Tremoli E, Giorgi C, Pinton P (2023) Comprehensive analysis of mitochondrial dynamics alterations in heart diseases. Int J Mol Sci 24(4):3414. https://doi.org/10.3390/ijms24043414

Mosconi L (2013) Glucose metabolism in normal aging and Alzheimer's disease: methodological and physiological considerations for PET studies. Clin Translat Imaging 1(4). https://doi.org/10.1007/s40336-013-0026-y

Muddapu VR, Dharshini SAP, Chakravarthy VS, Gromiha MM (2020) Neurodegenerative diseases – is metabolic deficiency the root cause? Front Neurosci 14:213. https://doi.org/10.3389/fnins.2020.00213

Murphy E, Liu JC (2022) Mitochondrial calcium and reactive oxygen species in cardiovascular disease. Cardiovasc Res 119(5):1105–1116. https://doi.org/10.1093/cvr/cvac134

Murphy E, Ardehali H, Balaban RS, DiLisa F, Dorn GW, Kitsis RN, Otsu K, Ping P, Rizzuto R, Sack MN, Wallace D, Youle RJ (2016) Mitochondrial function, biology, and role in disease. Circ Res 118(12):1960–1991. https://doi.org/10.1161/RES.0000000000000104

Napolitano G, Fasciolo G, Venditti P (2021) Mitochondrial management of reactive oxygen species. Antioxidants 10(11):1824. https://doi.org/10.3390/antiox10111824

Nayab DE, Din FU, Ali H, Kausar WA, Urooj S, Zafar M, Khan I, Shabbir K, Khan GM (2023) Nano biomaterials based strategies for enhanced brain targeting in the treatment of neurodegenerative diseases: an up-to-date perspective. J Nanobiotechnol 21(1):477. https://doi.org/10.1186/s12951-023-02250-1

Needs HI, Protasoni M, Henley JM, Prudent J, Collinson I, Pereira GC (2021) Interplay between mitochondrial protein import and respiratory complexes assembly in neuronal health and degeneration. Life (Basel) 11(5):5. https://doi.org/10.3390/life11050432

Nelson PT, Braak H, Markesbery WR (2009) Neuropathology and cognitive impairment in Alzheimer disease: a complex but coherent relationship. J Neuropathol Exp Neurol 68(1):1–14. https://doi.org/10.1097/NEN.0b013e3181919a48

Nicoletti V, Palermo G, Del Prete E, Mancuso M, Ceravolo R (2021) Understanding the multiple role of mitochondria in Parkinson's disease and related disorders: lesson from genetics and protein–interaction network. Front Cell Dev Biol 9. https://www.frontiersin.org/articles/10.3389/fcell.2021.636506

Norat P, Soldozy S, Sokolowski JD, Gorick CM, Kumar JS, Chae Y, Yağmurlu K, Prada F, Walker M, Levitt MR, Price RJ, Tvrdik P, Kalani MYS (2020) Mitochondrial dysfunction in neurological disorders: exploring mitochondrial transplantation. NPJ Regen Med 5(1):1. https://doi.org/10.1038/s41536-020-00107-x

Nuytemans K, Theuns J, Cruts M, Van Broeckhoven C (2010) Genetic etiology of parkinson disease associated with mutations in the SNCA, PARK2, PINK1, PARK7, and LRRK2 genes: a mutation update. Hum Mutat 31(7):763–780. https://doi.org/10.1002/humu.21277

Olesen MA, Villavicencio-Tejo F, Quintanilla RA (2022) The use of fibroblasts as a valuable strategy for studying mitochondrial impairment in neurological disorders. Transl Neurodegener 11(1):36. https://doi.org/10.1186/s40035-022-00308-y

Oliver D, Reddy PH (2019) Dynamics of dynamin-related protein 1 in Alzheimer's disease and other neurodegenerative diseases. Cells 8(9):961. https://doi.org/10.3390/cells8090961

Onyango IG, Dennis J, Khan SM (2016) Mitochondrial dysfunction in Alzheimer's disease and the rationale for bioenergetics based therapies. Aging Dis 7(2):201–214. https://doi.org/10.14336/AD.2015.1007

Osellame LD, Blacker TS, Duchen MR (2012) Cellular and molecular mechanisms of mitochondrial function. Best Pract Res Clin Endocrinol Metab 26(6):711–723. https://doi.org/10.1016/j.beem.2012.05.003

Pagano G, Niccolini F, Politis M (2016) Current status of PET imaging in Huntington's disease. Eur J Nucl Med Mol Imaging 43:1171–1182. https://doi.org/10.1007/s00259-016-3324-6

Pallotti F, Bergamini C, Lamperti C, Fato R (2021) The roles of coenzyme Q in disease: direct and indirect involvement in cellular functions. Int J Mol Sci 23(1):128. https://doi.org/10.3390/ijms23010128

Pandey S, Srivanitchapoom P (2017) Levodopa-induced dyskinesia: clinical features, pathophysiology, and medical management. Ann Indian Acad Neurol 20(3):190–198. https://doi.org/10.4103/aian.AIAN_239_17

Papa S, Martino PL, Capitanio G, Gaballo A, De Rasmo D, Signorile A, Petruzzella V (2012) The oxidative phosphorylation system in mammalian mitochondria. In: Scatena R, Bottoni P, Giardina B (eds) Advances in mitochondrial medicine. Springer, Dordrecht, pp 3–37. https://doi.org/10.1007/978-94-007-2869-1_1

Park J-S, Davis RL, Sue CM (2018) Mitochondrial dysfunction in Parkinson's disease: new mechanistic insights and therapeutic perspectives. Curr Neurol Neurosci Rep 18(5):21. https://doi.org/10.1007/s11910-018-0829-3

Pauwels EKJ, Coumou AW, Kostkiewicz M, Kairemo K (2013) [18F]Fluoro-2-Deoxy-D-Glucose positron emission tomography/computed tomography imaging in oncology: initial staging and evaluation of cancer therapy. Med Princ Pract 22(5):427–437. https://doi.org/10.1159/000346303

Penke B, Szűcs M, Bogár F (2023) New pathways identify novel drug targets for the prevention and treatment of Alzheimer's disease. Int J Mol Sci 24(6):6. https://doi.org/10.3390/ijms24065383

Perez Ortiz JM, Swerdlow RH (2019) Mitochondrial dysfunction in Alzheimer's disease: role in pathogenesis and novel therapeutic opportunities. Br J Pharmacol 176(18):3489–3507. https://doi.org/10.1111/bph.14585

Pérez MJ, Jara C, Quintanilla RA (2018) Contribution of tau pathology to mitochondrial impairment in neurodegeneration. Front Neurosci 12. https://doi.org/10.3389/fnins.2018.00441

Picard M, McEwen BS, Epel ES, Sandi C (2018) An energetic view of stress: Focus on mitochondria. Front Neuroendocrinol 49:72–85. https://doi.org/10.1016/j.yfrne.2018.01.001

Pizzino G, Irrera N, Cucinotta M, Pallio G, Mannino F, Arcoraci V, Squadrito F, Altavilla D, Bitto A (2017) Oxidative stress: harms and benefits for human health. Oxidative Med Cell Longev 2017:8416763. https://doi.org/10.1155/2017/8416763

Pollard PJ, Brière JJ, Alam NA, Barwell J, Barclay E, Wortham NC, Hunt T, Mitchell M, Olpin S, Moat SJ, Hargreaves IP, Heales SJ, Chung YL, Griffiths JR, Dalgleish A, McGrath JA, Gleeson MJ, Hodgson SV, Poulsom R et al (2005) Accumulation of Krebs cycle intermediates and overexpression of HIF1α in tumours which result from germline FH and SDH mutations. Hum Mol Genet 14(15):2231–2239. https://doi.org/10.1093/hmg/ddi227

Porstmann T, Santos CR, Griffiths B, Cully M, Wu M, Leevers S, Griffiths JR, Chung Y-L, Schulze A (2008) SREBP activity is regulated by mTORC1 and contributes to Akt-dependent cell growth. Cell Metab 8(3–3):224–236. https://doi.org/10.1016/j.cmet.2008.07.007

Quiles JM, Gustafsson ÅB (2020) Mitochondrial quality control and cellular proteostasis: two sides of the same coin. Front Physiol 11:515. https://doi.org/10.3389/fphys.2020.00515

Quintanilla RA, Matthews-Roberson TA, Dolan PJ, Johnson GVW (2009) Caspase-cleaved tau expression induces mitochondrial dysfunction in immortalized cortical neurons: implications for the pathogenesis of Alzheimer disease*. J Biol Chem 284(28):18754–18766. https://doi.org/10.1074/jbc.M808908200

Quntanilla RA, Tapia-Monsalves C (2020) The role of mitochondrial impairment in Alzheimer´s Disease neurodegeneration: the tau connection. Curr Neuropharmacol 18(11):1076–1091. https://doi.org/10.2174/1570159X18666200525020259

Raimundo N, Baysal BE, Shadel GS (2011) Revisiting the TCA cycle: signaling to tumor formation. Trends Mol Med 17(11):641–649. https://doi.org/10.1016/j.molmed.2011.06.001

Ralhan I, Chang C-L, Lippincott-Schwartz J, Ioannou MS (2021) Lipid droplets in the nervous system. J Cell Biol 220(7):e202102136. https://doi.org/10.1083/jcb.202102136

Ramesh S, Arachchige ASPM (2023) Depletion of dopamine in Parkinson's disease and relevant therapeutic options: a review of the literature. AIMS Neurosci 10(3):200–231. https://doi.org/10.3934/Neuroscience.2023017

Rangaraju V, Lewis TL, Hirabayashi Y, Bergami M, Motori E, Cartoni R, Kwon S-K, Courchet J (2019) Pleiotropic mitochondria: the influence of mitochondria on neuronal development and disease. J Neurosci 39(42):8200–8208. https://doi.org/10.1523/JNEUROSCI.1157-19.2019

Rao VK, Carlson EA, Yan SS (2014) Mitochondrial permeability transition pore is a potential drug target for neurodegeneration. Biochim Biophys Acta 1842(8):1267–1272. https://doi.org/10.1016/j.bbadis.2013.09.003

Read AD, Bentley RET, Archer SL, Dunham-Snary KJ (2021) Mitochondrial iron–sulfur clusters: structure, function, and an emerging role in vascular biology. Redox Biol 47:102164. https://doi.org/10.1016/j.redox.2021.102164
Reddy PH, Shirendeb UP (2012) Mutant huntingtin, abnormal mitochondrial dynamics, defective axonal transport of mitochondria, and selective synaptic degeneration in Huntington's disease. Biochim Biophys Acta 1822(2):101–110. https://doi.org/10.1016/j.bbadis.2011.10.016
Rehman MU, Sehar N, Dar NJ, Khan A, Arafah A, Rashid S, Rashid SM, Ganaie MA (2023) Mitochondrial dysfunctions, oxidative stress and neuroinflammation as therapeutic targets for neurodegenerative diseases: an update on current advances and impediments. Neurosci Biobehav Rev 144:104961. https://doi.org/10.1016/j.neubiorev.2022.104961
Ren L, Chen X, Chen X, Li J, Cheng B, Xia J (2020) Mitochondrial dynamics: fission and fusion in fate determination of mesenchymal stem cells. Front Cell Dev Biol 8:580070. https://doi.org/10.3389/fcell.2020.580070
Ronowska A, Szutowicz A, Bielarczyk H, Gul-Hinc S, Klimaszewska-Łata J, Dyś A, Zyśk M, Jankowska-Kulawy A (2018) The regulatory effects of Acetyl-CoA distribution in the healthy and diseased brain. Front Cell Neurosci 12:169. https://doi.org/10.3389/fncel.2018.00169
Rossi A, Pizzo P, Filadi R (2019) Calcium, mitochondria and cell metabolism: a functional triangle in bioenergetics. Biochim Biophys Acta Mol Cell Res 1866(7):1068–1078. https://doi.org/10.1016/j.bbamcr.2018.10.016
Rossmann MP, Dubois SM, Agarwal S, Zon LI (2021) Mitochondrial function in development and disease. Dis Model Mech 14(6):dmm048912. https://doi.org/10.1242/dmm.048912
Roux PP, Topisirovic I (2018) Signaling pathways involved in the regulation of mRNA translation. Mol Cell Biol 38(12):e00070–e00018. https://doi.org/10.1128/MCB.00070-18
Roy RG, Mandal PK, Maroon JC (2023) Oxidative stress occurs prior to amyloid Aβ plaque formation and tau phosphorylation in Alzheimer's disease: role of glutathione and metal ions. ACS Chem Neurosci 14(17):2944–2954. https://doi.org/10.1021/acschemneuro.3c00486
Schenkel LC, Bakovic M (2014) Formation and regulation of mitochondrial membranes. Int J Cell Biol 2014:709828. https://doi.org/10.1155/2014/709828
Schieber M, Chandel NS (2014) ROS function in redox signaling and oxidative stress. Curr Biol 24(10):R453–R462. https://doi.org/10.1016/j.cub.2014.03.034
Schulte J, Littleton JT (2011) The biological function of the Huntingtin protein and its relevance to Huntington's disease pathology. Curr Trends Neurol 5:65–78
Scott I, Youle RJ (2010) Mitochondrial fission and fusion. Essays Biochem 47:85–98. https://doi.org/10.1042/bse0470085
Shankar GM, Walsh DM (2009) Alzheimer's disease: synaptic dysfunction and Aβ. Mol Neurodegener 4(1):48. https://doi.org/10.1186/1750-1326-4-48
Sharifi-Rad M, Anil Kumar NV, Zucca P, Varoni EM, Dini L, Panzarini E, Rajkovic J, Tsouh Fokou PV, Azzini E, Peluso I, Prakash Mishra A, Nigam M, El Rayess Y, Beyrouthy ME, Polito L, Iriti M, Martins N, Martorell M, Docea AO et al (2020) Lifestyle, oxidative stress, and antioxidants: back and forth in the pathophysiology of chronic diseases. Front Physiol 11. https://www.frontiersin.org/articles/10.3389/fphys.2020.00694
Sharma P, Jha AB, Dubey RS, Pessarakli M (2012) Reactive oxygen species, oxidative damage, and antioxidative defense mechanism in plants under stressful conditions. J Bot 2012:e217037. https://doi.org/10.1155/2012/217037
Sharma C, Kim S, Nam Y, Jung UJ, Kim SR (2021) Mitochondrial dysfunction as a driver of cognitive impairment in Alzheimer's Disease. Int J Mol Sci 22(9):9. https://doi.org/10.3390/ijms22094850
Shen Y, Jiang W-L, Li X, Cao A-L, Li D, Li S-Z, Yang J, Qian J (2023) Mitochondrial dynamics in neurological diseases: a narrative review. Ann Transl Med 11(6):264. https://doi.org/10.21037/atm-22-2401
Shi Q, Gibson GE (2011) Up-regulation of the mitochondrial malate dehydrogenase by oxidative stress is mediated by miR-743a. J Neurochem 118(3):440–448. https://doi.org/10.1111/j.1471-4159.2011.07333.x

Shibuya M (2011) Vascular endothelial growth factor (VEGF) and its receptor (VEGFR) signaling in angiogenesis. Genes Cancer 2(12):1097–1105. https://doi.org/10.1177/1947601911423031

Simmons EC, Scholpa NE, Schnellmann RG (2020) Mitochondrial biogenesis as a therapeutic target for traumatic and neurodegenerative CNS diseases. Exp Neurol 329:113309. https://doi.org/10.1016/j.expneurol.2020.113309

Smith KM, Matson S, Matson WR, Cormier K, Del Signore SJ, Hagerty SW, Stack EC, Ryu H, Ferrante RJ (2006) Dose ranging and efficacy study of high-dose coenzyme Q10 formulations in Huntington's disease mice. Biochim Biophys Acta 1762(6):616–626. https://doi.org/10.1016/j.bbadis.2006.03.004

Soto C, Satani N (2011) The intricate mechanisms of neurodegeneration in prion diseases. Trends Mol Med 17(1):14–24. https://doi.org/10.1016/j.molmed.2010.09.001

Stillwell W (2016) Chapter 18 – membrane-associated processes. In: Stillwell W (ed) An introduction to biological membranes, 2nd edn. Elsevier, pp 381–421. https://doi.org/10.1016/B978-0-444-63772-7.00018-X

Su B, Wang X, Bonda D, Perry G, Smith M, Zhu X (2010) Abnormal mitochondrial dynamics – a novel therapeutic target for Alzheimer's disease? Mol Neurobiol 41(2–3):87–96. https://doi.org/10.1007/s12035-009-8095-7

Sugden MC, Holness MJ (2011) The pyruvate carboxylase-pyruvate dehydrogenase axis in islet pyruvate metabolism. Islets 3(6):302–319. https://doi.org/10.4161/isl.3.6.17806

Sun F, Zhou Q, Pang X, Xu Y, Rao Z (2013) Revealing various coupling of electron transfer and proton pumping in mitochondrial respiratory chain. Curr Opin Struct Biol 23. https://doi.org/10.1016/j.sbi.2013.06.013

Swerdlow RH (2020) The mitochondrial hypothesis: dysfunction, bioenergetic defects, and the metabolic link to Alzheimer's disease. Int Rev Neurobiol 154:207–233. https://doi.org/10.1016/bs.irn.2020.01.008

Tondo M, Marin-Valencia I, Ma Q, Pascual JM (2015) Chapter 25 – pyruvate dehydrogenase, pyruvate carboxylase, Krebs cycle and mitochondrial transport disorders. In: Rosenberg RN, Pascual JM (eds) Rosenberg's molecular and genetic basis of neurological and psychiatric disease, 5th edn. Academic Press, pp 291–297. https://doi.org/10.1016/B978-0-12-410529-4.00025-5

Tönnies E, Trushina E (n.d.) Oxidative stress, synaptic dysfunction, and Alzheimer's disease. J Alzheimers Dis 57(4):1105–1121. https://doi.org/10.3233/JAD-161088

Trancikova A, Tsika E, Moore DJ (2012) Mitochondrial Dysfunction in Genetic Animal Models of Parkinson's Disease. Antioxid Redox Signal 16(9):896–919. https://doi.org/10.1089/ars.2011.4200

Triarhou LC (2013) Dopamine and parkinson's disease. In: Madame Curie bioscience database [Internet]. Landes Bioscience. https://www.ncbi.nlm.nih.gov/books/NBK6271/

Trimmer PA, Bennett JP (2009) The cybrid model of sporadic Parkinson's disease. Exp Neurol 218(2):320–325. https://doi.org/10.1016/j.expneurol.2009.03.016

Trimmer PA, Keeney PM, Borland MK, Simon FA, Almeida J, Swerdlow RH, Parks JP, Parker WD, Bennett JP (2004) Mitochondrial abnormalities in cybrid cell models of sporadic Alzheimer's disease worsen with passage in culture. Neurobiol Dis 15(1):29–39. https://doi.org/10.1016/j.nbd.2003.09.011

Troutwine BR, Strope TA, Franczak E, Lysaker CR, Hamid L, Mansel C, Stopperan JA, Gouvion CM, Haeri M, Swerdlow RH, Wilkins HM (2022) Mitochondrial function and Aβ in Alzheimer's disease postmortem brain. Neurobiol Dis 171:105781. https://doi.org/10.1016/j.nbd.2022.105781

Túnez I, Santamaría A (2009) Model of Huntington's disease induced with 3-nitropropionic acid. Rev Neurol 48(8):430–434

Túnez I, Tasset I, Pérez-De La Cruz V, Santamaría A (2010) 3-Nitropropionic acid as a tool to study the mechanisms involved in Huntington's disease: past, present and future. Molecules 15(2):878–916. https://doi.org/10.3390/molecules15020878

Uddin TM, Chakraborty AJ, Khusro A, Zidan BRM, Mitra S, Emran TB, Dhama K, Ripon MKH, Gajdács M, Sahibzada MUK, Hossain MJ, Koirala N (2021) Antibiotic resistance in microbes:

History, mechanisms, therapeutic strategies and future prospects. J Infect Public Health 14(12):1750–1766. https://doi.org/10.1016/j.jiph.2021.10.020

Vamecq J, Dessein A-F, Fontaine M, Briand G, Porchet N, Latruffe N, Andreolotti P, Cherkaoui-Malki M (2012) Mitochondrial dysfunction and lipid homeostasis. Curr Drug Metab 13(10):1388–1400. https://doi.org/10.2174/138920012803762792

Van Coster R, Seneca S, Smet J, Van Hecke R, Gerlo E, Devreese B, Van Beeumen J, Leroy JG, De Meirleir L, Lissens W (2003) Homozygous Gly555Glu mutation in the nuclear-encoded 70 kDa flavoprotein gene causes instability of the respiratory chain complex II. Am J Med Genet A 120A(1):13–18. https://doi.org/10.1002/ajmg.a.10202

Van Hove JLK, Saenz MS, Thomas JA, Gallagher R, Lovell MA, Fenton LZ, Shanske S, Myers SM, Wanders RJA, Ruiter J, Turkenburg M, Waterham HR (2010) Succinyl-CoA ligase deficiency: a mitochondrial hepatoencephalomyopathy. Pediatr Res 68(2):159–164. https://doi.org/10.1203/PDR.0b013e3181e5c3a4

Varela-López A, Giampieri F, Battino M, Quiles JL (2016) Coenzyme Q and its role in the dietary therapy against aging. Molecules 21(3):373. https://doi.org/10.3390/molecules21030373

Veech RL (2013) Tricarboxylic acid cycle. In: Lennarz WJ, Lane MD (eds) Encyclopedia of biological chemistry, 2nd edn. Academic Press, pp 436–440. https://doi.org/10.1016/B978-0-12-378630-2.00608-3

Verma M, Lizama BN, Chu CT (2022) Excitotoxicity, calcium and mitochondria: a triad in synaptic neurodegeneration. Transl Neurodegener 11:3. https://doi.org/10.1186/s40035-021-00278-7

Vilas-Boas EA, Cabral-Costa JV, Ramos VM, da Silva CCC, Kowaltowski AJ (2023) Goldilocks calcium concentrations and the regulation of oxidative phosphorylation: too much, too little, or just right. J Biol Chem 299(3). https://doi.org/10.1016/j.jbc.2023.102904

Walters GC, Usachev YM (2023) Mitochondrial calcium cycling in neuronal function and neurodegeneration. Front Cell Dev Biol 11:1094356. https://doi.org/10.3389/fcell.2023.1094356

Wang X, Wang W, Li L, Perry G, Lee H, Zhu X (2014) Oxidative stress and mitochondrial dysfunction in Alzheimer's disease. Biochim Biophys Acta 1842(8):1240–1247. https://doi.org/10.1016/j.bbadis.2013.10.015

Wang Q, Liu Y, Zhou J (2015) Neuroinflammation in Parkinson's disease and its potential as therapeutic target. Transl Neurodegener 4:19. https://doi.org/10.1186/s40035-015-0042-0

Wang H, Naghavi M, Allen C, Barber RM, Bhutta ZA, Carter A, Casey DC, Charlson FJ, Chen AZ, Coates MM, Coggeshall M, Dandona L, Dicker DJ, Erskine HE, Ferrari AJ, Fitzmaurice C, Foreman K, Forouzanfar MH, Fraser MS et al (2016) Global, regional, and national life expectancy, all-cause mortality, and cause-specific mortality for 249 causes of death, 1980–2015: a systematic analysis for the Global Burden of Disease Study 2015. Lancet 388(10053):1459–1544. https://doi.org/10.1016/S0140-6736(16)31012-1

Wang W, Zhao F, Ma X, Perry G, Zhu X (2020) Mitochondria dysfunction in the pathogenesis of Alzheimer's disease: recent advances. Mol Neurodegener 15:30. https://doi.org/10.1186/s13024-020-00376-6

Wang B, Huang M, Shang D, Yan X, Zhao B, Zhang X (2021) Mitochondrial behavior in axon degeneration and regeneration. Front Aging Neurosci 13. https://www.frontiersin.org/articles/10.3389/fnagi.2021.650038

Wang F, Xie X, Xing X, Sun X (2022a) Excitatory synaptic transmission in ischemic stroke: a new outlet for classical neuroprotective strategies. Int J Mol Sci 23(16):9381. https://doi.org/10.3390/ijms23169381

Wang Y, Hong F, Yang S (2022b) Roles of nitric oxide in brain ischemia and reperfusion. Int J Mol Sci 23(8):4243. https://doi.org/10.3390/ijms23084243

Wang Q, Li M, Zeng N, Zhou Y, Yan J (2023) Succinate dehydrogenase complex subunit C: Role in cellular physiology and disease. Exp Biol Med 248(3):263–270. https://doi.org/10.1177/15353702221147567

Wareham LK, Liddelow SA, Temple S, Benowitz LI, Di Polo A, Wellington C, Goldberg JL, He Z, Duan X, Bu G, Davis AA, Shekhar K, Torre AL, Chan DC, Canto-Soler MV, Flanagan JG, Subramanian P, Rossi S, Brunner T et al (2022) Solving neurodegeneration: common mechanisms and strategies for new treatments. Mol Neurodegener 17(1):23. https://doi.org/10.1186/s13024-022-00524-0

Williams GSB, Boyman L, Lederer WJ (2015) Mitochondrial calcium and the regulation of metabolism in the heart. J Mol Cell Cardiol 78:35–45. https://doi.org/10.1016/j.yjmcc.2014.10.019

Wilson DM, Cookson MR, Van Den Bosch L, Zetterberg H, Holtzman DM, Dewachter I (2023) Hallmarks of neurodegenerative diseases. Cell 186(4):693–714. https://doi.org/10.1016/j.cell.2022.12.032

Winklhofer KF, Haass C (2010) Mitochondrial dysfunction in Parkinson's disease. Biochim Biophys Acta 1802(1):29–44. https://doi.org/10.1016/j.bbadis.2009.08.013

Xiang M, Lu Y, Xin L, Gao J, Shang C, Jiang Z, Lin H, Fang X, Qu Y, Wang Y, Shen Z, Zhao M, Cui X (2021) Role of oxidative stress in reperfusion following myocardial ischemia and its treatments. Oxidative Med Cell Longev 2021:6614009. https://doi.org/10.1155/2021/6614009

Yan X, Hu Y, Wang B, Wang S, Zhang X (2020) Metabolic DYSREGULATION COntributes to the progression of Alzheimer's disease. Front Neurosci 14:530219. https://doi.org/10.3389/fnins.2020.530219

Yang L, Calingasan NY, Wille EJ, Cormier K, Smith K, Ferrante RJ, Beal MF (2009) Combination therapy with coenzyme Q10 and creatine produces additive neuroprotective effects in models of Parkinson's and Huntington's diseases. J Neurochem 109(5):1427–1439. https://doi.org/10.1111/j.1471-4159.2009.06074.x

Yang D, Wang X, Zhang L, Fang Y, Zheng Q, Liu X, Yu W, Chen S, Ying J, Hua F (2022) Lipid metabolism and storage in neuroglia: role in brain development and neurodegenerative diseases. Cell Biosci 12:106. https://doi.org/10.1186/s13578-022-00828-0

Yang M, Ali O, Bjørås M, Wang J (2023) Identifying functional regulatory mutation blocks by integrating genome sequencing and transcriptome data. iScience 26(8):107266. https://doi.org/10.1016/j.isci.2023.107266

Zadoorian A, Du X, Yang H (2023) Lipid droplet biogenesis and functions in health and disease. Nat Rev Endocrinol 19(8):8. https://doi.org/10.1038/s41574-023-00845-0

Zahoor I, Shafi A, Haq E (2018) Pharmacological treatment of Parkinson's disease. In: Stoker TB, Greenland JC (eds) Parkinson's disease: pathogenesis and clinical aspects. Codon. http://www.ncbi.nlm.nih.gov/books/NBK536726/

Zhang X, Alshakhshir N, Zhao L (2021) Glycolytic METABOLISM, BRAIN Resilience, and Alzheimer's disease. Front Neurosci 15:662242. https://doi.org/10.3389/fnins.2021.662242

Zhang D, Wang F, Li P, Gao Y (2022) Mitochondrial Ca2+ homeostasis: emerging roles and clinical significance in cardiac remodeling. Int J Mol Sci 23(6):3025. https://doi.org/10.3390/ijms23063025

Zhao X, Bhattacharyya A (2018) Human models are needed for studying human neurodevelopmental disorders. Am J Hum Genet 103(6):829–857. https://doi.org/10.1016/j.ajhg.2018.10.009

Zhao X-Y, Lu M-H, Yuan D-J, Xu D-E, Yao P-P, Ji W-L, Chen H, Liu W-L, Yan C-X, Xia Y-Y, Li S, Tao J, Ma Q-H (2019) Mitochondrial dysfunction in neural injury. Front Neurosci 13:30. https://doi.org/10.3389/fnins.2019.00030

Zhu Z, Zhang J, Li K, Zhao S (2005) Cloning, mapping and association study with carcass traits of the porcine SDHD gene. Anim Genet 36:191–195. https://doi.org/10.1111/j.1365-2052.2005.01270.x

Zorova LD, Popkov VA, Plotnikov EY, Silachev DN, Pevzner IB, Jankauskas SS, Babenko VA, Zorov SD, Balakireva AV, Juhaszova M, Sollott SJ, Zorov DB (2018) Mitochondrial membrane potential. Anal Biochem 552:50–59. https://doi.org/10.1016/j.ab.2017.07.009

Fungicidal SDH Inhibitors in Agriculture: Regulation and Future Perspectives

10

Pragya Gawande, Souranava Jana, Akshay Kumar, Jaishree Biswas, Diksha Bharti, Surbhi Jyoti, Anjali Murmu, Balaji Wamanrao Matore, Jagadish Singh, and Partha Pratim Roy

Abstract

Fungal diseases pose significant threats to agriculture, impacting crop quality, productivity, and global food security. Succinate dehydrogenase Inhibitors (SDHIs) have played a pivotal role in the agrochemical industry due to their potent fungicidal activity, with widespread use for over 60 years. According to the Fungicide Resistance Action Committee (FRAC), the first SDHI was reported in 1966, and to date, 17 inhibitors have been approved and commercialized. This chapter provides a comprehensive overview of SDHIs, focusing on their classification, case studies, mechanism of action, role in disease control, structure–activity relationships (SAR), resistance mechanisms, synthetic approaches, and future challenges. It begins by defining and describing the structure of SDHIs and their critical roles in various metabolic pathways. It highlights case studies of fungal attacks on crops and their adverse effects on global crop yields. Fungal pathogens affect different plant parts—including roots, leaves, and fruits—leading to significant reductions in productivity. Novel SDHIs, particularly those incorporating carboxamide, nicotinamide, benzothiazole, and pyrazine-carboxamide-diphenyl-ether scaffolds, are essential in enhancing crop yields. However, resistance development due to genetic mutations in fungi and the environmental and health hazards associated with prolonged SDHI use remain sig-

P. Gawande · A. Kumar · J. Biswas · D. Bharti · S. Jyoti · A. Murmu · B. W. Matore · J. Singh · P. P. Roy (✉)
Laboratory of Drug Discovery and Ecotoxicology, Department of Pharmacy, Guru Ghasidas Vishwavidyalaya (A Central University), Bilaspur, Chhattisgarh, India
e-mail: parthapratim.roy@ggu.ac.in

S. Jana
Laboratory of Drug Discovery and Ecotoxicology, Department of Pharmacy, Guru Ghasidas Vishwavidyalaya (A Central University), Bilaspur, Chhattisgarh, India

Haldia Institute of Pharmacy, Haldia, West Bengal, India

V. M. Patil et al. (eds.), *Deciphering The Role of Succinate Dehydrogenase in Drug Discovery*, https://doi.org/10.1007/978-981-96-6831-1_10

nificant challenges. Recent research emphasizes sustainable approaches to overcome these issues, aiming to improve the future application of SDHIs in agriculture.

Keywords

Succinate dehydrogenase · Succinate dehydrogenase inhibitors · Kreb cycle · Agrochemical · SAR · Fungicides · Toxicity and resistance

Abbreviations

ADP	Adenosine diphosphate
ATP	Adenosine triphosphate
DMI	Demethylation inhibitor
DNA	Deoxyribonucleic acid
FAD	Flavin adenine dinucleotide
FRAC	Fungicide Resistance Action Committee
GISTs	Gastrointestinal stromal tumors
IPM	Integrated pest management
MDR	Multidrug-resistant
MRC	Mitochondrial respiratory chain
NEt_3	Triethylamine
PGL	Paragangliomas
QH2	Ubiquinol
QoIs	Quinone-outside inhibitors
Qp	Ubiquinone-binding
ROS	Reactive oxygen species
SDH	Succinate dehydrogenase
SDHI	Succinate dehydrogenase inhibitors
SNP_S	Single-nucleotide polymorphisms
TBTU	N,N,N′,N′-Tetramethyluroniumtetrafluoroborate
TCA	Tricarboxylic acid
UQ	Ubiquinone
US FDA	United States Food and Drug Administration

10.1 Introduction

The control of fungal diseases is a crucial problem in the ever-changing field of modern agriculture, one that has an immediate effect on crop quality, yield, and overall food security (Avenot and Michailides 2010; Jain et al. 2019). With more than 99,000 distinct species, fungi encompass a diverse range, including polypore, molds, yeasts, and mushrooms (Barnes 2012). These adaptable organisms thrive on various organic materials, such as food, and can affect diverse species, including humans, animals, and plants (Naranjo-Ortiz and Gabaldón 2019). Flourishing in nearly every habitat—from rivers and seas to soil and air—fungi wield a dual nature, causing diseases and leaving a significant imprint on social, agricultural, and industrial realms. Their pervasive influence extends to agriculture, food production, drug manufacturing, and the degradation of organic matter (Duarte Hospital et al. 2023). Notably, fungi are the main cause of plant diseases; over 19,000 species have been identified globally as responsible for diseases in agricultural plants, which has led to significant crop losses and compromised agricultural commodities (Jain et al. 2019).

These microscopic organisms lack chlorophyll and manifest as threadlike filaments called hyphae, which form mycelium-resting structures including rhizomorphs and sclerotia (Jain et al. 2019). Timely detection and identification of fungal infections are vital, given their detrimental effects on crops by depleting nutrients, disrupting growth metabolism, and inhibiting growth or causing death (Peng et al. 2021). Fungi spread through spores, generating identifiable fruiting bodies known as pathogen indications. Leaves are particularly susceptible, and fungal spread can affect the entire plant, leading to issues like seed rot, seedling blights, and root concerns (Omotayo and Babalola 2023). Swift symptom identification is imperative for effective management. Varied rapid detection methods exist to control phytopathogenic fungi, safeguarding food, and agricultural outputs. Additionally, some fungi pose a dual threat by producing toxins in hosts, further jeopardizing food safety (El-Baky and Amara 2021).

In contemporary times, a primary approach to addressing phytopathogenic fungi involves the application of fungicides. Despite their widespread use, this extensive application has resulted in resistance development and environmental pollution caused by certain harmful fungi (Kilaru et al. 2015). Consequently, managing fungal diseases in plants require development of innovative and effective fungicides (Elkhairy et al. 2023).

Historically, since the 1940s, the control of phytopathogenic fungi has primarily relied on the use of chemical and synthetic fungicides (El-Baky and Amara 2021; Amara and El-Baky 2023). To manage plant diseases effectively, various approaches have been employed, encompassing regulatory, genetic, cultural, biological, and chemical controls (Pandit et al. 2022). Integrated pest management (IPM) plays a crucial role by strategically selecting, integrating, and utilizing pest management techniques, considering predicted economic, aesthetic, sociological, and ecological consequences (Chandler et al. 2011). However, the escalating emergence of resistant strains among fungal phytopathogens intensifies the challenges associated with treating plant fungal diseases (Hahn 2014). The Fungicide Resistance Action Committee (FRAC) has categorized 55 types of fungicides, with a particular emphasis on two categories—quinone-outside inhibitors (QoIs) and sterol biosynthesis inhibitors—which are especially relevant to disease control in commercial agriculture (Kiiker et al. 2021).

Over the years, the agricultural industry has witnessed significant advancements in fungicidal technologies, with succinate dehydrogenase inhibitors (SDHIs) emerging as a prominent class of fungicides (Yankovskaya et al. 2003; Avenot and Michailides 2010; Chai et al. 2023). Succinate dehydrogenase (SDH), a key enzyme in the cellular respiration pathway of fungi, has become a focal point in SDHI design (Avenot and Michailides 2010). By inhibiting this enzyme, these fungicides disrupt the energy production processes crucial for fungal growth, providing a targeted and efficient means of disease control (Yanicostas and Soussi-yanicostas 2021).

The term "SDHIs" refers to a group of newly introduced or announced fungicides that inhibit the SDH enzyme (Bénit et al. 2022). The current edition of FRAC (17) contains a list of 17 SDHIs compounds that fall under different chemical classes: benzovindiflupyr (Ishii et al. 2016), bixafen (D'Hose et al. 2021), boscalid (Vielba-Fernández et al. 2021), carbosin (Yanicostas and Soussi-Yanicostas 2021), fenfuram (Yanicostas and Soussi-Yanicostas 2021), fluopyram (Vielba-Fernández et al. 2021), flutolanil (Yanicostas and Soussi-Yanicostas 2021; Duarte Hospital et al. 2023), fluxapyroxad (Mao et al. 2024), furametpyr (Liu et al. 2022b), isopyrazam (Guo et al. 2024), mepronil (Yanicostas and Soussi-Yanicostas 2021), oxycarboxin (Avenot and Michailides 2010) (Sang and Lee 2020), penflufen (Sun et al. 2021), penthiopyrad (Fernández-Ortuño et al. 2017), sedaxane (Walter et al. 2015), and thifluzamide (Masiello et al. 2019). Except for oomycete activity, which is still absent, the "overall" spectrum of SDHI fungicides is currently very broad and comparable to the QoI spectrum. Some United States Food and Drug Administration (US FDA)-approved drugs also share chemical structures with SDH inhibitors (Fig. 10.1) (Bezawork-Geleta et al. 2016).

As we carry out more research, the real-world applications of SDHIs in agriculture become more apparent. The versatility of these inhibitors holds promise for a resilient and productive agricultural future, from sustaining farming practices to shielding staple crops from devastating fungal infections.

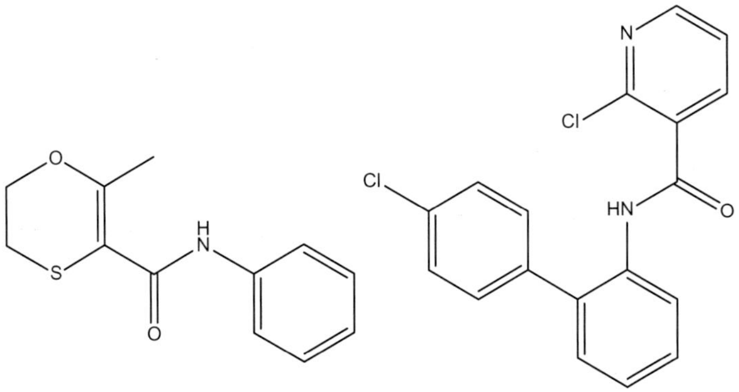

Carboxin Boscalid

Fig. 10.1 Chemical structures of US FDA-approved drugs for SDH inhibitors

10.2 Succinate Dehydrogenase (SDH)

The enzyme "SDH" was originally discovered by Sir *Hans Adolf Krebs* and *Fritz Albert Lipmann* in the 1930s. The citric acid cycle, sometimes referred to as the Krebs cycle, is an essential component of cellular respiration (Wilson et al. 2010). A key component of SDH is the investigation of cellular metabolism and the identification of significant biochemical pathways (Rutter et al. 2010).

SDH is part of complex II of the mitochondrial respiratory chain (MRC) and consists of four subunits: SDHA, SDHB, SDHC, and SDHD. It is susceptible to inhibition by SDH fungicides (Avenot and Michailides 2010). Situated at the convergence of two crucial mitochondrial pathways—the Krebs cycle and the mitochondrial respiratory chain (MRC)—SDH plays a pivotal role in catalyzing the conversion of succinate to fumarate and the reduction of ubiquinone (coenzyme Q), respectively. These functions are essential for adenosine triphosphate (ATP) synthesis and metabolic equilibrium, interlinked by electron transfers between processes (Zorov et al. 2014; Bénit et al. 2022). Consequently, inhibiting the SDH complex disrupts electron transfer, halts ATP synthesis, interrupts the Krebs cycle, and may elevate succinate levels, the substrate for SDH. Unlike other MRC complexes (I, III, and IV), whose inactivation primarily affects ATP synthesis and redox state, SDH's distinctive role underscores the significance of SDHI fungicides as mitotoxic herbicides capable of influencing various mitochondrial activities (Tang et al. 2022; Bouillaud 2023). According to Sierotzki and Scalliet (2013), SDHI fungicides bind tightly to the ubiquinone-binding site, formed by the SDHB, SDHC, and SDHD subunits, thereby physically obstructing ubiquinone access and preventing the catalysis of succinate (Sierotzki and Scalliet 2013; Bouillaud 2023). As such, SDHI fungicides are recognized as competitive inhibitors of the ubiquinone-binding site and noncompetitive inhibitors of the succinate binding site. Consequently, these fungicides operate by impeding the universal cellular respiration function in mitochondria (Iftikhar et al. 2022).

10.3 Structure of SDH

The SDH heterotetrameric protein is present in both prokaryotes' cytoplasmic membrane and eukaryotes inner mitochondrial membrane. It is made up of four subunits encoded by the nuclear genome (Gawryluk and Gray 2009; Kim et al. 2012; Bénit et al. 2022). SDHA and SDHB form the "head" of the complex, entering the mitochondria. The other two, SDHC and SDHD, attach the complex to the mitochondrial matrix and are located in the greasy portion (hydrophobic tail). Flavin adenine dinucleotide (FAD), which functions as a head helper molecule, is present in the SDHA subunit (Goulin et al. 2022). The 72,692 Da SDHA gene, which is located on chromosome 5p15.33 and consists of 16 sections, is a relatively large molecule with 664 amino acids. Its unique spot allows it to bind to the succinate molecule (Sun et al. 2022a). Conversely, SDHB serves as a sort of catalyst and transfers electrons, which is a fundamental function. It weighs approximately

31,630 Da and is composed of 280 amino acid genes, which is divided into eight sections, and is located on chromosome 1p35-36. The SDHB subunit houses three iron–sulfur clusters ([2Fe–2S], [4Fe–4S], and [3Fe–4S]) linked to the substrate-binding and ubiquinone-binding sites through a series of redox centers (Fig. 10.2). Scientists have discovered that SDHB functions in our bodies like a superhero, preventing the growth of tumors (Wilson et al. 2010; Huang and Millar 2013; Olamiju et al. 2021; Jin et al. 2021).

Made up of SDHC and SDHD, the portion of SDH that protrudes into the cell membrane functions as ubiquinone's docking site. The gene for SDHC is located on chromosome 1q21, has six sections composed of 169 amino acids with 18,610 Da. The gene for SDHD, located on chromosome 11q23, is even smaller—weighs about 17,043 Da, containing only 159 amino acids, and has four sections (Schiavi 2005; Her and Maher 2015), which include a heme b group and a ubiquinone-binding site, making SDH a target for fungicides acting on complex II of respiration. Crucially, SDH plays a dual role in aerobic respiration—oxidizing succinate to fumarate in the mitochondrial matrix and participating in ubiquinone reduction in the mitochondrial membrane. Notably, a single point mutation near the Q-site decreases the

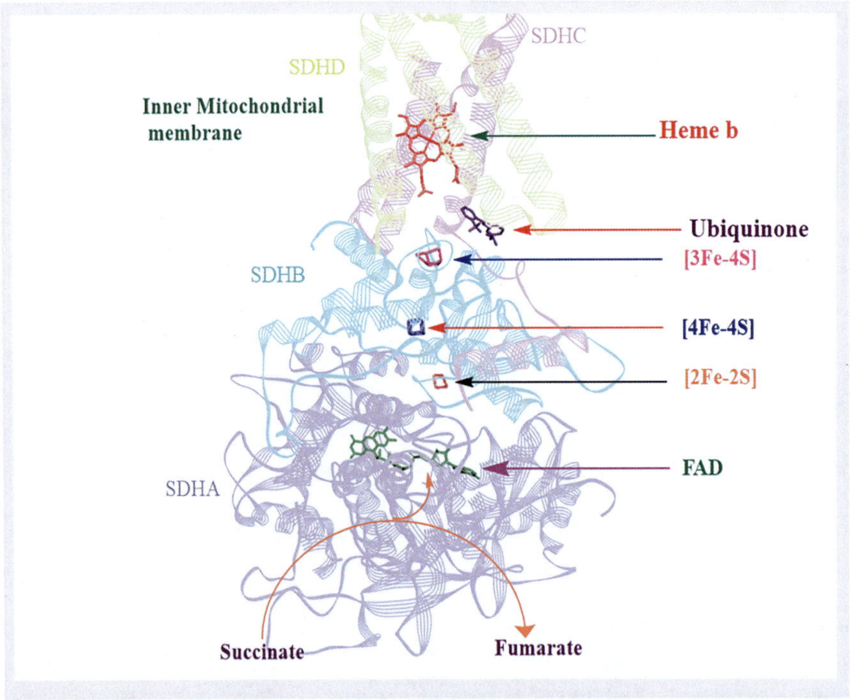

Fig. 10.2 Succinate dehydrogenase (PDB ID: 4YXD) embedded in the mitochondrial inner membrane: SDHA (purple ribbon), SDHB (blue ribbon), SDHC (pink ribbon), SDHD (yellow ribbon), FAD (green stick), FeS clusters —[2Fe–2S], [4Fe–4S], [3Fe–4S]—from bottom (blue, pink, and orange sticks), and ubiquinone in the QP site (purple stick), heme b (red stick)

binding affinity to carboxamides fungicides, such as SDHIs (Rutter et al. 2010; Chen et al. 2021).

Adding to the original four members, scientists found two more members in the past ten years, SDHAF1 and SDHAF2. The process of flavination of SDHA, which aids in complex formation within the mitochondria, is carried out by SDHAF2, also referred to as SDH5. SDHAF1 aids in the insertion of a component known as the FeS residue into SDHB. SDHIs are specific molecules binding to the ubiquinone-binding site (Q-site), thereby inhibiting fungal respiration (Van Vranken et al. 2014; Bezawork-Geleta et al. 2016; Dwight et al. 2017).

10.4 Role of SDH in Mitochondrial Function and Cellular Energy Production

This enzyme, also referred to as respiratory chain complex II, succinate-ubiquinone oxidoreductase, or SDH, is a vital enzyme that plays a crucial role in the energy production process within mitochondria. It is unique because it takes part in the tricarboxylic acid (TCA) cycle as well as the respiratory chain (Huang and Millar 2013). SDH is necessary for mitochondrial metabolism in a variety of organisms, and its absence can have adverse effects (Brière et al. 2005). SDH is the target of a particular class of fungicides called SDHIs. It may play a role as a tumor suppressor because genetic suppression of the protein—whether by chemicals or mutations—has been connected to human cancers and mitochondrial disorders (Aspuria et al. 2014; Bouillaud 2023). Superoxide scavenging and oxygen sensing are regulated by SDH, an essential antioxidant enzyme in mitochondria that works in tandem with ubiquinone. SDH deficiency can lead to serious health problems due to its crucial role in preserving respiratory chain function and metabolic homeostasis, regardless of whether the deficiency is caused by chemicals or genetic mutations (Xiong et al. 2015; Fan et al. 2019).

In the TCA cycle, succinate is oxidized to fumarate through the catalytic action of succinate dehydrogenase (Kregiel 2012). The SDHA subunit of SDH is necessary for this process because it has a conserved histidine residue covalently attached to its flavin adenine dinucleotide (FAD) cofactor. This oxidation produces $FADH_2$, a high-energy electron carrier and participant in the citric acid cycle. Adenosine triphosphate (ATP), the energy source used in cellular respiration, cannot be produced without this cycle (Duarte Hospital et al. 2023).

SDH reduces ubiquinone to produce $FADH_2$, which then transports electrons to ubiquinone in the inner mitochondrial membrane. This process is closely related to the electron transport chain (ETC). A proton gradient is subsequently created across the inner mitochondrial membrane as electrons move through the electron transport chain (Huang and Millar 2013). The iron–sulfur centers in SDH, such as [2Fe–2S], [4Fe–4S], and [3Fe–4S], help electrons move from SDHA to membrane domains (Gawryluk and Gray 2009; Read et al. 2021). The reduction of ubiquinone and the subsequent production of ubiquinol (QH2) are facilitated by this electron transfer.

When the liberated ubiquinol binds to additional enzymes in the respiratory chain of the mitochondria, ATP synthesis results (Goulin et al. 2022).

During the Electron Transport Chain and ATP synthesis, protons (H^+) are pumped across the inner mitochondrial membrane. In the process of oxidative phosphorylation, this generates an electrochemical proton gradient that is essential. ATP synthase uses this proton motive force generated by the electron transport chain to convert adenosine diphosphate (ADP) and inorganic phosphate into ATP. SDH facilitates the flow of electrons through the electron transport chain, which helps cells produce ATP, the main energy unit (OMURA et al. 1988).

10.5 Role of SDH in Various Diseases

SDH is a multifunctional enzyme essential for many cellular functions, most notably the production of energy in mitochondria (Read et al. 2021). SDH deficiency is linked to a variety of conditions in humans, including leukoencephalopathies, cardiomyopathies, leukodystrophies, and encephalopathies—, which are uncommon mitochondrial disorders (Baysal et al. 2000; Duarte Hospital et al. 2023). Furthermore, certain uncommon types of cancer are associated with deficiencies in SDH, and the tumor suppressor genes SDHB, SDHC, and SDHD are responsible for this association. However, there is no evidence that SDHA suppresses tumors. Which SDH gene is mutated determines the exact type of disease, for example, Leigh syndrome is linked to the SDHA gene, whereas pheochromocytoma and hereditary paraganglioma are caused by mutations in the SDHB gene (Neumann 2004). Numerous benign and malignant tumors have also been linked to other SDH genes, including SDHC and SDHD (Andrews et al. 2018).

In mammals, including dogs, deficiencies in SDH can lead to the development of pheochromocytomas (PCC) and paragangliomas (PGL), with metastases identified in various organs (Fig. 10.3). Canine PGLs and PCCs have been reported to have mutations in the SDHB and SDHD genes, suggesting that the incidence of these conditions may be underestimated (Holt et al. 2014). Genetic inactivation of the SDH complex can result in oxidative stress, succinate accumulation, pseudohypoxia, and metabolic reprogramming in cellular and molecular processes linked to SDH impairment (Lemarie and Grimm 2011). These processes all contribute to the severity and evolution of cancer phenotypes. Although gastrointestinal stromal tumors (GISTs) rarely contain SDH mutations, when they do, SDH deficiency is linked to the mutation. The most frequent mutations are found in SDHB, and then SDHC and SDHD (Miettinen and Lasota 2014; Andrews et al. 2018).

SDH impairment is linked to cellular and molecular processes such as metabolic reprogramming, succinate accumulation, oxidative stress, and pseudohypoxia (Lemarie and Grimm 2011). Reactive oxygen species (ROS) can rise as a result of SDH inactivation, which can also cause oxidative stress and interfere with cellular respiration. SDH deficiency causes succinate accumulation, which inhibits the Krebs cycle and is linked to the development of cancer. SDH-related carcinomas grow as a result of pseudohypoxia brought on by succinate accumulation (Fig. 10.3)

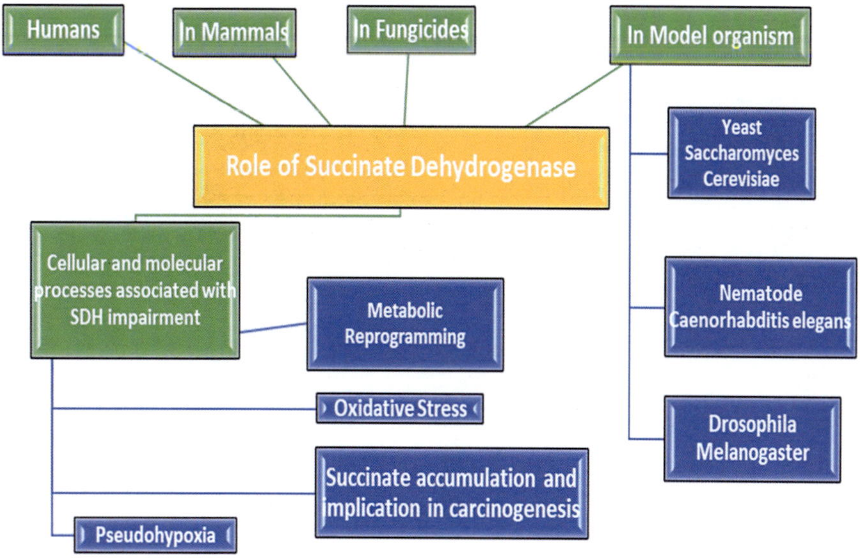

Fig. 10.3 Role of SDH in various diseases

(Zorov et al. 2014). Under the influence of a disturbed Krebs cycle, metabolic reprogramming promotes the growth and metastasis of cancer cells. Scientists investigate the effects of SDH inactivation using genetically modified organisms. It results in oxidative stress and modifications to cellular processes in yeast (Panizza et al. 2013). SDH mutations cause altered respiration, increased oxidative stress, and a shorter lifespan in nematodes. SDH inactivation causes a syndrome that mimics early mortality, behavioral abnormalities, and premature aging (Brière et al. 2005). By preventing fungal respiration, fungicides known as SDHIs are essential for managing plant diseases. By preventing the transfer of electrons, these inhibitors help promote sustainable agriculture by reducing fungicide resistance and minimizing environmental pollution caused by overuse of fungicides (Ding et al. 2022).

10.6 Mechanism of Action of SDHIs

Fungicides use diverse mechanisms to combat fungal growth. Historically, multisite fungicides like maneb (1955), mancozeb (1961), and captan (1952) targeted multiple cellular processes in fungi (Sierotzki and Scalliet 2013). Since the 1960s, most fungicides focus on single-site actions, disrupting specific fungal functions such as nucleic acid metabolism, respiration, and membrane integrity (Deising et al. 2008; Sierotzki and Scalliet 2013; Duarte Hospital et al. 2023). Around 25% of fungicides target cellular components like mitochondria, crucial for ATP production through oxidative phosphorylation. For instance, 60 fungicides obstruct various stages of the mitochondrial respiratory chain. Some inhibit complexes I–III, while others disrupt ATP transport or act as uncouplers (Duarte Hospital et al. 2023). Additionally,

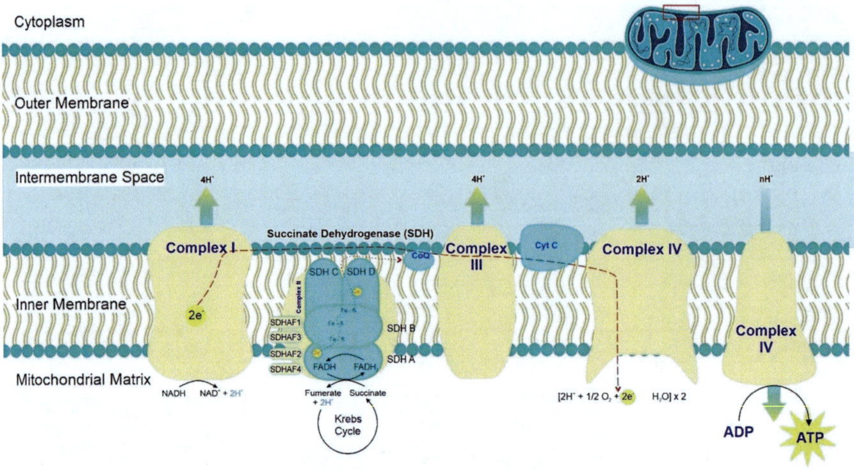

Fig. 10.4 Structure of the SDH and the mitochondrial respiration chain (MRC)

antifungals impact synthetic pathways, structural components of cell walls, and cellular replication by targeting ergosterol synthesis, β-(1,3)-glucan synthesis, and microtubule formation (Abe et al. 2018).

SDHIs specifically target SDH, a key enzyme in the electron transport chain. By inhibiting SDH, these fungicides prevent the conversion of succinate to fumarate, disrupting electron transfer to ubiquinone and ATP synthesis (Bouillaud 2023; Duarte Hospital et al. 2023). This accumulation of succinate hampers cellular energy production, affecting various biological processes (Fig. 10.4). SDHIs, introduced in the early twenty-first century, remain crucial in agriculture due to their broad-spectrum activity. However, resistance in fungal populations poses challenges, prompting ongoing research to enhance their effectiveness and reduce environmental impact (Gullino et al. 2000). Current efforts focus on improving SDHI formulations, application techniques, and ensuring sustainable use to balance crop protection with environmental and health considerations (Wu et al. 2018; Brischigliaro and Zeviani 2021; Duarte Hospital et al. 2023).

10.7 Classification of SDH Inhibitors

Various categories of SDHIs are summarized in Fig. 10.5.

10.8 On the Basis of Generation

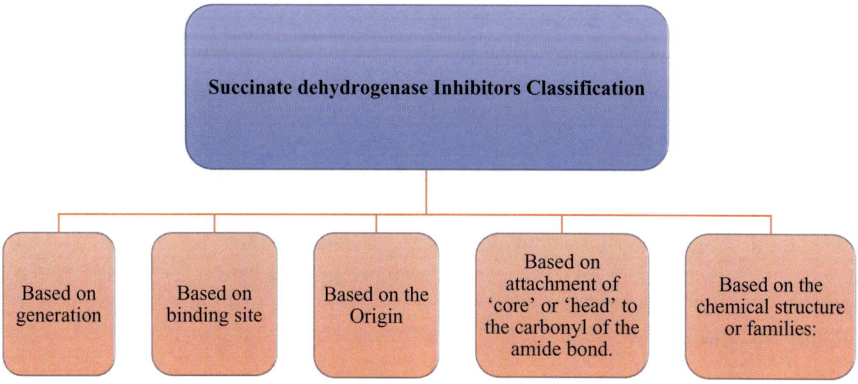

Fig. 10.5 Different classes of SDHIs

(i) **First Generation**

The first-generation molecules in this class carboxin, was commercialized in the late 1960s and exhibited a narrow spectrum of activity which restricted to basidiomycetes, especially *Rhizoctonia* species (Sierotzki and Scalliet 2013).

(ii) **Second Generation**

The second generation SDHIs emerged in the late 1990s to early 2000s, with key compounds Furametpyr and Thifluzamide developed in early 1990s to control cereal smut diseases, caused by the fungi *Ustilago nuda* and *Ustilago hordei*. (OMURA et al. 1988).

(iii) **Third Generation**

In this generation, new varieties of SDHIs have been developed and regularly introduced to the market with a broader spectrum of activity. For example, in 2002, boscalid obtained a marketing license in order to control the spread of *ascomycetes*, such as *Sclerotinia sclerotiorum* and *basidiomycetes* (OMURA et al. 1988).

10.9 Classification Based on Binding Site

According to the binding site, SDHIs fall into two main classifications:

(i) Those that bind to the pocket where ubiquinone binds (e.g., malonate).
(ii) Those that attach to the succinate-binding pocket (e.g., carboxamides).

SDHIs bind strongly to the ubiquinone-binding (Qp) site. After binding, they physically block the binding of substrate, which consequently prevents further cycling of succinate oxidation (Guo and Lemire 2003; Sierotzki and Scalliet 2013). Ubiquinone (UQ) is located in a pocket composed of residues derived from three subunits—SDHB, SDHC, and SDHD—, which are close to the [3Fe–4S] cluster (Xiong et al. 2015). Furthermore, ubiquinone binds to the entry site, the Q1-site, where just a carbonyl oxygen is involved. SDHIs are regarded as competitive inhibitors of the ubiquinone-binding site and noncompetitive inhibitors of the succinate binding site (Sierotzki and Scalliet 2013).

10.10 Classification Based on the Origin (Natural Products)

Natural products are small molecules produced by living organisms including plants, invertebrates, and microorganisms. While these substances are not essential to life, they are additionally referred to as secondary metabolites and play a role for defense and cell-to-cell communication (Dias et al. 2012; Omokhefe Bruce 2022).

For many years, natural products have been an important tool for pharmaceutical companies in discovering new bioactive compounds. Currently, over 50% of the molecules under investigation are either natural products or their semisynthetic counterparts, or they are inspired by such scaffolds (Xiong et al. 2015).

Macro and micro fungi have been part of human life for thousands of years. They have been used as food (mushrooms), in preparation of alcoholic beverages (yeasts), in traditional medicine, and for cultural purposes (Xiong et al. 2015). Currently with the advances in microbiology, their uses have extended to enzyme production, biological control, antibiotics, and other pharmacologically active products. Notably, one of the most famous natural product discoveries is penicillin, derived from *Penicillium notatum* and discovered by Fleming in 1929 (Bénit et al. 2022).

The natural SDHI in fungicide is atpenin, which is a novel antifungal antibiotic complex discovered in a soil isolate of Penicillium sp. FO-125 (OMURA et al. 1988).

Another natural SDHI is harzianopyridone, which is a secondary metabolite produced by certain fungi, particularly from the genus *Trichoderma*. These fungi are well known for their role in biocontrol and plant disease suppression in agriculture (Khan et al. 2020). Harzianopyridone is one of the bioactive compounds produced by these fungi and contributes to their biocontrol properties (OMURA et al. 1988; Yao et al. 2023).

10.11 Classification Based on Attachment of "Core" or "Head" to the Carbonyl of the Amide Bond

While the structures of SDHIs differ greatly, they all possess an amide bond as an essential characteristic. Six distinct chemical forms have been identified based on the "core" or "head" of the molecule (Fig. 10.6) that is connected to the carbonyl of the amide bond. The core component of the molecule extensively reaches the SDH active site and plays an essential role for binding and in vivo potency. All SDHIs have an amide bond, which defines the orientation of the molecule and is strictly conserved (Yan et al. 2019).

A "linker" is attached to the amine side of the amide bond. This linker may be a simple, non-substituted phenyl ring (as in carboxin), or it may carry substituents at different places. In most cases, the linker is a phenyl group. The last component of the molecule is known as the "hydrophobic rest," which is present in the ortho configuration (Sierotzki and Scalliet 2013; Yan et al. 2019). This

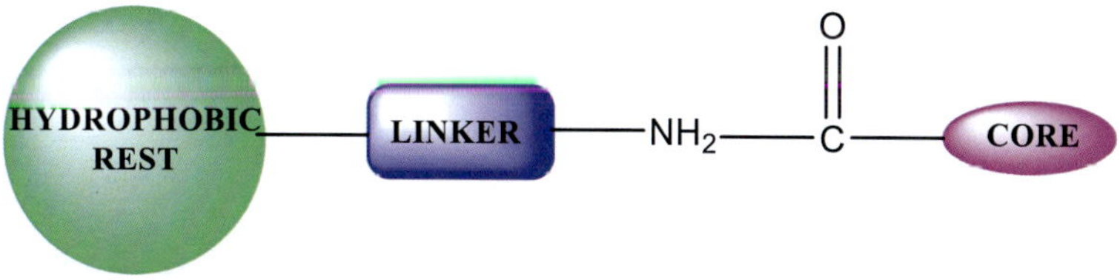

Fig. 10.6 Pharmacophoric feature of SDHI

Table 10.1 Structures of SDHIs

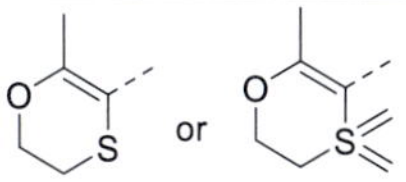 Oxathiin Carboxamide	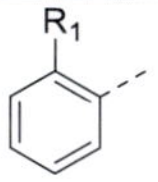 Benzamides	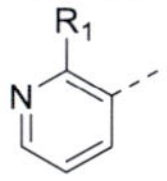 Pyridine Carboxamides
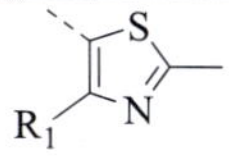 Thiazole Carboxamides	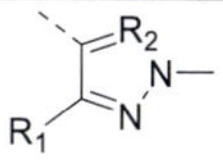 Pyrazole Carboxamides	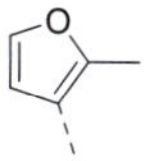 Furan Carboxamides

Six core ring systems define the current chemical classification for SDHIs

classification is based on the nature of the substituted five- or six-membered rings, as shown in Table 10.1.

In terms of easily accessible compounds, pyrazole carboxamides are the most represented class, followed closely by benzamides and oxathiin carboxamides (Xiong et al. 2015).

10.12 Classification According to Chemical Structure or Families

This classification is based on the FRAC codes. The FRAC issues a number and/or letter combination to recognize active components that display signs of cross-resistance. Fungicides with the same FRAC code carry the risk of establishing cross-resistance since they share the same target site. In addition to the chemical compound's FRAC group, the pathogen's response and relationship to the chemical compound are also taken into account in order to establish the fungicide resistance risk (low, medium, or high. The FRAC group or code can usually be found in the upper corner of a fungicide label (Xiong et al. 2015).

- Despite notable differences in chemical structure, all SDHI conserve the amide bond.
- SDHIs are categorized into 12 different chemical families, as listed in Table 10.2:

Table 10.2 SDHIs and their chemical structures

Pyrazole-4-carboxamides			
Fluxapyroxad	Penflufen	Penthiopyrad	Furametpyr
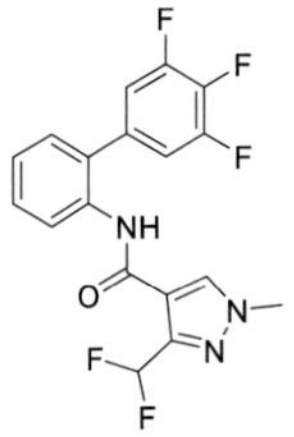	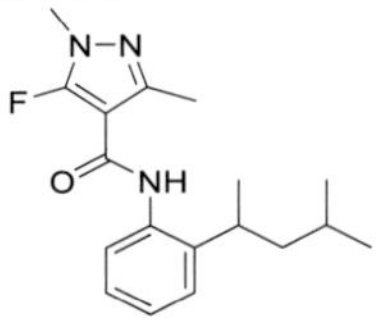	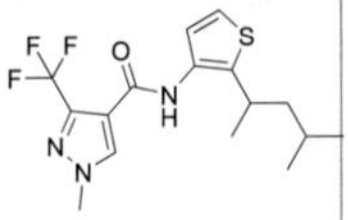	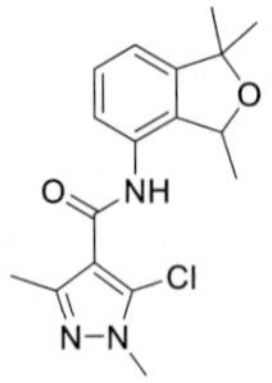
Bixafen	Benzovindiflupyr	Isopyrazam	Sedaxane
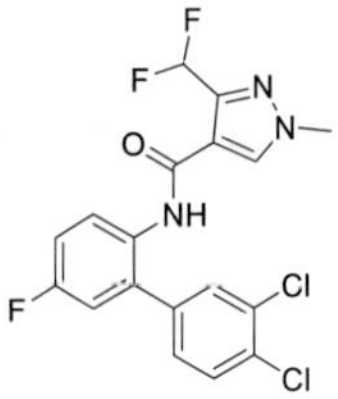	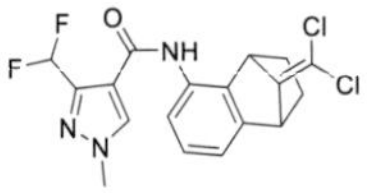	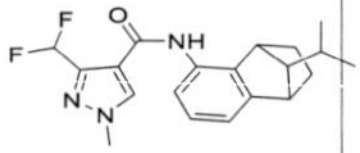	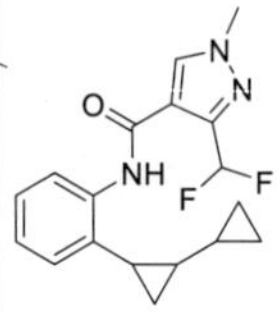
Fluindapyr		Inpyrfluxam	
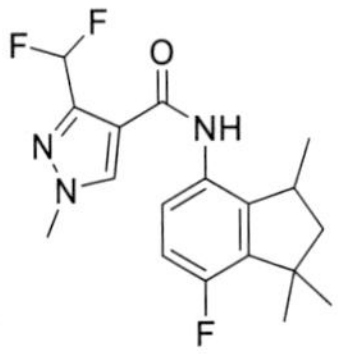		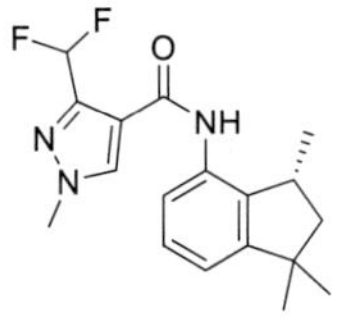	
Thiazole carboxamides	**Pyrazinecarboxamides**	**Furan carboxamides**	**N-methoxy-(phenylethyl)-pyrazole-carboxamides**
Thifluzamide	Pyraziflumid	Fenfuram	Pydiflumetofen
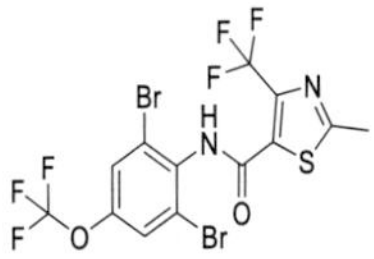	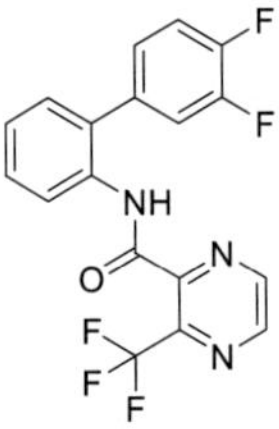	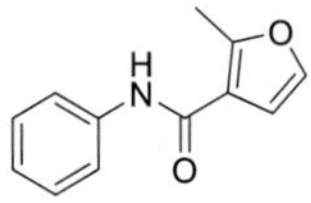	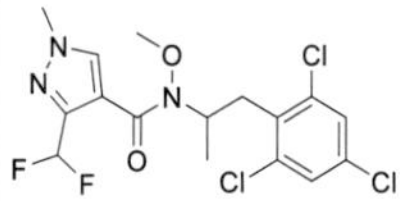
N-cyclopropyl-N-benzyl-pyrazole carboxamides	**Pyridine-carboxamides**	**Pyridinyl-ethyl-benzamides**	**Phenyl-cyclobutyl pyridine amides**
Isoflucypram	Boscalid	Fluopyram	Cyclobutrifluram

(continued)

Table 10.2 (continued)

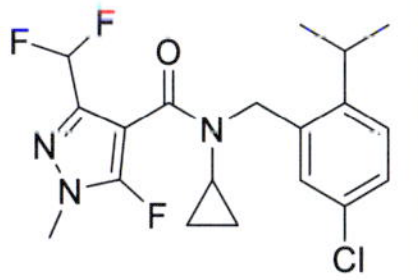	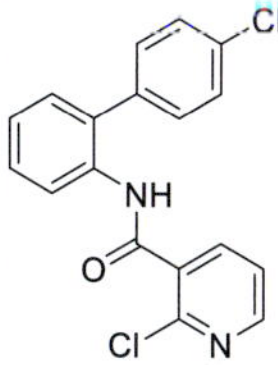	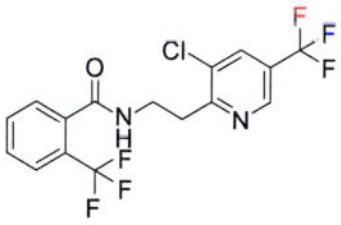	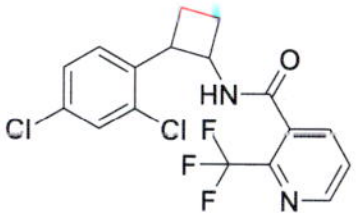
Phenyl-oxo-ethyl thiophene amides			
Isofetamid			
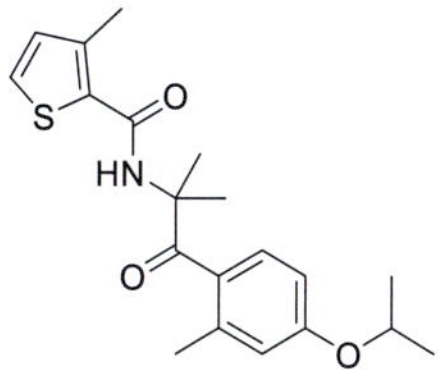			
Phenylbenzamides			
Flutolanil	Mepronil	Benodanil	
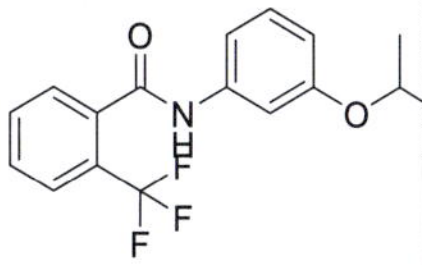		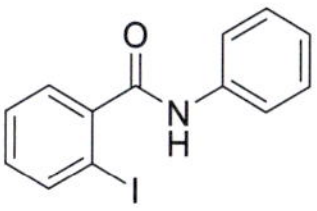	
Oxathiin carboxamides			
Carboxin		Oxycarboxin	
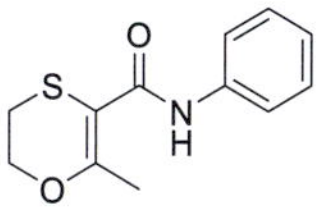		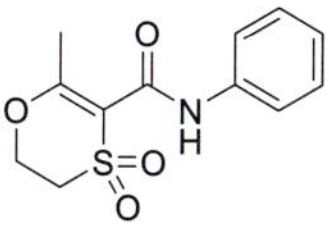	

10.13 Synthetic Approaches and Chemical Synthesis of SDH Inhibitors

10.13.1 Synthesis of Atypical SDH Inhibiting Fungicides

10.13.1.1 Synthesis of Pyrazole-4-Acetohydrazide Derivative

Inhibitors of succinate dehydrogenase are an emerging class of fungicides that are widely employed in the agricultural field as pesticides. Recently, several classes of nuclei have been reported as SDHIs. Wang et al. developed a new series of pyrazole-4-acetohydrazide derivatives as fungicides in agriculture for protection. The synthetic strategy based on the reaction between substituted trifluoromethyl pyrazole derivatives and substituted phenylhydrazine in a dichloromethane solution (Scheme 10.1). A solution of triethylamine (NEt_3) and O-(benzotriazol-1-yl)-N,N,N′,N′-tetramethyluronium tetrafluoroborate (TBTU) was required to facilitate the reaction and achieve the desired compound.

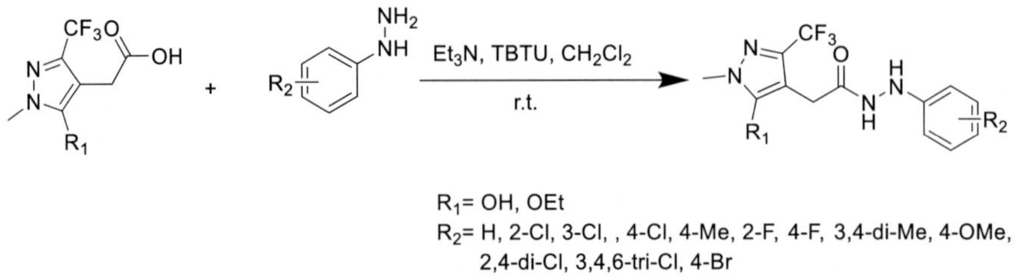

Scheme 10.1

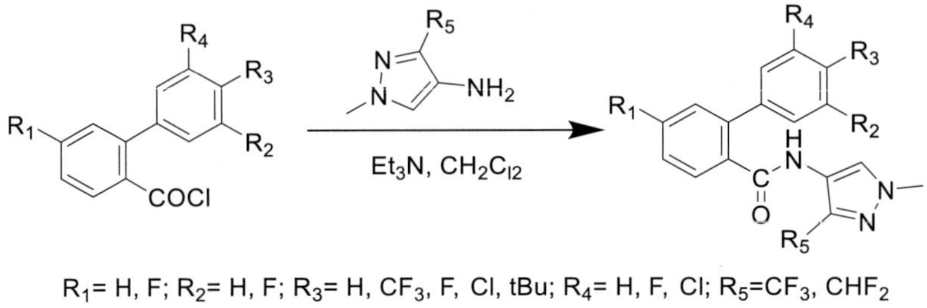

Scheme 10.2

The final compounds were insoluble in water, and all the derivatives were obtained in good yields. Ethanol was used for recrystallization. The authors synthesized 23 derivatives of pyrazole-4-acetohydrazide and evaluated their desired activity against *Rhizoctonia solani*, *Fusarium graminearum*, and *Botrytis cinerea* (Wang et al. 2021b).

10.13.1.2 Synthesis of Pyrazole Carboxamide Derivative with Transposal of Amide Linkage

Another novel SDHI derivative was introduced by Yang et al. by the transposal of the carbonyl and amide groups of fluxapyroxad, a broad-spectrum pyrazole carboxamide fungicide. A tandem reaction between substituted biphenyl carbonyl chloride and substituted 1-methyl-pyrazole-4-amines in the presence of dry dichloromethane is shown in Scheme 10.2. All analogs were obtained in moderate to good yields and evaluated for in vitro activity against *Sclerotinia sclerotiorum*, *Rhizoctonia cerealis*, *Puccinia sorghi*, and *Rhizoctonia solani* (Yang et al. 2019).

10.13.1.3 Synthesis of Pyrrolidine Derivative

A different moiety, pyrrolidine, was introduced as an SDHI fungicide by Luo et al., with activity established against five fungi: *Monilinia fructicola, Rhizoctonia solani, Fusarium graminearum Schw., Fusarium oxysporum*, and *Phytophthora infestans*. The final compounds were designed by modifying fluopyram, a representative of SDHI fungicide as shown in Scheme 10.3. A base-catalyzed reaction between substituted pyrrolidine and substituted pyridine has been reported for the synthesis of target compounds. The researchers have synthesized 25 title compounds and all the analogues obtained with general yield of 55–70% (Luo et al. 2024).

R= 2-OCH_3-phenyl, 2-F-phenyl, 3-(difluromethyl)-1-methyl-1H-pyrazole-4-yl, Thiphene-2-yl, 2-CF_3-pyridine-3-yl, Furan-2-yl
R_1= Cl, Br; R_2= CF_3, Cl, Br;

Scheme 10.3

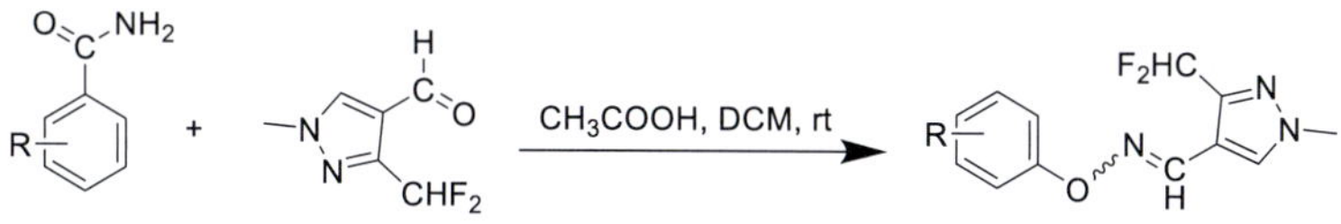

(E- and Z-isiomer)

R= H, 2-CH_3, 3-CH_3, 4-CH_3, 2-F, 3-F, 4-F, 2-Cl, 3-Cl, 4-Cl, 2-Br, 3-Br, 4-Br, 2-I, 3-I, 4-I, 2CF_3, 3CF_3, 4-CF_3

Scheme 10.4

10.13.1.4 Synthesis of Pyrazolyl Oxime Ethers

The synthesis of pyrazolyl oxime ethers has been reported as novel SDHI fungicides. The scheme is based on the formation of a novel bridge instead of the typical amide bridge between the aromatic and aryl moiety. Nineteen pairs of *E*- and *Z*-isomers of pyrazolyl oxime were synthesized via the reaction of substituted benzamide with the free aldehyde of a pyrazole derivative in the presence of dichloromethane for the formation of oxime as shown in Scheme 10.4 (Wang et al. 2021a).

10.13.1.5 Synthesis of Thiazolamide Derivative Appended Diphenyl Ester

A novel class of thiazolamide derivatives with appended diphenyl ester moiety was developed as SDHIs for fungicidal action in agriculture. The development of target compounds was based on replacement of non-heterocycles aryl group by diphenyl ester moiety pertaining antifungal activity. The scheme 10.5 illustrates the condensation of a thiazolidine moiety with a diphenyl ester in the presence of methyl cyanide, resulting in the formation of a thiazolamide derivative appended to a diphenyl ester through an amide linkage (Yan et al. 2023).

10.13.1.6 Synthesis of Pyrazole Carboxamide Appended Thiazole Derivatives

Among existing SDHIs, the pyrazole-4-carboxamide scaffold is the most common in fungicide chemistry. A recent study synthesized novel pyrazole carboxamide-appended thiazole derivatives and confirmed their activity against plant fungi pathogens. The synthetic route involved condensation of a thiazole moiety with carboxamide via a substituted methylene bridge. The reaction strategy based on

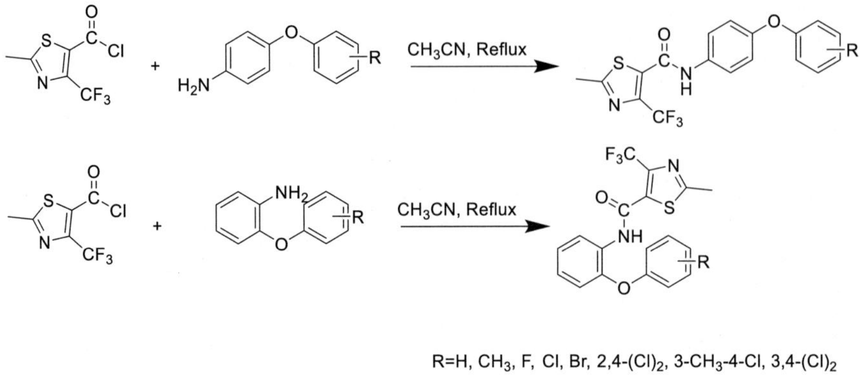

Scheme 10.5

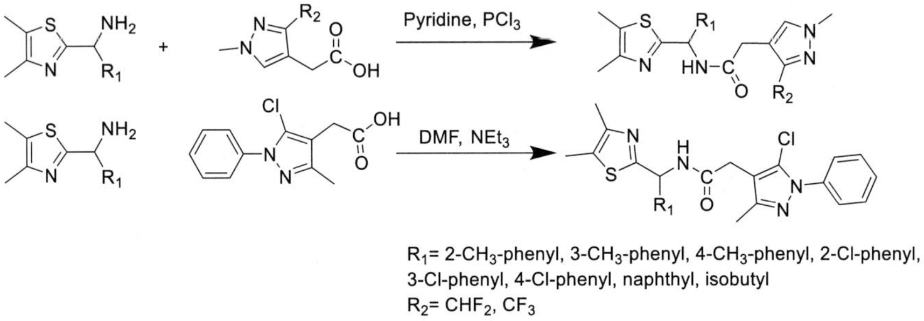

Scheme 10.6

reaction between substituted thiazole-2-methanamine and pyrazole-4-carboxylic acid with substitution in the presence of pyridine and phosphorous trichloride yield the title fungicides as shown in Scheme 10.6. They also offer another route where substituted thiazole-2-methanamine react with phenyl substituted pyrazole in the presence of triethylamine and N,N-dimethylformamide as the solvent yield the final derivatives (Li et al. 2023).

10.13.1.7 Synthesis of 1,3-Benzodioxole-Pyrimidine Derivative

A novel series of 1,3-benzodioxole-pyrimidine derivatives was introduced by Y. Sun et al. in the field of agriculture to provide cidal action on fungi. The activity as fungicide was established by using in vitro trailed by in vivo experiment on tomato fruits and leaves. A scaffold was developed wherein the benzodioxole moiety is directly connected to a substituted pyridine moiety at the 5-position. The synthesis involves a cyclization reaction where the aromatic substituted benzodioxole moiety with a free carbonyl side chain is treated with an amino group to form the title compound as shown in Scheme 10.7. Thiourea and guanylurea in the presence of sodium hydroxide formed the title compounds. The substituted benzodioxole moiety may be obtained by the treatment of substituted aldehyde and 3′,4′-(methylenedioxy) acetophenone (Sun et al. 2022b).

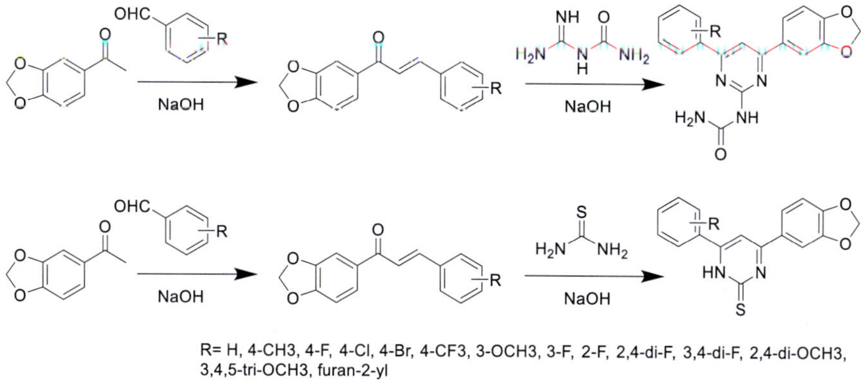

Scheme 10.7

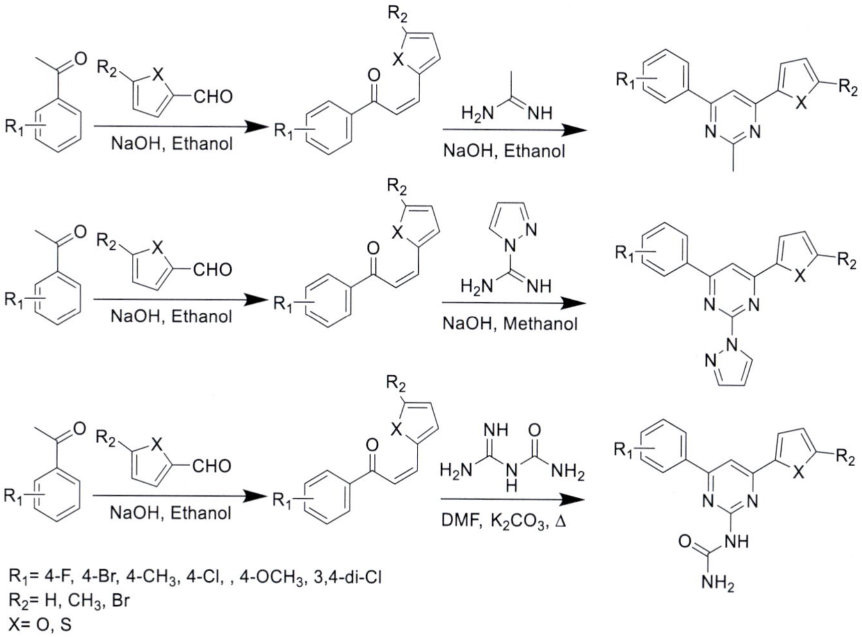

Scheme 10.8

10.13.1.8 Synthesis of 2-6-Diarypyrimidne Derivative

The design of 2,6-diarylpyrimidine derivatives was based on a synthesis involving the condensation of aromatic benzaldehydes with heterocyclic rings containing a carbaldehyde group in the presence of sodium hydroxide and methanol, followed by reaction with a hydrazine derivative. This approach led to the development of several pyrimidine analogs as potent succinate dehydrogenase inhibitors (SDHIs). The final compound is composed of two moieties, one is six-membered substituted

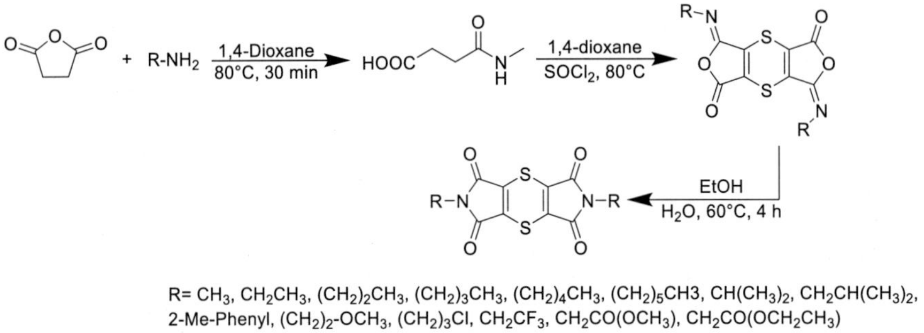

Scheme 10.9

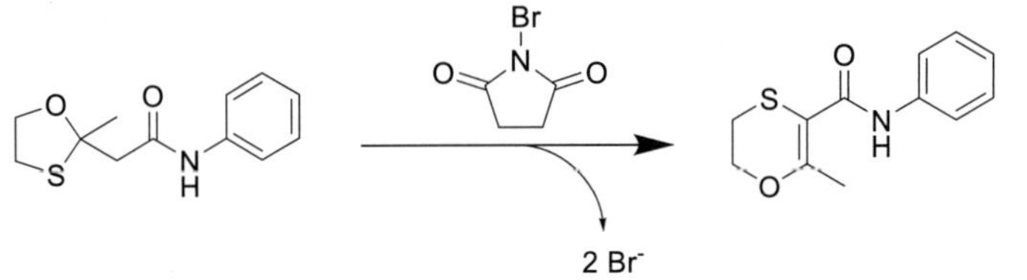

Scheme 10.10

phenyl ring and another is five-membered heterocycles as shown in Scheme 10.8. Both moieties are connected to a pyrimidine base nucleus via a cyclization reaction (Yan Yingkun 2020).

10.13.1.9 Synthesis of Dithiin Tetracarboximide Derivatives

A different class of novel dithiin tetracarboximide derivatives was synthesized by a reaction between maleic anhydride and different primary amines to form intermediates based on ring-opening reaction as shown in Scheme 10.9. Further, the intermediates were treated with thionyl chloride as an oxidant to yield a heterocyclic moiety, which further tautomerized to give the final compoundsas described by Ma et al. (2023).

10.14 Synthesis of Typical SDH Inhibiting Fungicides

10.14.1 Synthesis of Carboxin

Carboxin or Vitavax, 2-methyl-5,6-dihydro-1,4-oxathiine-3-carboxanilide, has been prepared by N-bromosuccinimide (NBS)-promoted oxidative rearrangement of acetoacetanilide 1,3-oxathiolane in chloroform solvent, as shown in Scheme 10.10. The reaction was carried out under mild conditions to form above 90% yield (Caputo et al. 1995).

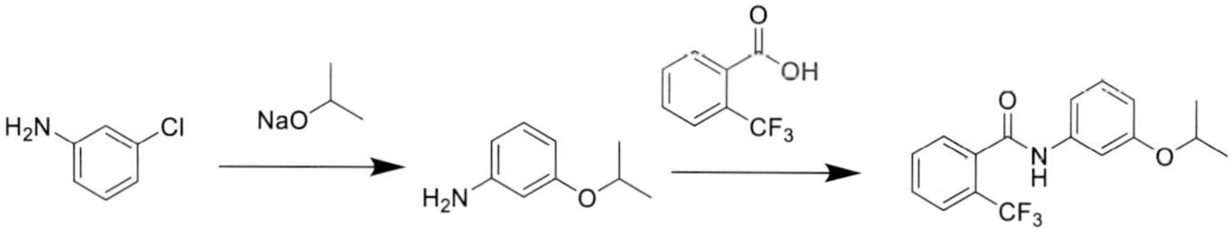

Scheme 10.11

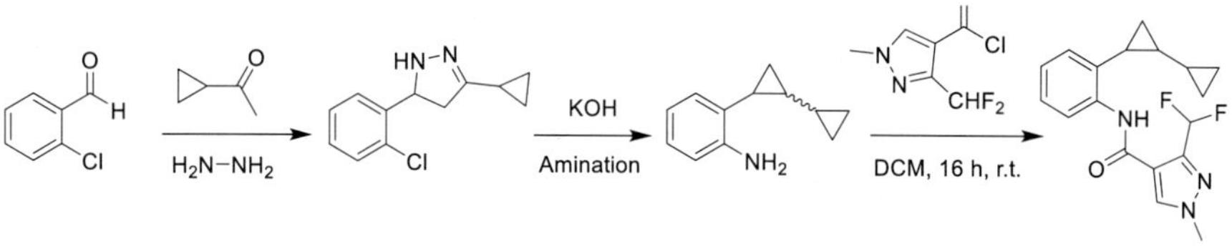

Scheme 10.12

10.14.2 Synthesis of Flutolanil

Flutolanil, N-(3-propan-2-yloxyphenyl)-2-(trifluoromethyl)benzamide, is a member of the class of benzamides, synthesized by a condensation reaction between the carboxy group of 2-(trifluoromethyl)benzoic acid and the amino group of 3-(isprop-pyloxy)aniline. The latter was obtained by reacting 3-chloroaniline with sodium Sodium isopropoxide, as shown in Scheme 10.11 (Müller et al. 2011).

10.14.3 Synthesis of Sedaxane

Sedaxane, N-(2-[1,1′-bi(cyclopropyl)-2-yl]phenyl)-3-(difluoromethyl)-1-methyl-1pyrazole-4-carboxamide, is obtained by a base-catalyzed crossed aldol condensation reaction between 2-chlorobenzaldehyde and cyclopropyl methyl ketone to form an adduct, which reacts with hydrazine to give a dihydropyrazole derivative, as shown in Scheme 10.12. Upon heating in the presence of a base, the dihydropyrazole forms an intermediate, 2-bicyclopropyl-1-chlorobenzene, which is converted into aniline by a Pd-catalyzed benzyl-amination reaction followed by metal-catalyzed deprotection yield, 2-bicyclopropylaniline. Coupling this with pyrazole acid chloride forms the desired compound (German Patent No. EP1480955B, 2011).

10.14.4 Synthesis of Boscalid

Boscalid, 2-chloro-N-[2-(4-chlorophenyl)phenyl]pyridine-3-carboxamide, consists of two separate parts in which pyridyl heterocycle is connected to biphenyl ring. Boscalid is prepared by the based-on condensation reaction between 2-chloronicotinic acid and the amino group of 4′-chlorobiphenyl-2-amine, as shown in Scheme

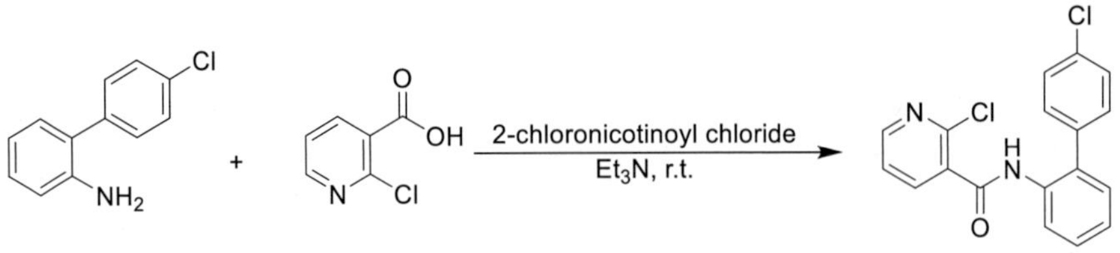

Scheme 10.13

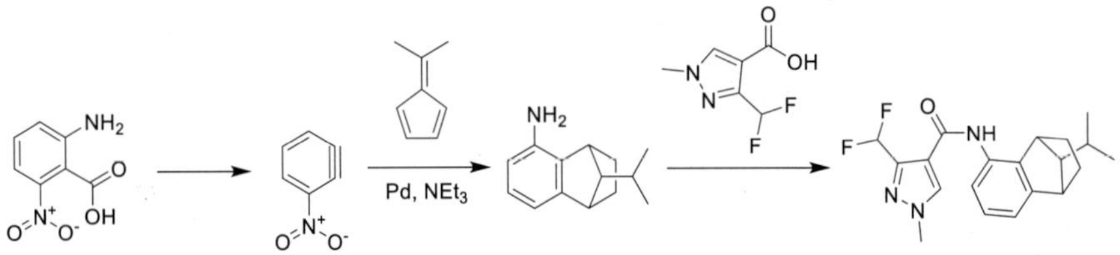

Scheme 10.14

10.13. In generally Biphenyl compound synthesized using the well-established Suzuki coupling reaction (Xiong et al. 2015).

10.14.5 Synthesis of Isopyrazam

Isopyrazam, 3-(difluoromethyl)-N-(9-isopropyl-1,2,3,4-tetrahydro-1,4-methan onaphthalen-5-yl)-1-methylpyrazole-4-carboxamide, is an aromatic amide with organofluoride compound prepared by condensation reaction of carboxy group of 3-(difluoromethyl)-1-methylpyrazole-4-carboxylic acid with the amino part of 9-isopropyl-1,2,3,4-tetrahydro-1,4-methanonaphthalen-5-amine. The amino part synthesized from 6-nitroanthranilic acid forms nitrobenzene, which further reacts with dimethylfulvene to yield nitrobenzonorbornadiene by cycloaddition reaction, as shown in Scheme 10.14. The nitrobenzonorbornadiene on palladium reduction in the presence of triethylamine and high pressure yield the title compound (Walter 2016).

10.14.6 Synthesis of Benzovindiflupyr

Benzovindiflupyr, N-[9-(dichloromethylene)-1,2,3,4-tetrahydro-1,4-methano-naphthalen-5-yl]-3-(difluoromethyl)-1-methylpyrazole-4-carboxamide, is obtained by a synthetic method similar to isopyrazam. In the preparation of benzovindiflupyr the amine component is generated using dichlorofulvene instead of dimethylfulvene, yielding a nitrobenzonorbornadiene intermediate. This intermediate is reduced and then condensed with acid to form the title compound, as shown in Scheme 10.15 (Walter 2016).

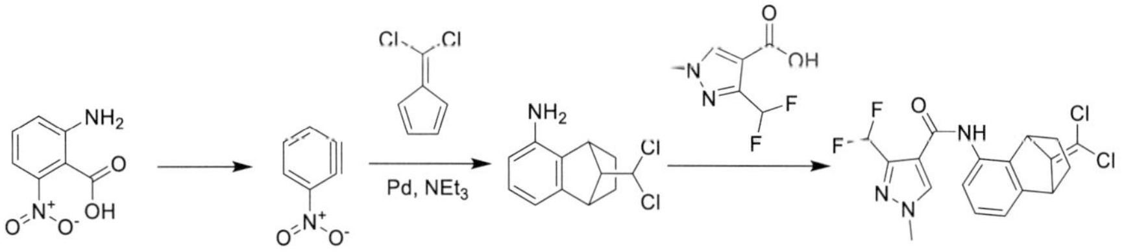

Scheme 10.15

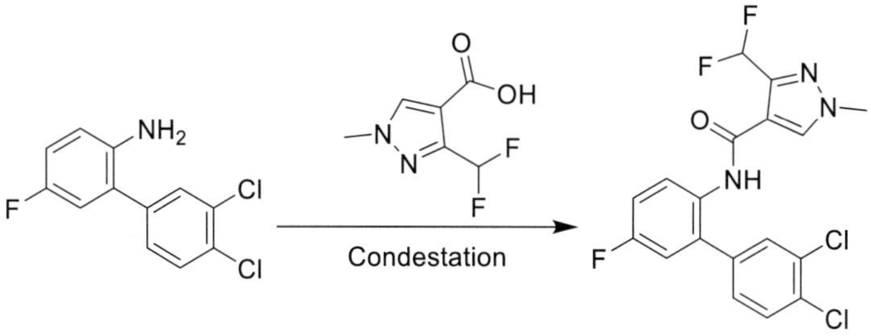

Scheme 10.16

10.14.7 Synthesis of Bixafen

Bixafen, N-[2-(3,4-dichlorophenyl)-4-fluorophenyl]-3-(difluoromethyl)-1-methylpyrazole-4-carboxamide, is obtained by condensation of carboxy group of 3-(difluoromethyl)-1-methylpyrazole-4-carboxylic acid with the amino group of 3′,4′-dichloro-5-fluorobiphenyl-2-amine, as shown in Scheme 10.16. The synthetic strategy is similar to the synthesis of boscalid, as shown in Scheme 10.13, irrespective of reactants for the formation of amide linkage (Gooßen et al. 2008).

10.15 Structure–Activity Relationship (SAR) of SDH Inhibitors

Yu et al. (2021) synthesized a series of 43 novel pyrazole-aromatic carboxamide compounds and tested their antifungal activity. At 50 μg/mL, these compounds exhibited strong in vitro activity against *Botrytis cinerea*, *Rhizoctonia cerealis*, and *Sclerotinia sclerotiorum*, and in vivo activity against *R. solani*. Notably, Compound 10.1 demonstrated significant in vitro activity against *R. cerealis*, with an EC_{50} value of 0.93 μg/mL, making it about twice as potent as the previously reported lead compound and approximately 11 times more effective than the commercial SDH inhibitor thifluzamide. The difluoromethyl pyrazole-(*m*-benzene) carboxamide scaffold was identified as crucial for enhancing antifungal activity. Molecular docking simulations using Autodock Tools-1.5.6, with a grid size of 20 × 20 × 20 and grid space of 0.375 Å, predicted that Compound 10.1 binds to the RcSDH enzyme by forming hydrogen bonds with B-Trp203 and stacking π-π interactions with B-His246,

Compound 10.1

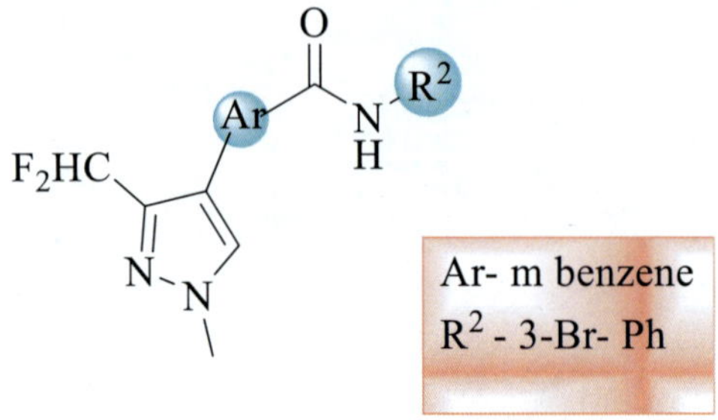

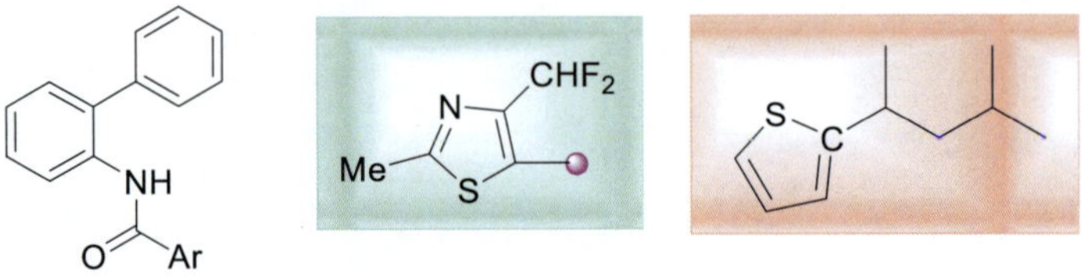

Compound 10.2

similar to thifluzamide. These interactions stabilize Compound 10.1 in the enzyme's active site, enhancing its efficacy (Yu et al. 2021).

Yoshikawa Yukihiro et al. studied the SAR of carboxin-related carboxamides in fungicide applications. They examined how various aromatic rings and N-substituents affect fungicidal activity. They found that compounds with hydrophobic branched alkyl groups at the ortho-position of the N-phenyl group showed broad-spectrum, high activity. The development of penthiopyrad, featuring enhanced alkyl groups on the thiophene ring, was especially effective against multiple diseases. Key structural elements contributing to strong fungicidal activity included CHF_2 or CF_3 groups and the 2-(1,3-dimethylbutyl)-2-thienyl group. This research underscores the significance of N-substituents and aromatic rings in optimizing fungicidal efficacy (Yoshikawa et al. 2011) (Compound 10.2).

Oda et al. presented two synthesis routes using commercially available reagents and established techniques to produce pyraziflumid and its derivatives. The key intermediate, methyl 3-(trifluoromethyl)pyrazine-2-carboxylate, was created via trifluoromethylation. Pyraziflumid was produced through reactions with pyrazine carboxylic acids and biphenyl-2-amines. The SAR showed that fluorine atoms at the 3′ and 4′ positions enhanced fungicidal efficacy. Pyraziflumid, particularly Compound 10.3, demonstrated superior activity against cucumber gray mold, wheat brown rust, and barley powdery mildew, outperforming the previous SDHI fungicide BC723 (Oda et al. 2017).

Luo et al. (2024) designed and synthesized amide derivatives with a pyrrolidine moiety as potential SDHI fungicides. A mycelial growth rate method was used to evaluate the antifungal activity of 25 compounds against five fungi. Compound 10.4 showed broad-spectrum activity, outperforming boscalid and fluopyram. It efficiently disrupted cell membranes, inhibited SDH activity, and prevented Monilinia fructicola zoospore germination. Substituents like furan, thiophene, and thiazole

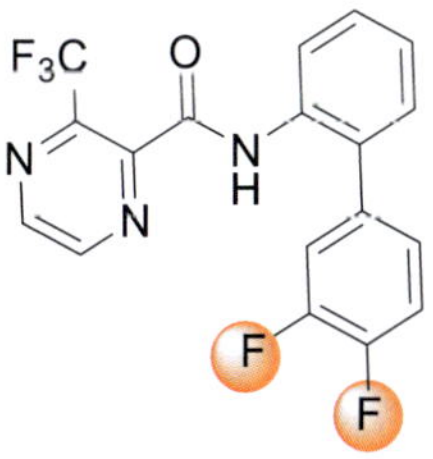

N-(3',4'-difluoro-[1,1'-biphenyl]-2-yl)-3-(trifluoromethyl)pyrazine-2-carboxamide

Compound 10.3

Compound 10.4

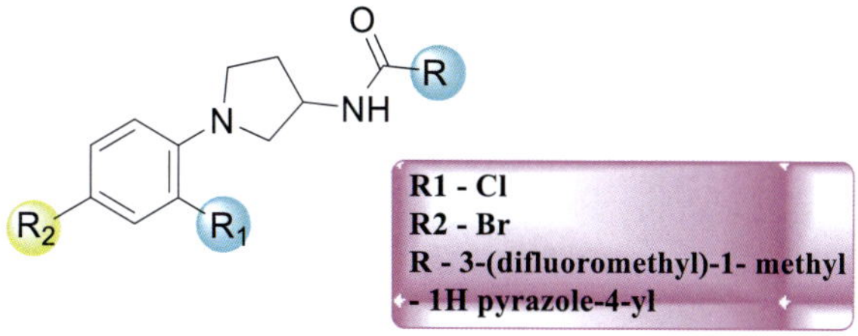

Compound 10.5

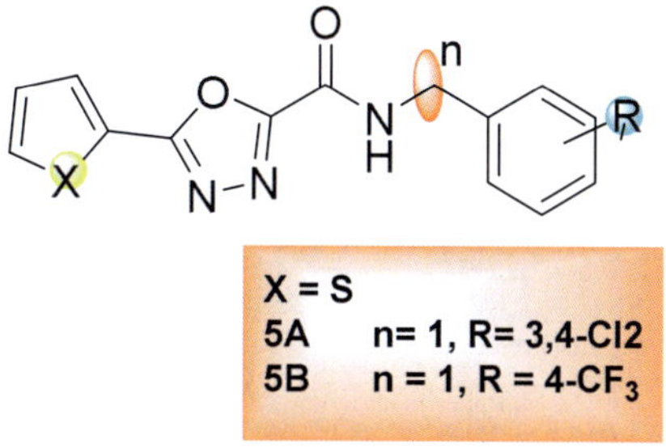

enhanced activity, while benzene rings reduced it. The study highlighted that Br or Cl atoms further improved efficacy. The pyrrolidine moiety in Compound 10.4, forming strong hydrogen bonds with the target protein, was key to its potent antifungal action (Luo et al. 2024).

Yang et al. (2021) synthesized 30 novel thiophene/furan-1,3,4-oxadiazole carboxamide derivatives as potential SDHI antifungal inhibitors. These compounds exhibited moderate to strong antifungal activity against seven phytopathogenic fungi, with Compound 10.5a showing superior potency against *Sclerotinia sclerotiorum*, surpassing boscalid. SAR analysis revealed that thiophene rings and electron-withdrawing groups on the benzylamine moiety enhanced antifungal efficacy, particularly against *S. sclerotiorum*. Compounds with fluorine groups also demonstrated significant inhibitory effects. Molecular docking and dynamics simulations confirmed Compound 10.5a's potential as an SDHI, offering a promising alternative to boscalid for crop protection (Yang et al. 2021).

Jiang (2022) synthesized 36 furan/thiophene carboxamides with 4,5-dihydropyrazole rings, assessing their antifungal activity against pathogens like

Compound 10.6

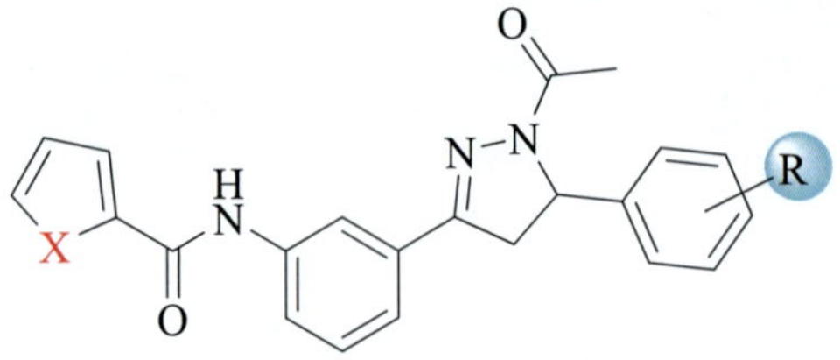

6a, 6b, 6c	**X = O**
6a	**R = 3,4-F2**
6b	**R = 3-Br, 4-F**
6c	**R = 3,4,5 - F3**

Penicillium digitatum, Physalospora piricola, Erysiphe graminis, and *Botrytis cinerea.* Compound 10.6a–c showed exceptional inhibitory effects, outperforming fluxapyroxad, especially against *B. cinerea.* Molecular dynamics simulations confirmed Compound 10.6c's higher affinity for SDH than fluxapyroxad. SAR analysis revealed that thiophene rings generally enhanced antifungal activity, while $-OCF_3$ substitutions on phenyl rings negatively impacted performance. Electron-withdrawing groups like –F and – Br also significantly influenced activity (Jiang et al. 2022).

10.16 Structure–Activity Relationship (SAR for SDH Inhibitors)

Aromatic rings, such as substituted benzene or biphenyl, boost SDH inhibition, with specific substituents enhancing antifungal activity. The pyrrolidine moiety plays a key role in increasing biological activity, while adding fluorine atoms at the 3′- and 4′-positions significantly improves fungicidal effectiveness (Jiang et al. 2022; Luo et al. 2024). Thiophene rings show stronger antifungal action compared to furan rings. Electron-withdrawing groups like fluoro or trifluoromethyl enhance inhibition, especially against *Sclerotinia sclerotiorum* (Yang et al., 2021). The antifungal potency of a compound depends on the location of substituents on the benzene ring. Comparing *para*-substituted compounds to their ortho- or meta-substituted counterparts exhibits greater inhibitory activity frequently. The substituents' location on the benzene ring and the length of the linker between the amide and benzene also impact potency. Length of linker: Antifungal activity is influenced by the length of the linker that connects the amide moiety and benzene ring. In comparison to longer linkers (e.g., ethylene linkers, n = 2), shorter linkers such as methylene linkers (n = 1) generally contribute to higher antifungal efficacy, as shown in Fig. 10.7. Effective inhibition is facilitated by hydrogen bonding with the SDH target, revealed through molecular docking, further aids inhibition (Yang et al. 2021). Region-specific effects, such as fluorine placement, contribute to broad-spectrum activity, and compounds outperforming established SDHIs like fluopyram and boscalid offer strong potential for future development (Vielba-Fernández et al. 2021; Lee et al. 2022).

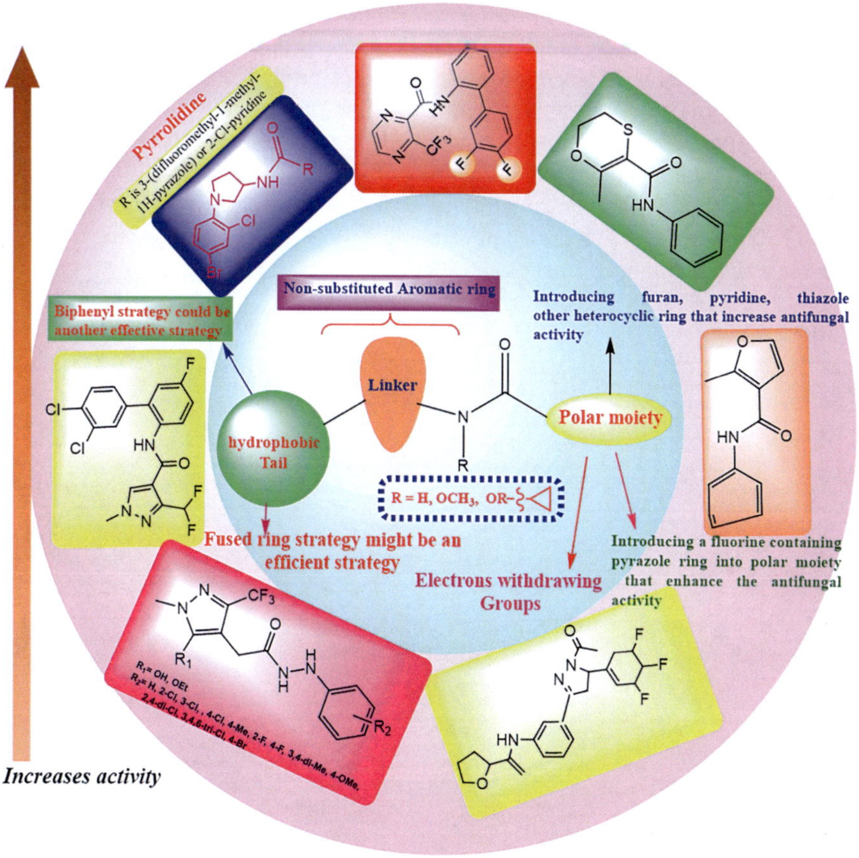

Fig. 10.7 Common SAR for succinate dehydrogenase inhibitors

10.17 Reported Scaffolds as Fungicidal SDH Inhibitor in Agriculture

Carboxamides

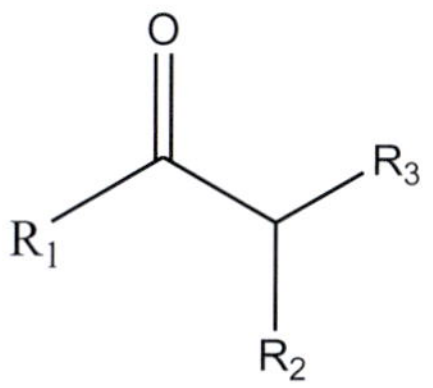

Carboxamide fungicides are crucial in global agriculture, combating plant diseases like *Botrytis Cinerea* and *Sclerotinia* spp. (Yan et al. 2022b; Zhang et al. 2022). The development began with carboxin—the first of its kind—and has since

evolved with compounds like thifluzamide, furametpyr, mepronil, and flutolanil, which target specific fungal infections. Notably, N-(biphenyl-2-yl)-2-methyl-1,4-thioxolane-3-carboxamide offers broad-spectrum control. These fungicides target SDH, a key enzyme in fungal energy metabolism, disrupting ATP synthesis and the electron transport chain, leading to fungal cell death. The carboxamide scaffold, with its flexible amide chain, underpins their efficacy. Recent innovations include pyrazole carboxamide fungicides with a diarylamine scaffold, which have shown significant antifungal activity in both laboratory and field settings, particularly against rice sheath blight (Brauer et al. 2019).

Nicotinamide

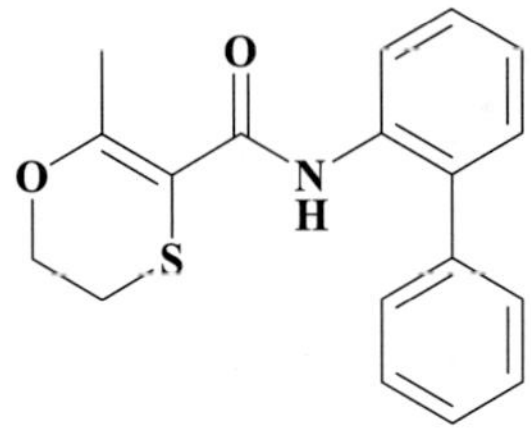

N-(biphenyl-2-yl)-2-methyl-1,4-thioxolane-3-carboxamide

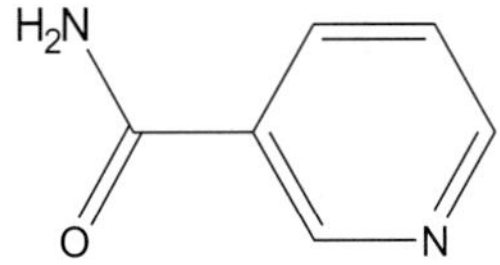

Nicotinamide

Recent research has revealed that nicotinamide (NAM), best known as a precursor to vitamin B3, also possesses unexpected antifungal properties. Studies indicate that nicotinamide suppresses the growth of several fungal pathogens, including *Cryptococcus neoformans, Fusarium oxysporum*, and *Candida albicans* (Yan et al. 2022a). Nicotinamide also helps antifungal drugs that are already marketed to work better, like amphotericin B and nystatin potentially reducing dosage requirements and associated side effects. Traditional antifungal treatments are not effective against the protective communities known as fungal biofilms. By preventing these communities from formation, nicotinamide may help. A crucial energy source for the growth of fungi, nicotinamide initiates the salvage pathway for NAD^+ production. However, elevated NAD^+ levels inside cells may result in oxidative stress, which would be detrimental to fungal cells. The histone deacetylase enzymes, which control the expression of particular genes, may be inhibited by it. This could disrupt essential fungal metabolic pathways and the cell cycle. It might also interact with fungal membranes, changing their permeability and causing cell death (Xing et al. 2019).

Benzothiazole

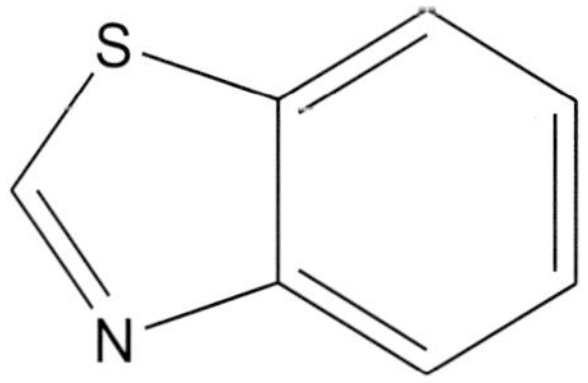

Benzothiazole

Current antifungal agents like azoles and antimetabolites, while effective, face issues such as low bioavailability, drug resistance, and limited spectrum. To address these, there is a need for new antifungals with high potency, broad activity, and minimal resistance. Benzothiazoles present a promising solution, showing strong efficacy against a range of plant diseases. They work by selectively inhibiting SDH, a crucial enzyme in fungal energy metabolism. By mimicking succinate, benzothiazoles competitively bind to the SDH active site, blocking succinate and disrupting the electron transport chain, which halts ATP production and kills fungal cells. Their broad-spectrum activity against various fungal groups and long-lasting protection makes benzothiazoles a valuable addition to fungicide options (Teixeira et al. 2022; Christophe 2023; Zou et al. 2023).

Pyrazine-Carboxamide-Diphenyl-Ethers

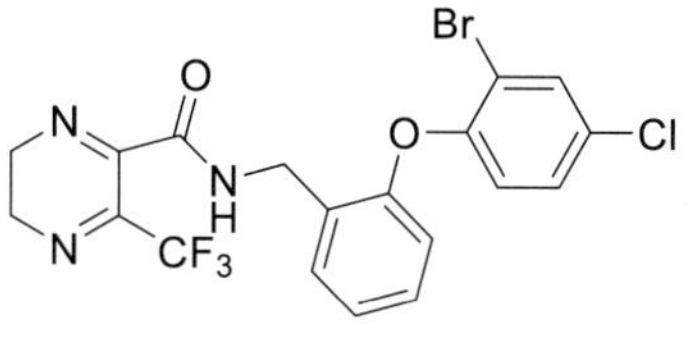

Pyrazine-carboxamides, valued for their potent antifungal activity and broad-spectrum efficacy as SDH inhibitors, have faced resistance issues due to prolonged use. This highlights the need for new SDHIs. Advances in drug discovery, including artificial intelligence and high-performance computing, have led to the development of the pyrazine-carboxamide-diphenyl-ethers scaffold, a novel combination of active fragments from pyraziflumid, flubeneteram, pydiflumetofen, and fluopyram. These compounds exhibit potent antifungal activity by binding to an allosteric site near the ubiquinone-binding pocket, blocking electron transport and ATP synthesis, which kills fungal cells. Their potent fusion with SDH enhances their potent fungicidal activity against a variety of fungal infections, such as Ascomycetes, Basidiomycetes, and Deuteromycetes. They offer effective and long-lasting protection against a range of fungal infections while being safe for most plants (Matsa et al. 2022; Liu et al. 2022a).

10.18 Resistance to SDH Inhibitors in Fungicides

Mutations, gene flow, and selection all contribute to the evolution of resistance to SDHIs in fungicides. Identifying baseline sensitivity, comprehending genetic adaptability, testing in the absence of naturally resistant strains, elucidating inheritance patterns, evaluating cross-resistance, modeling evolutionary potential, investigating mitigation strategies, and putting counter-resistance tactics into practice are the steps involved in addressing this. The effect of SDHI fungicides on fungal infections is revealed by an investigation spanning more than 50 years. With nine chemical groups and nineteen active ingredients, the SDHI group demonstrates cross-resistance. TCA cycle is inhibited by SDHIs, which impacts fungal mitochondrial respiration (Sierotzki and Scalliet 2013).

Penthiopyrad and bixafen are two examples of a chemical group of fungicides that have exerted high selection pressure on fungal pathogens on various plant hosts; however, certain SDHIs, such as fluopyram, have shown to exhibit greater activity against pathogens on broad-leafed plants as opposed to graminaceous hosts (Sang and Lee 2020). Mutations in the SDHI target genes, *SDHB*, *SDHC*, and *SDHD* (Fig. 10.8), which reduce the binding affinity for SDHI fungicides, are the primary

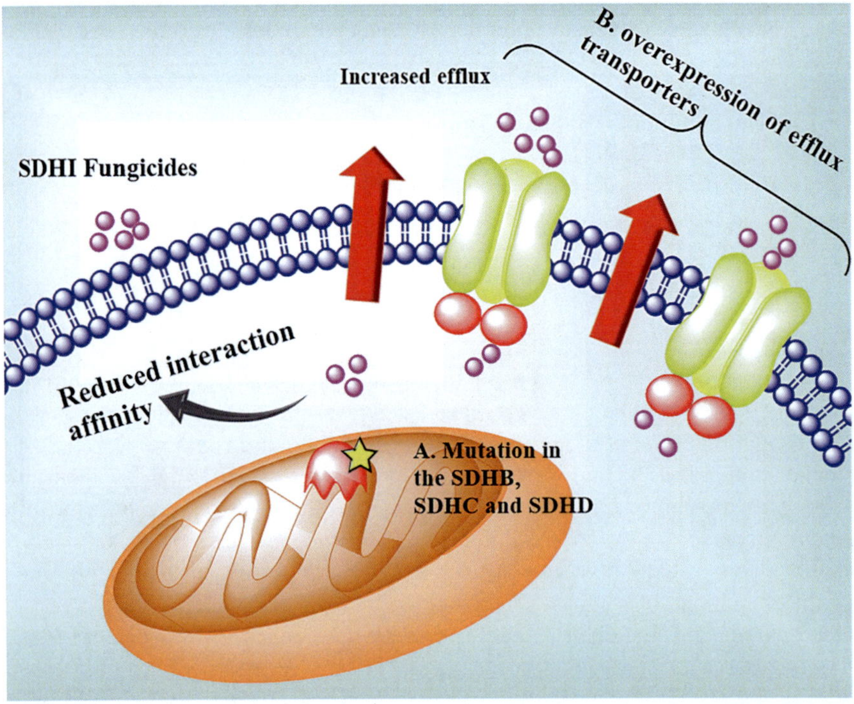

Fig. 10.8 Molecular mechanisms underlying plant pathogenic fungus' resistance to SDHI. (**a**) A mutation in any one of the *SDHB*, *SDHC*, or *SDHD* genes causes the binding affinity of SDHI fungicides to be reduced. (**b**) Reduced intracellular accumulation of SDHI fungicides is brought about by overexpressing efflux transporters

molecular mechanism of SDHI resistance in fungi (Cecchini 2003). Furthermore, lower intracellular accumulation of SDHI fungicides is brought about by overexpression of efflux transporter(s) in fungal species, resulting in decreased sensitivity to SDHI fungicides (Sierotzki and Scalliet 2013).

10.19 Exploring Resistance Mutations in the *SDH* Gene

The main cause of fungicide resistance to SDHIs is gene mutations (Fig. 10.9) related to SDH. When exposed to UV light, mutants of *Aspergillus nidulans* and *Ustilago maydis* showed resistance, leading to the first cases that were recorded in the 1970s (Georgopoulos et al. 1972; Tyul 1975; Georgopoulos and Ziogas 1977). According to recent observations, SDH gene mutations cause resistance in species such as *Alternaria alternata* and *Corynespora cassiicola* (Yang et al. 2015). Field populations have generated reports of over 27 resistance-related mutations, including SDH gene mutations. Multidrug resistance is also caused by mechanisms that entail the overexpression of efflux pump transporters like ShatrD and ShPDR1 (Popko et al. 2012, 2018). Three types of multidrug-resistant strains have been identified:

- MDR1 type strains (overexpression of ABC transporters AtrB)
- MDR2 type strains (overexpression of MFS transporters MFS2)
- MDR3 type strains (overexpressing both transporters)

Resistance to SDHIs in various fungi is a growing concern. Resistance to eight fungicides was brought about by mutations in SDHB (P225F) in *Botrytis cinerea* and resistance in *B. cinerea* has apple orchards, vineyard, and strawberry farms. *Zymoseptoria tritici* showed resistance to SDHB and SDHC mutations, indicating resistance to SDHIs. Many mutations were found to be responsible for the

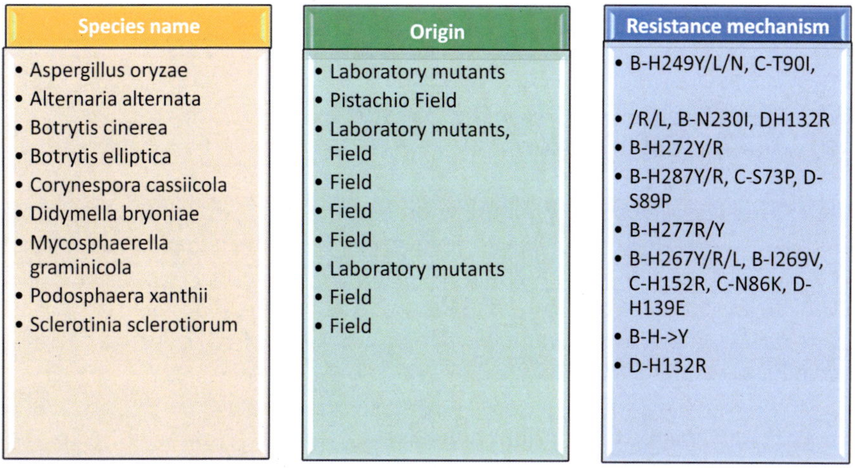

Fig. 10.9 Resistance mutation, origin, and resistant mechanism

resistance that other phytopathogenic fungi, such as *Corynespora cassiicola*, *Podosphaera xanthii*, and *Didymella bryoniae*, developed (Leroux et al. 2010; Leroch et al. 2011; Yin et al. 2011).

10.20 Resistance Detection, Transmission, and Surveillance

In Fig. 10.10, the ability of molecular diagnostic techniques to identify drug-resistant fungal species and genetic markers associated with antifungal resistance has been demonstrated, albeit underutilized.

- Clinical specimens can be directly applied to them due to their sensitivity, which expedites the process and removes the need for a culture.
- Direct sequencing of genes encoding drug target proteins (e.g., CYP51A in *A. fumigatus* or ERG11 in Candida spp.) can be used to identify putative resistance-associated mutations.
- Fungal samples can be obtained in a variety of ways from the environment or from the clinic. Typically, a well-known microbiology technique can be used to culture and isolate these samples, preparing them for the extraction of genomic DNA.
- These DNA fragments are used to build a sequencing library for whole-genome sequencing (WGS). There are many sequencing technologies available, offering both long- and short-read sequence data. Utilizing cloud computing or local processing, a quality check must be performed prior to mapping raw sequencing data to a reference genome. The identification of high-confidence single-nucleotide polymorphisms (SNPs) can help predict both drug resistance-related alleles and their evolutionary backgrounds.
- Phylodynamic inference and the development of interactive web portals (such as Micro react or Next strain), that are available to researchers and medical professionals, enable the tracking of transmission episodes (Hadfield et al. 2018; Fisher et al. 2022).

10.21 Case Studies on Diseases Caused by Fungicidal Infection

Case 1: Fungicidal Effect in Philippines on Banana

This case study was focused on Banana producer in Southern Philippines in the year 2014–2017 against a disease Fusarium wilt caused by *Fusarium oxysporum f.* sp. *cubense* tropical race4 (Foc TR4). Foc TR4 widely affects the wide range of banana variety mainly grown in tropical and subtropical regions of Asia, Africa, and the Middle East (Montiflor et al. 2019).

Case 2: Pyrenophora teres (Barley Pathogen)

This case study investigates the effects of the *Pyrenophora teres* fungus, which can cause a 10–40% reduction in yield when it causes Net Blotch disease on barley crop. Fungicide resistance is a medium risk associated with this fungus. Many

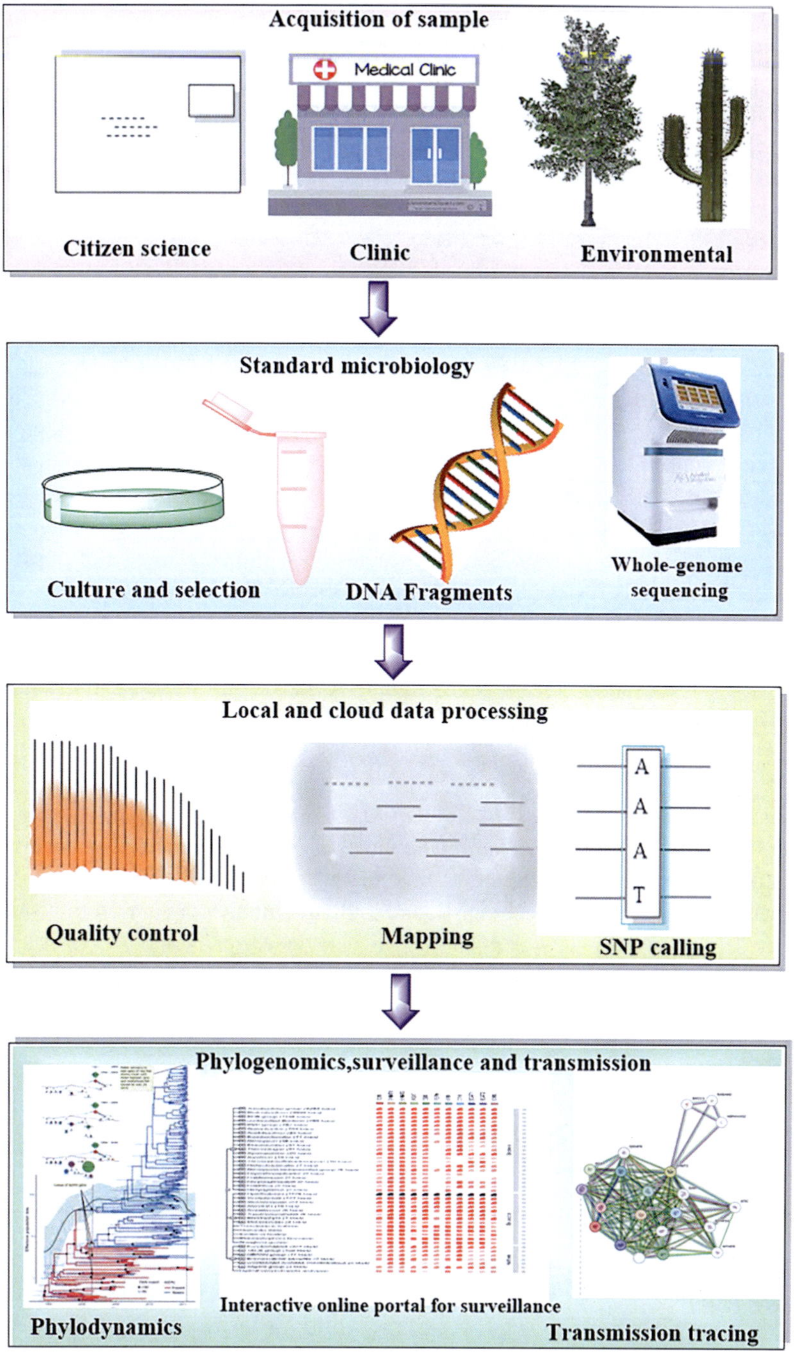

Fig. 10.10 Resistance detection, transmission, and surveillance

fungicides are used, but resistance to them is a concern, especially with QoI fungicides associated with the F129L mutation, which have less resistance than G143A in other pathogens. In Europe, F129L frequency is typically about 25%; in 2015, it peaked in the United Kingdom and Northern Germany. While resistance to SDHI increased in Northern Germany and France between 2012 and 2014 before stabilizing by 2015, resistance to DMIs and APs is generally low and stable. Particular mutations (like G79R) linked to moderate resistance were found (Minaříková and Polišenská 1999; Huang and Millar 2013).

Case 3: Zymoseptoria tritici (Wheat Pathogen)
The study highlights the medium risk of fungicide resistance in *Zymoseptoria tritici*, the cause of Septoria blotch in wheat. European cereal production depends on the control of this disease, resulting in an annual fungicide expenditure of $1.2 billion. SDHIs, DMIs, and multisites are widely used in the management of Septoria leaf blotch, while QoIs continue to play a role in Eastern and Southern Europe (O'Driscoll et al. 2014). After testing over 5000 isolates of *Z. tritici* over an 11-year period, stable DMI sensitivity was found in 2015. Since 2004, the SDHI sensitivity has been observed, and forward genetics studies have shown evidence of possible resistance. Moderately resistant mutations such as sdhC-H152R were found in 2014 and continued to be found at low frequencies in the following years. The data suggests that *Z. tritici* has evolved SDHI resistance more slowly than other pathogens. Other fungal species have homologous R-alleles to sdhC-H152R, suggesting a more extensive context for this mutation (Huang and Millar 2013; Gutiérrez-Alonso et al. 2017).

Case 4: Ramularia collo-cygni (Barley Pathogen)
The pathogen *Ramularia collo-cygni*, which causes Ramularia Leaf Spot in barley, is considered a high-risk pathogen for fungicide resistance by FRAC. The pathogen is currently controlled with SDHI, DMI, and multisite fungicides such as chlorothalonil. In 2014, UV mutants were generated to investigate potential SDHI resistance in *R. collo-cygni*. In 2015, Germany identified the first genotypes with significantly reduced sensitivity to SDHI, carrying H142R or H149R mutations. The H142R mutation is similar to those in other pathogens, while the H149R mutation is homologous to mutations in *Z. tritici* and *Venturia inaequalis*. This study sheds light on the evolving resistance of *R. collo-cygni* to SDHI fungicides, underscoring the importance of effective management strategies (Gutiérrez-Alonso et al. 2017; Kiiker et al. 2021).

10.22 Challenges and Future Directions

Fungicides have been widely used around the world for many years and decades to protect and improve the quality of crop yields. There is still a worldwide revolution dealing with a number of provocations, including issues concerning health, the environment, and fungal genetic evolution. Air, water, and land pollution are among the environmental issues brought on by the use of fungicides, whereas overuse of fungicides can contaminate groundwater by absorbing into the soil, inadequate dispersion can result in air pollution. Research suggests that there are microparticles of pesticides in groundwater.

Pesticides pose a health risk because they are persistent in the soil, easily absorbed by plants through different crop plant parts, and they enter in food chain. Long-term consumption of crops grown in high fungicide-use areas has been associated with common health problems, most notably cancer, from exposure to the environment and pesticide-contaminated food. A common occurrence for both farmers and consumers who eat crops grown with high fungicide levels is allergic reactions.

The emergence of drug-resistant fungal strains within particular fungal kingdoms, resulting from prolonged exposure to fungicides, presents a significant challenge. The effectiveness of current fungicides is significantly hampered by the evolution of fungal genes. Progress in research and development is essential to comprehending the dynamic behavior of fungal drug resistance, which is the cause of the destruction of crop plants and their fruits.

Fungicides can be used wisely and sustainably to mitigate these fundamental challenges, even though they may not be fully overcome. To address these issues, strict guidelines and appropriate usage procedures are essential. Looking ahead, the role of fungicides in improving environmental and health outcomes—as well as contributing to sustainable development—remains vital. While the potentially harmful consequences affecting both biological and non-biological systems cannot be ignored, researchers must concentrate on moving forward with their work.

From an outlook on the future, sustainable development becomes apparent as an imperative. In order to minimize negative effects, fungicidal investigations are necessary given the current challenges. A primary focus should be on biodegradable synthesis. Since fungicide use would severely reduce crop productivity and food supply, it is not feasible to completely stop using them given the size of the world's population. To shape future perspectives, it is therefore essential to investigate the factors that lead to environmental and toxicological effects on both aquatic and terrestrial ecosystems. In light of the increasing occurrence of biocide-resistant fungi to, the structural properties of the protein targets should form the foundation for designing promising fungicides.

10.23 Conclusion

In conclusion, SDHIs have emerged as a pivotal advancement in managing fungal diseases in agriculture by effectively targeting the SDH enzyme, which is integral to the Krebs cycle and mitochondrial respiratory chain. By disrupting SDH activity, these fungicides impair cellular respiration and energy production in fungi, providing broad-spectrum control over a wide range of pathogens. This targeted approach represents a significant improvement over traditional broad-spectrum fungicide. However, the rising prevalence of fungicide-resistant fungal strains, along with concerns about environmental pollution and health risks, highlights the urgent need for sustainable fungicide practices. The adverse effects of synthetic chemical fungicides have prompted the exploration of alternative strategies, such as biological soil disinfestation and bio fumigation. These methods, including the use of Trichoderma species, offer environmentally friendly solutions by enhancing soil health and providing natural pathogen control.

Researchers must focus on understanding fungal resistance mechanisms and developing innovative, eco-friendly fungicides to ensure long-term agricultural productivity and food safety. Future directions for SDHIs should focus on developing optimized formulations, refining application techniques, and incorporating biodegradable alternatives to reduce environmental and health impacts. A deeper understanding of fungal resistance mechanisms will be crucial in guiding the development of next-generation SDHIs with improved efficacy and specificity. Adopting a holistic approach that combines innovative fungicide technologies with sustainable agricultural practices will be essential for maintaining global food security. By balancing effective pathogen control with environmental stewardship and public health considerations, SDHIs can continue to play a vital role in agriculture. This strategy will help ensure the resilience and sustainability of agricultural systems in the face of evolving fungal threats and environmental challenges.

Acknowledgments The authors are thankful to The Department of Pharmacy and the authority of Guru Ghasidas Vishwavidyalaya, Bilaspur, Chhattisgarh, India, 495009, for providing the necessary facilities.

Conflict of Interest The authors declare no conflict of interest.

References

Abe M, Tsushima K, Sakayori M, Suzuki K, Ikari J, Terada J, Tatsumi K (2018) Utility of nintedanib for severe idiopathic pulmonary fibrosis: a single-center retrospective study. Drug Des Devel Ther 12:3369–3375. https://doi.org/10.2147/DDDT.S179427

Amara AA, El-Baky NA (2023) Fungi as a source of edible proteins and animal feed. J Fungi 9. https://doi.org/10.3390/jof9010073

Andrews KA, Ascher DB, Pires DEV, Barnes DR, Vialard L, Casey RT, Bradshaw N, Adlard J, Aylwin S, Brennan P, Brewer C, Cole T, Cook JA, Davidson R, Donaldson A, Fryer A, Greenhalgh L, Hodgson SV, Irving R, Lalloo F, McConachie M, McConnell VPM, Morrison PJ, Murday V, Park SM, Simpson HL, Snape K, Stewart S, Tomkins SE, Wallis Y, Izatt L, Goudie D, Lindsay RS, Perry CG, Woodward ER, Antoniou AC, Maher ER (2018) Tumour risks and genotype–phenotype correlations associated with germline variants in succinate dehydrogenase subunit genes SDHB, SDHC and SDHD. J Med Genet 55:384–394. https://doi.org/10.1136/jmedgenet-2017-105127

Aspuria P-JP, Lunt SY, Väremo L, Vergnes L, Gozo M, Beach JA, Salumbides B, Reue K, Wiedemeyer WR, Nielsen J, Karlan BY, Orsulic S (2014) Succinate dehydrogenase inhibition leads to epithelial-mesenchymal transition and reprogrammed carbon metabolism. Cancer Metab 2:21. https://doi.org/10.1186/2049-3002-2-21

Avenot HF, Michailides TJ (2010) Progress in understanding molecular mechanisms and evolution of resistance to succinate dehydrogenase inhibiting (SDHI) fungicides in phytopathogenic fungi. Crop Prot 29:643–651. https://doi.org/10.1016/j.cropro.2010.02.019

Barnes EH (2012) Atlas and manual of plant pathology, 1st edn. Springer, New York

Baysal BE, Ferrell RE, Willett-Brozick JE, Lawrence EC, Myssiorek D, Bosch A, van der Mey A, Taschner PE, Rubinstein WS, Myers EN, Richard CW 3rd, Cornelisse CJ, Devilee P, Devlin B (2000) Mutations in SDHD, a mitochondrial complex II gene, in hereditary paraganglioma. Science 80(287):848–851. https://doi.org/10.1126/science.287.5454.848

Bénit P, Goncalves J, El Khoury R, Rak M, Favier J, Gimenez-Roqueplo AP, Rustin P (2022) Succinate dehydrogenase, succinate, and superoxides: a genetic, epigenetic, metabolic, environmental explosive crossroad. Biomedicines 10:1–21. https://doi.org/10.3390/biomedicines10081788

Bezawork-Geleta A, Dong L, Rohlena J, Neuzil J (2016) The assembly factor SDHAF2 is dispensable for flavination of the catalytic subunit of mitochondrial complex II in breast cancer cells. J Biol Chem 291:21414–21420. https://doi.org/10.1074/jbc.C116.755017

Bouillaud F (2023) Inhibition of succinate dehydrogenase by pesticides (SDHIs) and energy metabolism. Int J Mol Sci 24:4045. https://doi.org/10.3390/ijms24044045

Brauer VS, Rezende CP, Pessoni AM, De Paula RG, Rangappa KS, Nayaka SC, Gupta VK, Almeida F (2019) Antifungal agents in agriculture: friends and foes of public health. Biomol Ther 9(10):521

Brière J-J, Favier J, Ghouzzi VE, Djouadi F, Bénit P, Gimenez AP, Rustin P (2005) Succinate dehydrogenase deficiency in human. Cell Mol Life Sci 62:2317–2324. https://doi.org/10.1007/s00018-005-5237-6

Brischigliaro M, Zeviani M (2021) Cytochrome c oxidase deficiency. Biochim Biophys Acta Bioenerg 1862:148335. https://doi.org/10.1016/j.bbabio.2020.148335

Caputo R, Ferreri C, Guaragna A, Palumbo G, Pedatella S (1995) New synthesis of carboxin and oxycarboxin pesticides: application to the preparation of their new analogues substituted at the C-2 methyl group. J Chem Soc Perkin Trans 1:1971-1973. https://doi.org/10.1039/p19950001971

Cecchini G (2003) Function and structure of complex II of the respiratory chain. Annu Rev Biochem 72:77–109. https://doi.org/10.1146/annurev.biochem.72.121801.161700

Chai JQ, Mei YD, Tai L, Wang XB, Chen M, Kong XY, Lu AM, Li GH, Yang CL (2023) Potential succinate dehydrogenase inhibitors bearing a novel Pyrazole-4-sulfonohydrazide scaffold: molecular design, antifungal evaluation, and action mechanism. J Agric Food Chem 71:9266–9279. https://doi.org/10.1021/acs.jafc.3c00126

Chandler D, Bailey AS, Mark Tatchell G, Davidson G, Greaves J, Grant WP (2011) The development, regulation and use of biopesticides for integrated pest management. Philos Trans R Soc B Biol Sci 366:1987–1998. https://doi.org/10.1098/rstb.2010.0390

Chen WM, Olson P, Arcot R, Nguyen H, Quereshi F, Kokenakes C, Cher ML (2021) Large retroperitoneal paraganglioma associated with germline mutation of the succinate dehydrogenase gene. J Kidney Cancer VHL 8:12–18. https://doi.org/10.15586/jkcvhl.v8i1.129

Christophe T (2023) Benzothiazole as a promising scaffold for the development of antifungal agents. Curr Top Med Chem 23:491–519

D'Hose D, Isenborghs P, Brusa D, Jordan BF, Gallez B (2021) The short-term exposure to SDHI fungicides Boscalid and Bixafen induces a mitochondrial dysfunction in selective human cell lines. Molecules 26:5842. https://doi.org/10.3390/molecules26195842

Deising HB, Reimann S, Pascholati SF (2008) Mechanisms and significance of fungicide resistance. Braz J Microbiol 39:286–295. https://doi.org/10.1590/S1517-83822008000200017

Dias DA, Urban S, Roessner U (2012) A historical overview of natural products in drug discovery. Meta 2:303–336. https://doi.org/10.3390/metabo2020303

Ding J-L, Li X-H, Lei J-H, Feng MG, Ying SH (2022) Succinate dehydrogenase subunit C contributes to mycelial growth and development, stress response, and virulence in the insect parasitic fungus Beauveria bassiana. Microbiol Spectr 10(5):e0289122 https://doi.org/10.1128/spectrum.02891-22

Duarte Hospital C, Tête A, Debizet K, Imler J, Tomkiewicz-Raulet C, Blanc EB, Barouki R, Coumoul X, Bortoli S (2023) SDHi fungicides: an example of mitotoxic pesticides targeting the succinate dehydrogenase complex. Environ Int 180:108219. https://doi.org/10.1016/j.envint.2023.108219

Dwight T, Na U, Kim E, Zhu Y, Richardson AL, Robinson BG, Tucker KM, Gill AJ, Benn DE, Clifton-Bligh RJ, Winge DR (2017) Analysis of SDHAF3 in familial and sporadic pheochromocytoma and paraganglioma. BMC Cancer 17:497. https://doi.org/10.1186/s12885-017-3486-z

El-Baky NA, Amara AAAF (2021) Recent approaches towards control of fungal diseases in plants: an updated review. J Fungi 7. https://doi.org/10.3390/jof7110900

Elkhairy BM, Salama NM, Desouki AM, Abdelrazek AB, Soliman KA, Ibrahim SA, Khalil HB (2023) Towards unlocking the biocontrol potential of Pichia kudriavzevii for plant fungal diseases: in vitro and in vivo assessments with candidate secreted protein prediction. BMC Microbiol 23:1–18. https://doi.org/10.1186/s12866-023-03047-w

Fan F, Sam R, Ryan E, Alvarado K, Villa-Cuesta E (2019) Rapamycin as a potential treatment for succinate dehydrogenase deficiency. Heliyon 5:e01217. https://doi.org/10.1016/j.heliyon.2019.e01217

Fernández-Ortuño D, Pérez-García A, Chamorro M, de la Peña E, de Vicente A, Torés JA (2017) Resistance to the SDHI fungicides Boscalid, Fluopyram, Fluxapyroxad, and penthiopyrad in Botrytis cinerea from commercial strawberry fields in Spain. Plant Dis 101:1306–1313. https://doi.org/10.1094/PDIS-01-17-0067-RE

Fisher MC, Alastruey-Izquierdo A, Berman J, Bicanic T, Bignell EM, Bowyer P, Bromley M, Brüggemann R, Garber G, Cornely OA, Gurr SJ, Harrison TS, Kuijper E, Rhodes J, Sheppard DC, Warris A, White PL, Xu J, Zwaan B, Verweij PE (2022) Tackling the emerging threat of antifungal resistance to human health. Nat Rev Microbiol 20:557–571. https://doi.org/10.1038/s41579-022-00720-1

Gawryluk RM, Gray MW (2009) A split and rearranged nuclear gene encoding the iron-sulfur subunit of mitochondrial succinate dehydrogenase in Euglenozoa. BMC Res Notes 2:16. https://doi.org/10.1186/1756-0500-2-16

Georgopoulos SG, Ziogas BN (1977) A new class of carboxin-resistant mutants of Ustilago maydis. Neth J Plant Pathol 83:235–242. https://doi.org/10.1007/BF03041438

Georgopoulos SG, Alexandri E, Chrysayi M (1972) Genetic evidence for the action of Oxathiin and Thiazole derivatives on the succinic dehydrogenase system of Ustilago maydis mitochondria. J Bacteriol 110:809–817. https://doi.org/10.1128/jb.110.3.809-817.1972

Gooßen LJ, Zimmermann B, Knauber T (2008) Palladium/copper-catalyzed Decarboxylative cross-coupling of aryl chlorides with potassium carboxylates. Angew Chem Int Ed 47:7103–7106. https://doi.org/10.1002/anie.200800728

Goulin EH, de Lima TA, dos Santos PJC, Machado MA (2022) RNAi-induced silencing of the succinate dehydrogenase subunits gene in Colletotrichum abscissum, the causal agent of postbloom fruit drop (PFD) in citrus. Microbiol Res 260. https://doi.org/10.1016/j.micres.2021.126938

Gullino ML, Leroux P, Smith CM (2000) Uses and challenges of novel compounds for plant disease control. Crop Prot 19:1–11. https://doi.org/10.1016/S0261-2194(99)00095-2

Guo J, Lemire BD (2003) The ubiquinone-binding site of the saccharomyces cerevisiae succinate-ubiquinone oxidoreductase is a source of superoxide. J Biol Chem 278:47629–47635. https://doi.org/10.1074/jbc.M306312200

Guo P, Ren Y, Pan X, Xu J, Wu X, Zheng Y, Du F, Dong F (2024) Stereoselective bioactivity and action mechanism of the fungicide isopyrazam. J Agric Food Chem 72:18909–18917. https://doi.org/10.1021/acs.jafc.4c06270

Gutiérrez-Alonso O, Hawkins NJ, Cools HJ, Shaw MW, Fraaije BA (2017) Dose-dependent selection drives lineage replacement during the experimental evolution of SDHI fungicide resistance in Zymoseptoria tritici. Evol Appl 10:1055–1066. https://doi.org/10.1111/eva.12511

Hadfield J, Megill C, Bell SM, Huddleston J, Potter B, Callender C, Sagulenko P, Bedford T, Neher RA (2018) Nextstrain: real-time tracking of pathogen evolution. Bioinformatics 34:4121–4123. https://doi.org/10.1093/bioinformatics/bty407

Hahn M (2014) The rising threat of fungicide resistance in plant pathogenic fungi: botrytis as a case study. J Chem Biol 7:133–141. https://doi.org/10.1007/s12154-014-0113-1

Her YF, Maher LJ (2015) Succinate dehydrogenase loss in familial paraganglioma: biochemistry, genetics, and epigenetics. Int J Endocrinol 2015:1–14. https://doi.org/10.1155/2015/296167

Holt DE, Henthorn P, Howell VM, Robinson BG, Benn DE (2014) Succinate dehydrogenase subunit D and succinate dehydrogenase subunit B mutation analysis in canine phaeochromocytoma and paraganglioma. J Comp Pathol 151:25–34. https://doi.org/10.1016/j.jcpa.2014.03.007
Huang S, Millar AH (2013) Succinate dehydrogenase: the complex roles of a simple enzyme. Curr Opin Plant Biol 16:344–349. https://doi.org/10.1016/j.pbi.2013.02.007
Iftikhar S, Bengyella L, Shahid AA, Nawaz K, Anwar W, Khan AA (2022) Discovery of succinate dehydrogenase candidate fungicides via lead optimization for effective resistance management of *Fusarium oxysporum* f. sp. *capsici*. 3 Biotech 12:102. https://doi.org/10.1007/s13205-022-03157-8
Ishii H, Zhen F, Hu M, Li X, Schnabel G (2016) Efficacy of SDHI fungicides, including benzovindiflupyr, against Colletotrichum species. Pest Manag Sci 72:1844–1853. https://doi.org/10.1002/ps.4216
Jain A, Sarsaiya S, Wu Q, Lu Y, Shi J (2019) A review of plant leaf fungal diseases and its environment speciation. In: Hammer ND, Kline KA (eds) Bioengineered. Springer, Boston, MA, pp 409–424
Jiang W, Zhang T, Wang J et al (2022) Design, synthesis, inhibitory activity, and molecular modeling of novel pyrazole-furan/thiophene carboxamide hybrids as potential fungicides targeting succinate dehydrogenase. J Agric Food Chem. https://doi.org/10.1021/acs.jafc.2c05054
Jin S, Hu Y, Fu H et al (2021) Identification and characterization of the succinate dehydrogenase complex iron sulfur subunit B gene in the oriental river prawn, macrobrachium nipponense. Front Genet 12. https://doi.org/10.3389/fgene.2021.698318
Khan RAA, Najeeb S, Hussain S, Xie B, Li Y (2020) Bioactive secondary metabolites from Trichoderma spp. against Phytopathogenic fungi. Microorganisms 8:817. https://doi.org/10.3390/microorganisms8060817
Kiiker R, Juurik M, Mäe A (2021) Fungicide resistance evolving in Ramularia collo-cygni population in Estonia. Microorganisms 9:1514. https://doi.org/10.3390/microorganisms9071514
Kilaru S, Schuster M, Studholme D, Soanes D, Lin C, Talbot NJ, Steinberg G (2015) A codon-optimized green fluorescent protein for live cell imaging in Zymoseptoria tritici. Fungal Genet Biol 79:125–131. https://doi.org/10.1016/j.fgb.2015.03.022
Kim HJ, Khalimonchuk O, Smith PM, Winge DR (2012) Structure, function, and assembly of heme centers in mitochondrial respiratory complexes. Biochim Biophys Acta, Mol Cell Res 1823:1604–1616. https://doi.org/10.1016/j.bbamcr.2012.04.008
Kregiel D (2012) Succinate dehydrogenase of Saccharomyces cerevisiae – the unique enzyme of TCA cycle – current knowledge and new perspectives. In: Dehydrogenases. InTech
Lee S, Abed DA, Nguyen M-U, Verzi MP, Hu L (2022) Structure-activity relationships of 1,4-bis(arylsulfonamido)-benzene or naphthalene-N,N′-diacetic acids with varying C2-substituents as inhibitors of Keap1-Nrf2 protein-protein interaction. Eur J Med Chem 237:114380. https://doi.org/10.1016/j.ejmech.2022.114380
Lemarie A, Grimm S (2011) Mitochondrial respiratory chain complexes: apoptosis sensors mutated in cancer? Oncogene 30:3985–4003. https://doi.org/10.1038/onc.2011.167
Leroch M, Kretschmer M, Hahn M (2011) Fungicide resistance phenotypes of Botrytis cinerea isolates from commercial vineyards in south West Germany. J Phytopathol 159:63–65. https://doi.org/10.1111/j.1439-0434.2010.01719.x
Leroux P, Gredt M, Leroch M, Walker A-S (2010) Exploring mechanisms of resistance to respiratory inhibitors in field strains of Botrytis cinerea, the causal agent of gray mold. Appl Environ Microbiol 76:6615–6630. https://doi.org/10.1128/AEM.00931-10
Li M, Wang W, Cheng X, Wang Y, Chen Y, Gong J, Chang X, Lv X (2023) Design, synthesis, and evaluation of antifungal bioactivity of novel Pyrazole Carboxamide Thiazole derivatives as SDH inhibitors. J Agric Food Chem 71:11365–11372. https://doi.org/10.1021/acs.jafc.3c02671
Liu D, Luo L, Wang Z, Ma X, Gan X (2022a) Design, synthesis and antifungal/nematicidal activity of novel 1,2,4-oxadiazole derivatives containing amide fragments. Int J Mol Sci 23:1596. https://doi.org/10.3390/ijms23031596
Liu K, Wen Z, Ma Z, Shao W (2022b) Biological and molecular characterizations of fluxapyroxad-resistant isolates of Botrytis cinerea. Phytopathol Res 4:2. https://doi.org/10.1186/s42483-022-00107-3

Luo X, Chen Y, Wang Y, Xing Z, Peng J, Chen J (2024) Design, synthesis and antifungal activity of novel amide derivatives containing a pyrrolidine moiety as potential succinate dehydrogenase inhibitors. Mol Divers 28:805–816. https://doi.org/10.1007/s11030-023-10622-w

Ma Z, Qiu S, Zhang D, Guo X, Lu Y, Fan Y, Chen X (2023) Design, synthesis, and antifungal activity of novel dithiin tetracarboximide derivatives as potential succinate dehydrogenase inhibitors. Pest Manag Sci 79:1922–1930. https://doi.org/10.1002/ps.7369

Mao X, Wu Z, Zhao F, Yang X, Zhou M, Hou Y (2024) Bioactivity and resistance risk of Fluxapyroxad, a novel SDHI fungicide, in Didymella bryoniae. Plant Dis 108:658–665. https://doi.org/10.1094/PDIS-07-23-1374-RE

Masiello M, Somma S, Ghionna V, Logrieco AF, Moretti A (2019) In vitro and in field response of different fungicides against aspergillus flavus and fusarium species causing ear rot disease of maize. Toxins (Basel) 11(1):11 https://doi.org/10.3390/toxins11010011

Matsa R, Makam P, Sethi G, Thottasseri AA, Kizhakkandiyil AR, Ramadas K, Mariappan V, Pillai AB, Kannan T (2022) Pyridine appended 2-hydrazinylthiazole derivatives: design, synthesis, in vitro and in silico antimycobacterial studies. RSC Adv 12:18333–18346. https://doi.org/10.1039/D2RA02163C

Miettinen M, Lasota J (2014) Succinate dehydrogenase deficient gastrointestinal stromal tumors (GISTs) – a review. Int J Biochem Cell Biol 53:514–519. https://doi.org/10.1016/j.biocel.2014.05.033

Minařiková V, Polišenská I (1999) Analysis of populations of Pyrenophora teres on barley in The Czech Republic. Plant Prot Sci 35:115–120. https://doi.org/10.1016/j.tplants.2014.04.0111

Montiflor MO, Vellema S, Digal LN (2019) Coordination as management response to the spread of a global plant disease: a case study in a major Philippine Banana production area. Front Plant Sci 10:1–14. https://doi.org/10.3389/fpls.2019.01048

Müller F, Ackermann P, Margot P (2011) Fungicides, agricultural, 2. individual fungicides. In: Ullmann's encyclopedia of industrial chemistry. Wiley

Naranjo-Ortiz MA, Gabaldón T (2019) Fungal evolution: major ecological adaptations and evolutionary transitions. Biol Rev 94:1443–1476. https://doi.org/10.1111/brv.12510

Neumann HPH (2004) Distinct clinical features of paraganglioma syndromes associated with SDHB and SDHD gene mutations. JAMA 292:943. https://doi.org/10.1001/jama.292.8.943

O'Driscoll A, Kildea S, Doohan F, Spink J, Mullins E (2014) The wheat-septoria conflict: a new front opening up? Trends Plant Sci 19:602–610. https://doi.org/10.1016/j.tplants.2014.04.011

Oda M, Furuya T, Morishita Y, Matsuzaki Y, Hasebe M, Kuroki N (2017) Synthesis and biological activity of a novel fungicide, pyraziflumid. J Pestic Sci 42:151–157. https://doi.org/10.1584/jpestics.D17-041

Olamiju B, Bhullar S, Coleman EL, Leventhal JS (2021) Management of paronychia with pseudopyogenic granulomas secondary to epidermal growth factor receptor inhibitors: an assessment of topical timolol and the need for multiple medical and procedural therapies. J Am Acad Dermatol 84:806–808. https://doi.org/10.1016/j.jaad.2020.06.010

Omokhefe Bruce S (2022) Secondary metabolites from natural products. In: Secondary metabolites – trends and reviews. IntechOpen

Omotayo OP, Babalola OO (2023) Fusarium verticillioides of maize plant: potentials of propitious phytomicrobiome as biocontrol agents. Front Fungal Biol 4:113230. https://doi.org/10.3389/fmicb.2021.670135

Omura S, Tomoda H, Kimura K, Zhen DZ, Kumagai H, Igarashi K, Imamura N, Takahashi Y, Tanaka Y, Iwai Y (1988) Atpenins, new antifungal antibiotics produced by Penicillium sp. production, isolation, physico-chemical and biological properties. J Antibiot (Tokyo) 41:1769–1773. https://doi.org/10.7164/antibiotics.41.1769

Pandit MA, Kumar J, Gulati S, Bhandari N, Mehta P, Katyal R, Rawat CD, Mishra V, Kaur J (2022) Major biological control strategies for plant pathogens. Pathogens 11(2):273

Panizza E, Ercolino T, Mori L, Rapizzi E, Castellano M, Opocher G, Ferrero I, Neumann HP, Mannelli M, Goffrini P (2013) Yeast model for evaluating the pathogenic significance of SDHB, SDHC and SDHD mutations in PHEO-PGL syndrome. Hum Mol Genet 22:804–815. https://doi.org/10.1093/hmg/dds487

Peng Y, Li SJ, Yan J, Tang Y, Cheng JP, Gao AJ, Yao X, Ruan JJ, Xu BL (2021) Research progress on phytopathogenic fungi and their role as biocontrol agents. Front Microbiol 12:250. https://doi.org/10.1080/21655979.2019.1649520

Popko JT, Ok C-H, Campbell-Nelson K, Jung G (2012) The association between in vitro propiconazole sensitivity and field efficacy of five New England Sclerotinia homoeocarpa populations. Plant Dis 96:552–561. https://doi.org/10.1094/PDIS-06-11-0524

Popko JT Jr, Sang H, Lee J, Yamada T, Hoshino Y, Jung G (2018) Resistance of Sclerotinia homoeocarpa field isolates to succinate dehydrogenase inhibitor fungicides. Plant Dis 102:2625–2631. https://doi.org/10.1094/PDIS-12-17-2025-RE

Read AD, Bentley RE, Archer SL, Dunham-Snary KJ (2021) Mitochondrial iron–sulfur clusters: structure, function, and an emerging role in vascular biology. Redox Biol 47:102164. https://doi.org/10.1016/j.redox.2021.102164

Rutter J, Winge DR, Schiffman JD (2010) Succinate dehydrogenase – assembly, regulation and role in human disease. Mitochondrion 10:393–401. https://doi.org/10.1016/j.mito.2010.03.001

Sang H, Lee HB (2020) Molecular mechanisms of succinate dehydrogenase inhibitor resistance in phytopathogenic fungi. Res Plant Dis 26:1–7. https://doi.org/10.5423/RPD.2020.26.1.1

Schiavi F (2005) Predictors and prevalence of Paraganglioma syndrome associated with mutations of the SDHC gene. JAMA 294:2057. https://doi.org/10.1001/jama.294.16.2057

Sierotzki H, Scalliet G (2013) A review of current knowledge of resistance aspects for the next-generation succinate dehydrogenase inhibitor fungicides. Phytopathology 103:880–887. https://doi.org/10.1094/PHYTO-01-13-0009-RVW

Sun Y, Shi H, Mao C, Wu J, Zhang C (2021) Activity of a SDHI fungicide penflufen and the characterization of natural-resistance in Fusarium fujikuroi. Pestic Biochem Physiol 179:104960. https://doi.org/10.1016/j.pestbp.2021.104960

Sun M, Yan Q, Qiao Y, Zhao H, Wang Y, Shan C, Zhang S (2022a) PIKE-A modulates mitochondrial metabolism through increasing SDHA expression mediated by STAT3/FTO Axis. Int J Mol Sci 23:11304. https://doi.org/10.3390/ijms231911304

Sun Y, Yang Z, Liu Q, Sun X, Chen L, Sun L, Gu W (2022b) Design, synthesis, and fungicidal evaluation of novel 1,3-Benzodioxole-pyrimidine derivatives as potential succinate dehydrogenase inhibitors. J Agric Food Chem 70:7360–7374. https://doi.org/10.1021/acs.jafc.2c00734

Tang X, Wang Z, Hu S, Zhou B (2022) Assessing drug-induced mitochondrial toxicity in Cardiomyocytes: implications for preclinical cardiac safety evaluation. Pharmaceutics 14:1313. https://doi.org/10.3390/pharmaceutics14071313

Teixeira MM, Carvalho DT, Sousa E, Pinto E (2022) New antifungal agents with azole moieties. Pharmaceuticals 15:1427. https://doi.org/10.3390/ph15111427

Tyul JM (1975) Acquired resistance to carboxin in aspergillus nidulans. Neth J Plant Pathol 81:122–123. https://doi.org/10.1007/BF01999863

Van Vranken JG, Bricker DK, Dephoure N, Gygi SP, Cox JE, Thummel CS, Rutter J (2014) SDHAF4 promotes mitochondrial succinate dehydrogenase activity and prevents neurodegeneration. Cell Metab 20:241–252. https://doi.org/10.1016/j.cmet.2014.05.012

Vielba-Fernández A, Polonio Á, Ruiz-Jiménez L, de Vicente A, Pérez-García A, Fernández-Ortuño D (2021) Resistance to the SDHI fungicides Boscalid and Fluopyram in Podosphaera xanthii populations from commercial cucurbit fields in Spain. J Fungi 7(9):733. https://doi.org/10.3390/jof7090733

Walter H (2016) Fungicidal succinate-dehydrogenase-inhibiting Carboxamides. Bioact Carboxylic Compd Classes Pharm Agrochem:405–425. https://doi.org/10.1002/9783527693931.ch31

Walter H, Tobler H, Gribkov D, Corsi C (2015) Sedaxane, Isopyrazam and Solatenol™: novel broad-spectrum fungicides inhibiting succinate dehydrogenase (SDH) – synthesis challenges and biological aspects. Chimia (Aarau) 69:425. https://doi.org/10.2533/chimia.2015.425

Wang M, Du Y, Ling C, Yang Z, Jiang B, Duan H, An J, Li X, Yang X (2021a) Design, synthesis and antifungal anti-oomycete activity of pyrazolyl oxime ethers as novel potential succinate dehydrogenase inhibitors. Pest Manag Sci 77:3910–3920. https://doi.org/10.1002/ps.6418

Wang X, Wang M, Han L, Jin F, Jiao J, Chen M, Yang C, Xue W (2021b) Novel Pyrazole-4-acetohydrazide derivatives potentially targeting fungal succinate dehydrogenase: design, synthesis, three-dimensional quantitative structure-activity relationship, and molecular docking. J Agric Food Chem 69:9557–9570. https://doi.org/10.1021/acs.jafc.1c03399

Wilson BA, Schisler JC, Willis MS (2010) Sir Hans Adolf Krebs: architect of metabolic cycles. Lab Med 41:377–380. https://doi.org/10.1309/LMZ5ZLAC85GFMGHU

Wu S, Lei L, Liu M, Song Y, Lu S, Li D, Shi H, Raley-Susman KM, He D (2018) Single and mixture toxicity of strobilurin and SDHI fungicides to Xenopus tropicalis embryos. Ecotoxicol Environ Saf 153:8–15. https://doi.org/10.1016/j.ecoenv.2018.01.045

Xing X, Liao Z, Tan F, Zhu Z, Jiang Y, Cao Y (2019) Effect of nicotinamide against *Candida albicans*. Front Microbiol 10:595

Yin YM, Zhang XM, Shang XY, Gao ZH, Liang ZB, Wang DW, Xi Z (2015) Succinate dehydrogenase: an ideal target for fungicide discovery. ACS Symp Ser 1204:175–194

Yan W, Wang X, Li K, Li TX, Wang JJ, Yao KC, Cao LL, Zhao SS, Ye YH (2019) Design, synthesis, and antifungal activity of carboxamide derivatives possessing 1,2,3-triazole as potential succinate dehydrogenase inhibitors. Pestic Biochem Physiol 156:160–169

Yan Y, Cheng W, Xiao T, Zhang G, Zhang T, Lu T, Tang X (2020) Discovery of Novel 2,4,6-Trisubstituted Pyrimidine Derivatives as Succinate Dehydrogenase Inhibitors. Chinese J Org Chem 40:4237

Yan Y, Liao Z, Shen J, Zhu Z, Cao Y (2022a) Nicotinamide potentiates amphotericin B activity against *Candida albicans*. Virulence 13:1533–1542

Yan Z, Yang Z, Qiu L, Chen Y, Li A, Chang T, Niu X, Zhu J, Wu S, Jin F (2022b) Discovery of novel pyridine carboxamides with antifungal activity as potential succinate dehydrogenase inhibitors. J Pestic Sci 47:118–124

Yan Z, Gao Y, Yang Z, Yu B, Zhao D, Li Y, Cao L, Qin M, Li M, Jin F (2023) Thiazolamide derivatives containing diphenyl ethers moiety: design, synthesis, fungicidal activity and docking study. J Chem Sci 135:66

Yang JH, Brannen PM, Schnabel G (2015) Resistance in Alternaria alternata to SDHI fungicides causes rare disease outbreak in peach orchards. Plant Dis 99:65–70

Yang D, Zhao B, Fan Z, Yu B, Zhang N, Li Z, Zhu Y, Zhou J, Kalinina TA, Glukhareva TV (2019) Synthesis and biological activity of novel succinate dehydrogenase inhibitor derivatives as potent fungicide candidates. J Agric Food Chem 67:13185–13194

Yang Z, Sun Y, Liu Q, Li A, Wang W, Gu W (2021) Design, synthesis, and antifungal activity of novel Thiophene/Furan-1,3,4-Oxadiazole Carboxamides as potent succinate dehydrogenase inhibitors. J Agric Food Chem 69:13373–13385

Yanicostas C, Soussi-Yanicostas N (2021) SDHI fungicide toxicity and associated adverse outcome pathways: what can zebrafish tell us? Int J Mol Sci 22(22):12362

Yankovskaya V, Horsefield R, Törnroth S, Luna-Chavez C, Miyoshi H, Léger C, Byrne B, Cecchini G, Iwata S (2003) Architecture of succinate dehydrogenase and reactive oxygen species generation. Science 299(80):700

Yao X, Guo H, Zhang K, Zhao M, Ruan J, Chen J (2023) Trichoderma and its role in biological control of plant fungal and nematode disease. Front Microbiol 14:1160551

Yin YN, Kim YK, Xiao CL (2011) Molecular characterization of Boscalid resistance in field isolates of Botrytis cinerea from apple. Phytopathology 101:986–995

Yoshikawa Y, Katsuta H, Kishi J, Yanase Y (2011) Structure-activity relationship of carboxin-related carboxamides as fungicide. J Pestic Sci 36:347–356

Yu B, Zhao B, Hao Z, Chen L, Cao L, Guo X, Zhang N, Yang D, Tang L, Fan Z (2021) Design, synthesis and biological evaluation of pyrazole-aromatic containing carboxamides as potent SDH inhibitors. Eur J Med Chem 214:113230

Zhang S-G, Wan Y-Q, Wen Y, Zhang W-H (2022) Novel Coumarin 7-Carboxamide/Sulfonamide derivatives as potential fungicidal agents: design, synthesis, and biological evaluation. Molecules 27

Zorov DB, Juhaszova M, Sollott SJ (2014) Mitochondrial reactive oxygen species (ROS) and ROS-induced ROS release. Physiol Rev 94:909–950

Zou Y, Zhang Y, Liu X, Song H, Cai Q, Wang S, Yi C, Chen J (2023) Research progress of Benzothiazole and Benzoxazole derivatives in the discovery of agricultural chemicals. Int J Mol Sci 24:10807

Metabolic Disorders and Succinate Dehydrogenase

11

Deepa S. Mandlik and Satish K. Mandlik

Abstract

The individual enzyme implicated in both oxidative phosphorylation and the tricarboxylic acid cycle is mitochondrial succinate dehydrogenase (SDH), often stated as electron transport chain complex II. Such metabolic enzymes are recognized to play a critical part in metabolic diseases. Due to its critical function in regulating mitochondrial metabolism and human health, SDH has drawn more and more attention. The finding of its function in human disease has rekindled interest in this complex, which has received the least amount of research on the mitochondrial respiratory complexes. SDH is recognized to perform a significant part in metabolic reprogramming, energy balancing and biosynthetic pathways. The lack of pancreatic β-cells and SDH compromises glucose-encouraged ATP formation, insulin secretion and β-cell development by mitochondrial membrane potential failure and impairs glucose-induced oxidative phosphorylation. Under pathological circumstances, SDH may potentially be a substantial generator of reactive oxygen species (ROS). We have previously revealed that elevated reactive oxygen species production is essential for the development of obesity. During the bacterial fermentation of dietary fibre, SDH is also generated at significant levels. In individuals with inflammatory bowel disease and intestinal swelling, elevated SDH levels have been observed in the gut lumen. It is unclear how SDH affects immune system inflammation in the gut mucosa, though. By describing its changes and dysfunctionality, we address the biology, pathophysiology and importance of the SDH enzyme in metabolic disorders in this book chapter.

D. S. Mandlik (✉) · S. K. Mandlik
Department of Pharmacology, Bharti Vidyapeeth (Deemed to be University), Poona College of Pharmacy, Pune, Maharashtra, India

V. M. Patil et al. (eds.), *Deciphering The Role of Succinate Dehydrogenase in Drug Discovery*, https://doi.org/10.1007/978-981-96-6831-1_11

Keywords

Succinate dehydrogenase · Metabolic disorders · Diabetes mellitus · Inflammatory bowel disease · Atherosclerosis · Obesity · Gut microbiota

Abbreviations

APCs Antigen-presenting capacity
ATP Adenosine triphosphate
CII Complex II
CPT1A Carnitine palmitoyltransferase 1A
DCs Dendritic cells
EPK Extracellular regulated kinase
ETC Electron transport chain
GABA γ-aminobutyric acid
HIF-1α Hypoxia-inducible factor-1α
IBD Inflammatory bowel diseases
IL Interleukin
IR Ischemia/reperfusion
KGDH α-ketoglutarate Dehydrogenase
KGDHCα-ketoglutarate dehydrogenase complex
LPS Lipopolysaccharide
NADNicotinamide adenine dinucleotide
NF-κB Nuclear factor kappa beta
ROS Reactive oxygen species
SCFAs Short-chain fatty acids
SCS Succinyl-CoA synthetase
SDH Succinate dehydrogenase
SIRT1 Sirtuin 1
SMCs Smooth muscle cells
SUCNR1 Succinate receptor 1
T2D Type 2 diabetes
TCA Tricarboxylic acid
TMAO Trimethylamine N-oxide
TNF-α Tumour necrosis factor-α
UC Ulcerative colitis
UD Crohn's disease

11.1 Introduction

According to Murphy and O'Neill (2018), succinate dehydrogenase (SDH) and succinate are crucial parts of the tricarboxylic acid (TCA) cycle and stay involved in the synthesis of adenosine triphosphate (ATP) in mitochondria. Additionally, succinate plays a decisive part in several metabolic pathways, helping to control a wide range of anabolic and catabolic activities during various phases of the TCA cycle (Mills and O'Neill 2014). The ability of immune cells to modify their metabolic processes to preserve immunological balance in hypoxic situations has come to light more and more throughout the last 10 years. It has been discovered that these metabolic changes significantly affect how the immune system reacts to tumours, ischaemia/reperfusion (IR) injury and chronic inflammation (Murphy and Chouchani 2022). The regulation of SDH and the roles of succinate amount beyond their conventional involvement in metabolism have been more and more highlighted in recent studies. In specific tumours and active macrophages, succinate, for example, has been shown to activate dendritic cells (DCs) and activate hypoxia-inducible factor-1α (HIF-1α) through the succinate receptor (O'Neill and Pearce 2016; Dalla Pozza et al. 2020). Moreover, SDH gene mutations have been found in several forms of cancer, pointing to possible pathways that cause an excessive build-up of succinate (Peti-Peterdi 2010). Through succinylation, succinate can cause alterations of proteins and act as an oncometabolite in the development of tumours and cancer (Mills et al. 2021). This article provides a thorough knowledge of the functions of succinate as well as the alterations in SDH expression or activity associated with changes in immune cell function under hypoxic settings. In addition to providing an overview of the routes that contribute to the build-up of succinate, our goal is to clarify possible targets for therapy by examining succinate metabolism in metabolic syndrome disease.

11.2 Metabolic Disorders

Metabolic disorders represent a significant global health system concern, characterized by aberrant disruptions in metabolic processes. A pandemic-like prevalence of metabolic disease has been attained (Ng et al. 2021; Hoffman et al. 2021). As a result of redox imbalance and cellular dysfunction brought on by the dysregulation of lipids, sugar and energy-yielding mitochondria in metabolic diseases, dysfunctional cells and tissues result (Rani et al. 2016). The mitochondria fail to provide enough energy to maintain the usual operations of body cells and disorders associated with mitochondrial dysfunction (Wallace 1999). Metabolic disorders can cause a number of health difficulties, including heart disease, hypertension, cancer, ischaemia and abnormalities in reproduction (Maurya et al. 2023a, b). In the United States, the cost of treating these illnesses is predicted to reach $66 billion by 2030. According to estimates by Chu et al. (2018), the United States would lose $444 million and $74 million in productivity per year due to hospitalization and reduced life expectancy, respectively. According to data from the World Health Organization in 2018 (WHO 2018), obesity alone results in 2.8 million deaths yearly and around 35.8 million disability-adjusted life years. There has to be a quick fix for this problem that lasts. To manage and avoid metabolic diseases, several treatment approaches have been proposed. For long-term weight management and the prevention of other metabolic illnesses, changing one's lifestyle is one of the most effective forms of treatment. Positive lifestyle changes are challenging to implement and sustain, though (Leitner et al. 2017). Nearly overweight patients with a body mass index of

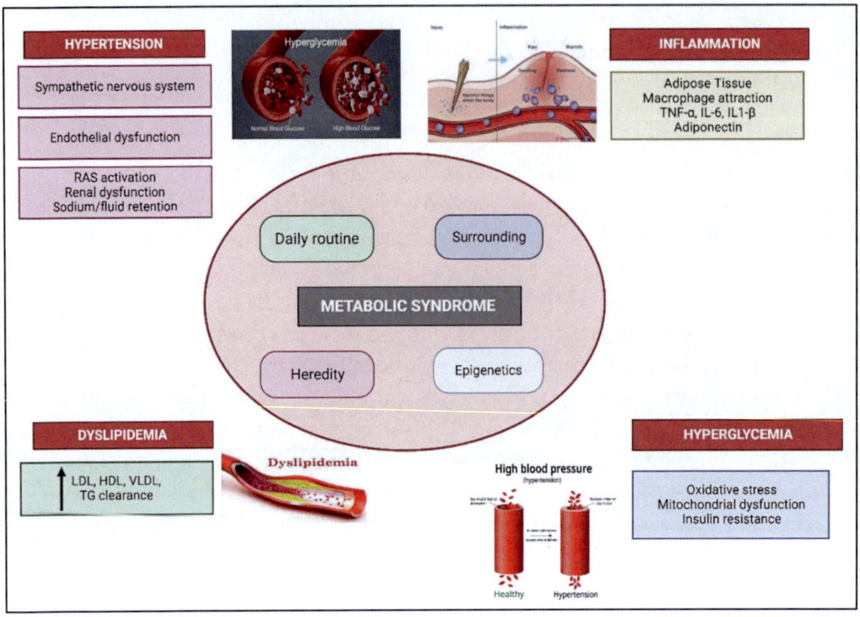

Fig. 11.1 Causative factors of metabolic disorders. Genetics, epigenetics, lifestyle and environment are well-known risk factors for metabolic syndrome. As for the aetiology of metabolic syndrome, it is still not entirely understood, although it is thought to be related to inflammation, dyslipidaemia, hyperglycaemia and hypertension

40 kg/m^2 or patients with type 2 diabetes (T2D) with a body mass index of 35 kg/m^2 prefer bariatric surgery as an alternative (Fried et al. 2013). However, without addressing the underlying causes of these conditions, drastic and quick weight reduction might result in malnourishment and osteoporosis (Leitner et al. 2017). Figure 11.1 has summarized the causative factors of metabolic disorders, covering genetics, epigenetics, lifestyle, environment, etc.

Drug therapy, such as metformin to treat type 2 diabetes, can be given to patients with metabolic problems; however, patients who are allergic to the chemicals in the drug or who have kidney impairment might not be able to benefit from it. Because they affect nearly every bodily component, some illnesses, such as mitochondrial abnormalities, are even more complicated (Gorman et al. 2016). As a result, research and development of new therapeutic approaches are still ongoing for diseases linked to mitochondrial dysfunction, including Parkinson's, Alzheimer's and muscular dystrophy (Avula et al. 2014; Wang et al. 2020; Malpartida et al. 2021; Budzinska et al. 2021; Maurya et al. 2023a, b).

11.3 Role of SDH in Diabetes Mellitus

A chronic disease of disturbed glucose homeostasis, T2D is considered to be continuous damage of β-cell mass and activity, commonly known as β-cell failure (Halban et al. 2014; Brereton et al. 2016). The pathophysiological mechanisms of β-cell mitochondrial dysfunction are still unknown; however, emerging evidence points to mitochondrial dysfunction as a major factor in T2D aetiology and β-cell damage (Ma et al. 2012; Haythorne et al. 2019). The tight control of glucose-activated insulin secretion and balanced β-cell growth is facilitated by mitochondria, which are essential for the connection of glucose metabolism in β-cell insulin release (Wollheim and Maechler 2002). Although the crucial function of the mitochondria in connecting the metabolism of glucose with the release of insulin is widely established, the molecular pathways connecting mitochondrial malfunction and increasing β-cell dysfunction remain poorly understood. Complex II (CII), also known as SDH, is one of the five mitochondrial complexes that is associated with the electron transport chain (ETC). According to Van Vranken et al. (2015), SDH is a heterotetrameric complex located in the inner mitochondrial membrane. Additionally, SDH catalyses the alteration of succinate to fumarate, which is a stage in the TCA cycle (Sun et al. 2005). Alongside this enzymatic activity, $FADH_2$ is produced. $FADH_2$ then uses CII to transfer electrons to ubiquinone for oxidative phosphorylation (Sun et al. 2005). SDH is crucial to both ATP synthesis and mitochondrial metabolism because of its simultaneous involvement in the TCA and ETC cycles. Accordingly, serious human metabolic diseases such as cardiomyopathy and Leigh syndrome are associated with reduced SDH/CII activity (Rutter et al. 2010). Remarkably, since it can be eschewed by other metabolic processes, the SDH complex is not necessary for oxidative phosphorylation (Rutter et al. 2010). As a result, the precise mechanism by which the absence of SDH severely compromises mitochondrial function remains unclear. The intracellular build-up of succinate, elevated formation of reactive oxygen species (ROS) and reduced ATP synthesis are hypothesized to be the disease-causing effects of SDH disruption (Rutter et al. 2010; Armstrong et al. 2022).

The cells that require constant energy, like neural, skeletal muscle and cardiomyocytes, are the most severely affected by SDH deficiency (Rutter et al. 2010). For this reason, β-cell—an extremely metabolic cell type that rumours mingling glucose amount through the production of ATP and insulin—is a perfect model to study the biological role of SDH. Inhibition of SDH/CII with 3-nitroprorionic acid was found to impede insulin secretion in recent research on mouse islets (Edalat et al. 2015), indicating an SDH role in controlling β-cell activity. Wojtovich et al. (2013) hypothesized that diabetes may be biologically related to decreased SDH action and the ensuing accumulation of succinate (Wojtovich et al. 2013). In this review paper, we investigate the theory that mitochondrial dysfunction in diabetes pathophysiology and β-cell failure is caused by SDH deficiency. T2D is primarily produced by mitochondrial dysfunction; however, the pathogenic processes in β-cells remain unclear. A mouse model of β-cell-specific SDH ablation (SDHBβKO) and samples from diabetic patients are used by Lee et al. (2022) to characterize the deficit of SDH as a driver of mitochondrial dysfunction in insulinopenic diabetes and β-cell failure. β-cell SDH deficit affects the glucose-induced respiratory oxidative phosphorylation and collapse of mitochondrial membrane potential, which compromises insulin release, β-cell development and glucose-stimulated ATP synthesis. The damage of SDH leads to an additional build-up of succinate that mis-activates the mTOR complex 1–regulated metabolic anabolism, counting enhanced SREBP-regulated synthesis of lipids. This is the mechanistic explanation supported by metabolomic and transcriptome investigations. Acute mTOR inhibition with rapamycin partially reverses these changes, which are like human β-cell dysfunction associated with diabetes. We recognize mTOR complex 1 inhibition as a possible mitigation technique and suggest SDH insufficiency as a contributive mechanism to the continuous failure of β-cell in diabetes (Lee et al. 2022).

11.4 Role of SDH in Inflammatory Bowel Diseases (IBDs)

Ulcerative colitis (UC) and Crohn's disease (CD) are two types of long-lasting, reverting gastrointestinal tract illnesses known as inflammatory bowel diseases (IBD) that are becoming more commonplace globally (Kaplan 2015). Although the exact cause of IBD is unidentified, it is believed to occur from complicated environmental and genetic variables, such as antibiotic usage and nutrition, causing the imbalance between the mucosal immune system and gut-resident microbiotic flora (Manichanh et al. 2012). Numerous interactions between the gut microbiotic flora and its metabolic products and the host affect homeostasis and disease. Variations in the structure and metabolism of the gut microbiotic flora are known risk factors for IBD, while the exact mechanisms are still unknown (Ni et al. 2017). A crucial metabolite in microbial and host activities is succinate. Despite being considered an intermediate, succinate has been shown to build up under specific pathophysiological conditions, particularly in places where there is metabolic stress and inflammation (Akram 2014). According to several studies (Ariza et al. 2012; Littlewood-Evans et al. 2016; Van Diepen et al. 2017), succinate is not just an inert by-product of metabolism but also actively participates in downstream cellular processes and could act as a proinflammatory mediator with systemic and tissue-specific effects. Extracellular succinate plays an important part in the control of colonic immunological responses, according to recent studies (Schneider et al. 2018; Nadjsombati et al. 2018). We will go over the data about succinate's effect on IBD in this section.

11.4.1 Succinate Accretion Is an Inflammatory Bowel Disease

Numerous research studies have reported that the concentrations of succinate in faeces are roughly four times higher in patients with IBD than in control patients (Hallert et al. 2003). Metabolomic research of the intestinal mucosa of patients with IBD has shown that succinate is amplified in inflammatory patches likened to healthy mucosal lesions (Ooi et al. 2011). Faecal succinate levels rise in mice with colitis caused by dextran sodium sulphate, and this is correlated with disease activity and severity (Osaka et al. 2017). It is most likely due to the acute inflammation and intestinal damage associated with IBD that mucosal tissue accumulates and releases succinate. The exact role that gut microbiota and dysbiosis—a recurring aspect of IBD—play in elevating luminal succinate is unknown, though. When compared to normal mice colonized with healthy samples, humanized gnotobiotic mice colonized with faecal specimens from UC and CD patients show a high amount of faecal succinate, indicating that accretion of succinate is a metabolic characteristic of IBD-linked imbalance in bacterial composition (Nagao-Kitamoto et al. 2016). This is reinforced by the discovery that the quantity of succinate-consuming (and acetate/propionate-producing) *Phascolarctobacterium* was suggestively lesser in IBD patients' faecal microbiomes when compared to controls in a metagenomic analysis of the microbiome (Morgan et al. 2012). In the faecal microorganism community of paediatric patients with CD compared to normal, our metagenomic investigation revealed improved wealth for enzymatic genes implicated in succinate synthesis. This difference was even more pronounced in patients who experienced a relapse soon after starting therapy (Dunn et al. 2016). Furthermore, it has been shown that Bacteroides that produce succinate are more prevalent in mice treated with DSS, and this is consistent with a higher colonic succinate content (Ariake et al. 2000). All these results point to a possible role for aberrant fermentation in IBD-related dysbiosis in the gut lumen's succinate build-up. Nonetheless, variations in the proportional amounts of succinate inside the intestinal lumen have been linked to variations in the course of diseases, indicating a connection between succinate and inflammation. In comparison to controls, germ-free mice that were non-colonized with *Bacteroides vulgatus* strains that produce succinate and were isolated from UC patients exhibited higher levels of caecal succinate and more severe colonic inflammatory reaction after dextran sulphate-prompted colitis (Setoyama et al. 2003). The association between succinate and inflammation has been noted in several studies that combine dietary interventions with mouse models of colitis. For example, in a DSS-induced colitis model, mice fed a pure fruit-oligosaccharide-complemented diet showed elevated caecal succinate and worsened diarrhoea and weight loss in comparison to mice on a normal diet (Goto et al. 2010). In contrast, interleukin (IL)-10-deficient cell transfer colitis mouse model mice supplemented with a diet that caused lower levels of succinate in the cecum showed noticeably less colonic inflammation (Kajiura et al. 2009; Paturi et al. 2012). These conclusions beg the question of whether mucosal inflammation is exacerbated by succinate build-up linked to IBD.

11.4.2 Stabilization of Hypoxia-Inducible Factor-1α (HIF-1α) in Inflammation

Remarkable study results suggest that succinate might be a new class of inflammatory regulators, providing important cues to adjust the inflammatory response (Tannahill et al. 2013; Grimolizzi and Arranz 2018). The importance of succinate as an indication in the immuno-inflammatory reaction is becoming more widely acknowledged. The stimulation of immune cells like dendritic cells (DCs) and macrophages, causes a metabolic swing in the direction of glycolysis in hypoxic inflammatory microenvironments (Mills and O'Neill 2014). It is believed that this metabolism switch in activated immune cells towards glycolysis has a major effect in low-oxygen environments, like hypoxic inflammatory areas. The build-up of succinate can significantly impact the immuno-inflammatory response and amplify inflammatory signalling. The build-up of succinate may originate from several possible places. First off, our investigations have confirmed that succinate build-up brought on by SDH mutations can alleviate HIF-1α in stimulated macrophages, especially in the presence of prolyl hydroxylase domain inhibition. The inflammatory cytokine interleukin-1β is subsequently produced as a result of this stabilization, thanks to HIF-1α signalling (O'Neill and Pearce 2016; Zhang et al. 2021; Peace and O'Neill 2022). Furthermore, it has been noted that the procedure of metabolism of glutamine that promotes the TCA cycle, can also lead to succinate build-up in lipopolysaccharide (LPS)-activated macrophages (Rubic et al. 2008). Following this, the "GABA shunt" route may cause succinate to build up because it is stimulated by raised γ-aminobutyric acid (GABA) and its transporters brought on by lipopolysaccharide (LPS). Vigabatrin can impede this route by blocking an important GABA shunt enzyme, which lowers the amount of succinate that accumulates. Furthermore, succinate build-up can be avoided by inhibiting GABA transaminase, an essential GABA shunt enzyme, with vigabatrin (Ren et al. 2019). Lastly, several different processes, including the defeat of SDH action during hypoxia, may result in succinate build-up. Apart from hypoxia, nicotinamide adenine dinucleotide (NAD) amount lowered by FAD as a result of LPS also inhibit SDH action (Wang et al. 2021). Through the sirtuins stimulation, a class of NAD^+-dependent deacetylases, NAD^+ displays a variation of anti-inflammatory characteristics. Sirtuin 1 (SIRT1) promotes oxidative metabolism and autophagy-related gene function while inhibiting glycolysis and the action of proinflammatory transcription factors including HIF-1α and nuclear factor-kappa B (NF-κB) (Misawa et al. 2013). On the other hand, it has been noted that the pyrin domain containing 3 (NLRP3) inflammasome and NOD-like receptor family is more stimulated when SIRT3 is inactivated (Schug et al. 2010). In addition, succinate normalizes immune cell functions, including those related to immune cell motility, cytokine synthesis and enhancement of antigen-presenting cells (APCs) capacity to trigger adaptive immunological reactions (Peace and O'Neill 2022). For example, succinate has been shown to serve as a chemokine and promote the relocation of monocyte-derived dendritic cells in vitro, hence boosting the migration of DCs treated with succinate to drain lymph nodes (Haffke et al. 2019). Via synergetic connections with specific ligands, such as tumour necrosis factor-α (TNF-α), succinate has been stated to activate the synthesis of cytokines (Rubic et al. 2008). Moreover, it has been established that activation of succinate with LPS increases the production of TNF-α and interleukin (IL)-1β in murine bone marrow–derived dendritic cells, in addition to TNF-α. Apart from encouraging the production of cytokines, succinate can also improve APCs' capacity to trigger immunological reactions that lead to inflammatory consequences. Numerous investigations have demonstrated succinate's ability to impede tumour growth; this will be discussed in more detail in the section that follows (Dalla Pozza et al. 2020). More importantly, altering succinate metabolism may be able to treat host-microbe interactions by microbiological means (Wei et al. 2023). These results imply that novel treatments for the avoidance and management of inflammation may be developed by modifying the metabolism of succinate.

11.4.3 Succinate and SUCNR1 Signalling During Inflammation

Outside its function as a metabolic signal in the inflammatory process via balancing HIF-1α, other studies have shown the communication between succinate and the ligand-receptor pair succinate receptor 1 (SUCNR1) (Ristic et al. 2017). Numerous G-protein coupled receptors, counting SUCNR1, were first identified by Wittenberger et al. (2001). Later, succinate was discovered as a specific SUCNR1 ligand (He et al. 2004). Following its discovery and investigation in the renal system, SUCNR1 was found to express significantly in the splenic, intestinal and hepatic tissues (de Castro et al. 2016; Gilissen et al. 2016). White adipocytes, haematopoietic progenitor cells and different blood and immune cell types (de Castro Fonseca et al. 2016; Krzak et al. 2021) have all been reported to express SUCNR1. Succinate causes SUCNR1 to become activated, which in turn starts signalling cascades across several channels. Like Gβ and Gγ signalling cascades, succinate stimulation causes extracellular regulated kinase (ERK) to be phosphorylated, mobilize calcium ions and accumulate inositol trisphosphate (Gilissen et al. 2016). Comprehending the signalling pathways that are sparked by the interaction between succinate and SUCNR1 in detail may afford novel insights for the growth of therapeutic methods that target disorders, especially those related to chronic inflammation. SUCNR1 is involved in several succinate-dependent inflammatory processes in vivo, according to recent investigations. To begin with, it was shown that mice lacking SUCNR1 had more DC migration than mice of the wild type. Second, the idea that succinate serves as a traditional signal to improve the antigen-presenting capacity (APCs) is supported by the fact that the lack of SUCNR1 in DCs prevented the elevation of cytokines. Finally, graft survival following organ replacement from SUCNR1-deficient animals was longer than that of wild-type mice (Rubic et al. 2008). Significant evidence has been shown by Peruzzotti-Jametti et al. showing that carrier-mediated succinate uptake is reduced when SUCNR1 is inhibited. This implies that SUCNR1 signalling stimulates the expression of Na^+-dependent dicarboxylic acid transporters in the plasma membrane, which help move succinate across the membrane. According to Peruzzotti-Jametti et al. (2018), these results raise interesting possibilities about succinate's role in the paracrine or autocrine regulation of disease progress. In summary, the above-described research has demonstrated that SUCNR1 is essential for the immune-inflammatory reaction that succinate induces. Thus, a probable beneficial method for the management of inflammatory diseases is the regulation of SUCNR1 signalling.

11.4.4 Succinate and Succinylation in Signal Transduction

A post-translational modification method called succinylation involves adding a succinyl group to a lysine residue. This changes how amino groups work (Alleyn et al. 2018; Sreedhar et al. 2020; Chinopoulos 2021). Current studies have focused on the alteration known as succinylation, in which protein lysine and succinyl CoA can create an amide bond (Zhao et al. 2022). The discovery of increased succinate concentrations in a multitude of proteins implicated in the control of various cellular and biological functions emphasizes the importance of succinylation as a vital cellular activity (Hirschey and Zhao 2015). There is still much to learn about the catalytic mechanisms underlying the multiple enzyme-mediated succinylation process. As a result, studies are being conducted to clarify these basic processes to make it easier to create novel therapeutic approaches and related medications (Shen et al. 2023). Several studies have shown that both enzymatic and non-enzymatic processes can lead to succinylation in organisms. As widely reported in the literature (Zhao et al. 2022), most of the known processes are non-enzymatic. This is especially true for the work of Hirschey et al., who showed that succinyl-CoA has a higher chemical reactivity than other acyl-CoA species (Wagner et al. 2017). While the catalysis of enzymatic succinylation might be controlled by succinyl CoA, non-enzymatic succinylation seems to be impacted by a few variables, including the concentrations of reactants in mitochondria. E1k [α-ketoglutarate Dehydrogenase (KGDH)] and α-ketoglutarate dehydrogenase complex (KGDHC) may facilitate enzymatic succinylation in addition to succinyl CoA (Yang and Gibson 2019). KGDHC regularly mediates succinylation in the nucleus and cytoplasm. Additionally, succinyl-CoA levels can be controlled by enzymatic TCA cycle processes such as succinyl-CoA synthetase (SCS), which subsequently non-enzymatically drives succinylation (Carrico et al. 2018; Gut et al. 2020). Whilst carnitine palmitoyltransferase 1A (CPT1A) enhances lysine succinylation without changing succinyl CoA levels in mammalian cells, the CPT1A G710E mutant has been shown to improve cell propagation under the metabolic stress (Kurmi et al. 2018; Wang et al. 2019a, b). According to Sreedhar et al. (2020), this data implies that CPT1A's lysine succinyl transferase activity can succinylate downstream substrate proteins, promoting propagation. The process supported by CPT1A is still debatable and has not been demonstrated in vivo, in distinction to the known mechanism of succinyl-CoA non-enzymatic lysine succinylation (Shen et al. 2023). This process would, at most, explain a small portion of succinylation (Sreedhar et al. 2020). It has been determined that acyl-CoA distribution may influence succinylation in specific tissues (Weinert et al. 2013). This study suggests that new treatment approaches for modifying succinylation in specific disorders may be available due to the tissue-specific regulation of acyl-CoA amount. Moreover, the modulation of de-succinylation is essential for regulating diverse catalytic activities in every cell compartment, while succinylation may demonstrate a mutually balanced association. Interestingly, it has been discovered that desuccinylase activity dynamic fluctuations are influenced by NAD^+-dependent SIRT5 levels. Research has demonstrated that the suppression of SIRT5 leads to an increase in succinylation in specific

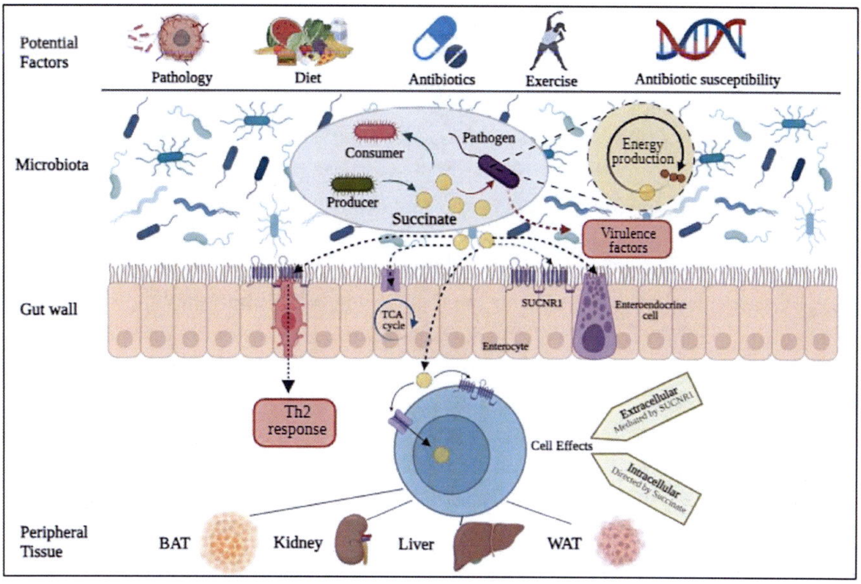

Fig. 11.2 Succinate as a potential intermediate in host-microbiota interactions

proteins, which in turn aids in the initiation and spread of cancer (Park et al. 2013). Comparatively, SIRT7 functions as a histone desuccinylase and is connected to both genome stability and chromatin compaction (Fig. 11.2). SIRT7 catalyses the desuccinylation process, which mostly takes place in the nucleus and is essential for both cell survival and the DNA damage response (Li et al. 2016).

As mentioned earlier, the control of physiological processes depends critically on the preservation of a fine balance between the succinylation and desuccinylation processes (Wang and Lin 2021). Many illnesses, like inflammatory disorders, IR damage and cancer, might arise when this homeostasis is upset (Sabari et al. 2017). Researchers are presently studying the complex role that SIRT5 plays in the carcinogenesis process (Yang et al. 2017). It has been suggested that SIRT5-mediated desuccinylation processes could be involved in the start and development of cancer. Hyper-succinylation may have comparable effects on tumour growth by modifying SIRT5, as the suppression of SIRT5 has been shown to reduce cell proliferation and tumour growth (Xiangyun et al. 2017). Moreover, a growing body of research indicates that excessively high lysine succinylation occurs during the development of cancer, suggesting that succinylation controls the energy metabolism of tumours (Song et al. 2017). Isocitrate dehydrogenase isoform 1 mutations that inhibit SDH lead to an increase in succinyl-CoA, which in turn promotes malignant metabolism (Li et al. 2015). The increase in succinylation in hypoxic environments could mitigate the impact of IR injury on the heart and brain. Effective succinylation control has the potential to reveal new treatment approaches for IR injury and even speed up transplant processes (Murphy and O'Neill 2018). In conclusion, understanding the tissue-specific mechanisms involved is critical to the development of pharmaceutical therapies that target de- and succinylation in hypoxia-related diseases like cancer and inflammation.

11.5 Role of SDH in Atherosclerosis

Atherosclerosis is the principal underlying pathology of many cardiovascular disorders, especially coronary artery diseases. It starts as fat strips on the vascular inner layer and develops into the formation of atherosclerotic plaques. The main factors causing atherosclerosis to advance are oxidation and inflammation. Atherosclerosis advances due to endothelial cell dysfunction caused by inflammatory cytokines linked to pro-atherosclerosis pathways (Beverly and Budoff 2020). After that, pro-inflammatory cytokines and growth factors are released by activated endothelial cells, which cause vascular smooth muscle cells (SMCs) to change and encourage the formation of fibrous plaques and atherosclerotic lesions (Tabas et al. 2015). The recruitment of inflammatory cells, like dendritic cells and macrophages that cling to activated endothelium cells and then penetrate the intima, is largely dependent on these cells. According to Foks and Bot (2017), the interaction between these cells advances atherosclerotic damage and plaque rupture, which may result in myocardial infarction and death.

11.5.1 Excessive Build-Up of Succinate in Atherosclerosis

An increasing amount of research has shown that excessive succinate is crucial to multiple pathological processes related to atherosclerosis and that levels are markedly elevated in the setting of hyperlipidaemia (Martínez-Reyes and Chandel 2020). High concentrations of glucose, fat and LPS impair the function of SDH and cause succinate to build up as a reaction substrate, which is why succinate is upregulated in atherosclerosis (Flores et al. 2018). Additionally, extra isocitrate is metabolized into succinate (Quartararo et al. 2013) and there is an enhanced change of glutamine to succinate (Jiang and Yan 2017).

11.5.2 Pathogenesis of Succinate in Different Cells Involved in the Progression of Atherosclerosis

The build-up of succinate influences smooth muscle cells, endothelial cells, lymphocytes and macrophages, among other cell types that are involved in the development of atherosclerosis (Ko et al. 2017).

11.5.3 Endothelial Cells

It has been demonstrated that succinate dramatically worsens endothelial cell dysfunction. It does this, in part, by encouraging the overexpression of reactive oxygen species (ROS) in the mitochondria. According to Ko et al. (2017), this is accomplished by the migration of cytosolic dynamin-related protein 1 to the outside of the mitochondrial membrane. Additionally, through a process known as reverse electron transfer, oxidized succinate can increase the formation of reactive oxygen species. This mechanism reduces nitric oxide's ability to dilate blood vessels in the endothelium, which promotes endothelial dysfunction (Ryan et al. 2019). Increased ROS can also interfere with the production of nitric oxide, which exacerbates endothelial dysfunction. Succinate has been shown to have an impact on ROS and activate the renin-angiotensin system through its communication with SUCNR1. Angiotensin II levels rise as a result of this interaction, encouraging vasoconstriction and inhibiting the synthesis of nitric oxide, both of which exacerbate endothelial dysfunction (Tang et al. 2020). Additionally, the interaction of succinate with GPR91 on platelets increases thromboxane A2 synthesis, which in turn initiates platelet activation. According to Tang et al. (2020), this activation is essential for the onset and progression of atherosclerosis. Damage to endothelial cells caused by platelet attachment to the endothelium can trigger a proinflammatory reaction and possibly result in thrombotic actions. Trimethylamine N-oxide (TMAO) is an additional significant metabolite that has been connected to elevated platelet reactivity. According to Zhu et al. (2016), this disorder promotes thrombus formation using an increased discharge of calcium ions from intracellular reserves, linking TMAO to atherosclerosis. Considering the effects of succinate and TMAO on atherosclerosis risk, a further comprehensive study of their collective effects is necessary.

11.5.4 Smooth Muscle Cells

Selak et al. (2005) have reported that the build-up of succinate triggers the expansion and penetration of smooth muscle cells (SMCs) into the inner layer of blood vessels, a process that is seen in atherosclerosis. Three main routes are responsible for the abnormal development of SMCs. First, angiotensin II can be released by stimulating the RAS, which is made possible by succinate. The SMAD pathways are then activated by this chemical, leading to fibrosis, SMC development and hypertrophy. Moreover, angiotensin II could increase the production of growth and proinflammatory substances, which will cause SMCs to change from a narrowing state to a synthesis one. After transforming, SMCs generate extracellular matrix components and proinflammatory cytokines, both of which worsen atherosclerosis. The second pathway is mediated by increased succinate, which oxidizes Fe^{2+} to promote the transcription and accumulation of HIF-1α (Selak et al. 2005). Through the promotion of several growth factors and mitochondrial division, the rise in HIF-1α drives the progress of SMC. The third path is typified by succinate acting as an agonist on NF-κB ligands to activate the nuclear factor kappa B (NF-κB) signalling path. Through the NF-κB and mammalian target of rapamycin pathways, this stimulation increases the production of proteins in SMCs and promotes their proliferation, differentiation and migration.

11.5.5 Immune Cells

Additionally, succinate can stimulate proinflammatory phenotype macrophage polarization, which exacerbates atherosclerosis (Komers 2013). Elevations in succinate cause HIF-1α to rise, which in turn causes IL-1β to be produced (Shio et al. 2015). Furthermore, proinflammatory factors like TNF-α, IL-1β and IL-18 are generated more easily by succinate-derived ROS and the communication among succinate and its receptor, which causes proinflammatory phenotypic macrophages to polarize (Rubic et al. 2008). According to Weber and Noels (2011), dendritic cells are also involved in the aetiology of atherosclerosis. Succinate can cause the release of proinflammatory cytokines from these cells by activating dendritic cell surface-expressed GPR91. DCs secrete such cytokines as interferon-γ and TNF-α, which could promote atherosclerosis and maintain macrophages with a proinflammatory phenotype (Mills and O'Neill 2014). According to recent research, lymphocytes—particularly B and T cells—play a critical part in the pathophysiology of atherosclerosis (Pattarabanjird et al. 2021; Chen et al. 2023). B cells have special abilities to respond to damage, stress and infection. These abilities include cytokine production, antigen presentation and intercellular interaction. According to Sage et al. (2019), these cells are linked to both systemic and local immunity, which accelerates the development of atherosclerosis. In particular, stimulated endothelial cells covering atherosclerotic plaques facilitate different immunoglobulins made by B cells from penetrating plaque regions and carrying out specific activities in the context of dyslipidaemia. Here, plasma cells are also able to manufacture immunoglobulins in situ. Furthermore, pro-atherosclerotic TNF-α and anti-atherosclerotic IL-10 are among the many substances that B cells can produce, and they can contribute to the disease's advancement (Pattarabanjird et al. 2021). Many T cell subtypes have been found to have different roles in the course of atherosclerotic disease in recent years. These consist of anti-inflammatory regulatory T cells, pluripotent CD4 T cells and proinflammatory CD8 T cells (Tabas and Lichtman 2017). According to the most recent research, the immunological milieu within atherosclerotic plaques has weakened multiple T cell peripheral immunity checkpoints, which may be a major factor accelerating the development of atherosclerosis disease (Wang et al. 2023). Growing data over the past few decades has indicated a potential link between the pathogenic processes and immune cells underlying atherosclerosis. Nevertheless, the connection between lymphocytes and atherosclerotic succinate is still unknown. Future study on the possible relationship between succinate and the development of atherosclerosis via the control of T- and B cell activities looks very promising.

11.5.6 Probable Treatment Outlooks of Succinate in Atherosclerosis

It has been demonstrated that treatment approaches that focus on the succinate-induced atherosclerosis pathway protect the cardiovascular system. By inhibiting NF-κB, one can prevent plaque rupture by reducing the hydrolysis of matrix metalloproteinase-9 and limiting smooth muscle cell migration. Moreover, the inflammatory reaction triggered by succinate in atherosclerosis can be lessened by cinnamaldehyde, which could inhibit the synthesis of HIF-1α and IL-1β by succinate (Liu et al. 2020).

11.5.7 Renin Angiotensin System and Hypertension

SUCNR1 can be activated by succinate. It is normally present in the kidney's vascular lumen, mainly in the glomerular vasculature and afferent arterioles (Vargas et al. 2009). GPR91 activation can increase cyclooxygenase-2 production, which leads to prostaglandin E2 synthesis and release. The cells of the juxtaglomerular apparatus are then stimulated to produce renin as a result of this interaction with their EP2/EP4 receptors (Toma et al. 2008). The afferent arteriole dilates as a result of this contact. By avoiding angiotensin II's negative feedback loop, succinate and GPR91 are associated, which may result in hypertension. Usually, the angiotensin II-mediated negative feedback loop regulates the renin release. Renin release via the calcium protein kinase C path is inhibited by the communication between angiotensin II and its receptor (Schweda et al. 2007). Nevertheless, in pathological situations characterized by hypertension, the GPR91 signal and GPR91-induced release of renin play a role in the production of angiotensin II, which can initiate the tubule's synthesis of (pro)renin (Steckelings et al. 2009). Tubular cell proliferation and tubulointerstitial fibrosis are facilitated by the interaction of (pro)renin and its receptor with activated GPR91, which increases activation of extracellular signal-regulated kinase 1/2 (ERK1/2; Peti-Peterdi et al. 2008).

11.6 Role of Gut Microbiota–Derived Succinate in Metabolic Syndrome

The billions of microbes that live in our gastrointestinal tract, known as the gut microbiota, play a complex ecosystem that greatly influences immune system function, food uptake and metabolism and gut maturation in healthy hosts (Sommer and Bäckhed 2013). It is commonly known that dysbiosis, or modifications in the structure and metabolism of the gut microbial flora, is related to a variety of clinical conditions. An increasing body of research proposes that dysbiosis of the gut microbiota plays a part in the aetiology of several diseases, including extra-intestinal conditions like non-alcoholic fatty liver disease, obesity, T2D and cancer, as well as intestinal conditions like inflammatory bowel disease (IBD) (Honda and Littman

2016; Canfora et al. 2019). Nonetheless, it is still unclear in humans whether microbiota dysbiosis and disease are directly causally related, or if the former is a result of the latter (Larsen et al. 2010; Gevers et al. 2014). Within this framework, gut health and illness have been associated with by-products of bacterial metabolism. In this book chapter, we concentrate on the TCA cycle metabolite succinate that is fast emerging as a model metabolite originating from the microbiota, playing significant roles in pathogen susceptibility, gut homeostasis and inflammatory disorders such as obesity and IBD. Interestingly, succinate is unique in that it is produced by both the host and the microbiota, which puts it at the centre of metabolic interactions between the gut microbiota and the host. This book chapter aims to give a few fundamental ideas and highlight open problems regarding succinate's role as a friend or foe in microbiota-associated health and disease, rather than to provide a thorough assessment of the research.

11.6.1 Succinate, a Metabolite with Pleiotropic Functions

Traditionally regarded as an energy substrate for oxidative phosphorylation in mitochondria, succinate is well recognized in host cells as an intermediary of the TCA cycle. Additionally, succinate has drawn more devotion as an extracellular and cytosolic signalling molecule. Research has shown that the accretion of succinate in the cytoplasm is directly connected to a growing number of processes, including the generation of reactive oxygen species, stabilization of the hypoxia-inducible factor family of transcription factors, stimulation of inflammatory programmes, epigenetic regulation and protein post-translational changes by succinylation (Chouchani et al. 2014; Murphy and O'Neill 2018). Additionally, it is widely believed that succinate regulates tissue injury, local stress and immunological reactions by acting as a pro-inflammatory stimulus (Littlewood-Evans et al. 2016; Toma et al. 2008). Furthermore, some data suggests that intracellular succinate has a positive impact on intestinal gluconeogenesis and thermogenesis (Wang et al. 2019a, b; Mills et al. 2018). SUCNR1, a G protein-coupled receptor (GPCR) expressed on the cell membrane of a variety of cells, is the cognate receptor for succinate that senses it extracellularly (Gilissen et al. 2016; De Castro et al. 2016). Similar to other GPCRs, SUCNR1 communicates with cells through a variety of channels (Ariza et al. 2012). Nevertheless, little is known about this receptor's signalling and functionality. According to Rubić-Schneider et al. (2017), succinate has been thought to be a stress-induced signalling mediator that enhances proinflammatory reactions for the best possible immune stimulation. On the other hand, SUCNR1's activation in neural stem cells has also lately been linked to anti-inflammatory properties (Peruzzotti-Jametti et al. 2018). The vigorous resolution of short-duration inflammation in the obesity situation is dependent on SUCNR1 signalling in macrophages, a previously unknown mechanism that our group has just discovered (Keiran et al. 2019). This finding represents a physiological circuit that is disrupted in human obesity. It is interesting to note that obesity is linked to decreased SUCNR1 signalling but increased circulating amounts of succinate, a condition we have dubbed succinate-resistant (Serena et al. 2018). Surprisingly, Lei et al. (2018) have reported that succinate-SUCNR1 signalling is a key mediator of helminth-stimulated type 2 immunity in the intestine, albeit it is still unclear where succinate comes from—dieting epithelial cells, the microbiota or the diet. It would appear obvious that the succinate/SUCNR1 axis may act as a mediator among metabolic trauma and immunological function, even though the indication assisting succinate as a damaging or useful signal is equivocal (van Diepen et al. 2017). Although others have reviewed the succinate role in the metabolic control of immune cells in great detail (Patil et al. 2019; Connors et al. 2018), more research is still required to fully understand the intracellular and extracellular role of succinate in immune cells as well as other organs and tissues outside of the immune system, despite recent significant advancements. This is crucial since SUCNR1 is a GPCR and may be strongly drugged with readily available small compounds (Rexen Ulven et al. 2018; Trauelsen et al. 2017). To assess the potential of extracellular succinate receptors as a therapeutic target, more research into the physiological and pathological roles of this substance is necessary. A plethora of data suggests that SUCNR1 is a factor in the aetiology of some disorders linked to obesity and type 2 diabetes. According to Regard et al. (2008), extracellular succinate inhibits the breakdown of lipids in adipose tissue, and SUCNR1-signalling appears to stay a key blood pressure regulator in T2D through a mechanism reliant on renin-angiotensin system activation. Furthermore, SUCNR1 signalling has pathological consequences in hypertrophic cardiomyopathy (Aguiar et al. 2014), steatohepatitis and liver fibrosis (Nguyen et al. 2018). It also plays a major part in diabetic retinopathy via the production of retinal neovascularization by vascular endothelial growth factor (Matsumoto et al. 2012). Numerous physiological situations, such as endurance exercise, and some inflammatory and metabolic disorders, such as ischaemic heart disease (Aguiar et al. 2014; De Castro et al. 2016; Gilissen et al. 2016), hypertension (Sadagopan et al. 2007), T2D and obesity, are related to elevated mingling succinate levels. However, the source of the circulating succinate is still unknown. Although tissue damage may have a role in the presence of succinate in the bloodstream in a pathological setting, a microbiological source for succinate should not be completely discounted. In keeping with this, we have shown for the first time that circulating succinate and gut microbiota in obese humans are closely related (Serena et al. 2018). The metabolite succinate, which is produced by the microbiota and may be important in intestinal and extraintestinal disorders linked to microbial dysbiosis, will be discussed in the following sections.

11.6.2 Succinate, a Common Microbiota By-Product

The gut microbiota breaks down various nutrients from the diet and the host into end products that the host can absorb, like lactate and succinate, and short-chain fatty acids (SCFAs). In particular, the primary sources of the SCFAs are butyrate, propionate and acetate, which are thought to be the greatest prevalent by-products of fermentation of microorganism dietary non-digestible carbohydrates (Wong et al. 2006). Despite having a variety of effects on host physiology, SCFAs principally provide several health-promoting roles by serving as important energy substrates for hepatocytes, enterocytes and colonocytes. Additionally, they act as signalling molecules that are recognized by particular GPCRs and primarily target immune and enteroendocrine cells in the gut mucosa's lamina propria (Koh et al. 2016; Husted et al. 2017). Although succinate is a metabolic by-product of certain bacteria, it has just historically remained recognized as a crucial step in the synthesis of propionate in microorganisms. Propionate is produced by two distinct microbiological processes (Louis and Flint 2017; Reichardt et al. 2014). Nevertheless, succinate is not only used by secondary fermenters as a food source, but it is also a precursor of propionate, which is frequently generated by primary fermenters like Bacteroides. Therefore, under certain development circumstances, such as great carbon dioxide proportions (Caspari and Macy 1983), a build-up of succinate has been reported in cultures of various *Bacteroides* spp. as well as in cultures of *Prevotella ruminicola* cultivated without vitamin B12 (Strobel 1992). Surprisingly, it was found that ordinary mice colonized with the succinate producer *Prevotella copri* had elevated levels of caecal succinate. Similar to this, various *Ruminococcaceae* have been identified as bacteria that produce succinate, including *Ruminococcus flavefaciens* (Macfarlane and Gibson 1997). However, some human colonic bacteria that are members of the Negativicutes class of Firmicutes can change succinate into propionate, including *Phascolarctobacterium succinatutens* (Watanabe et al. 2012; Flint et al. 2015). Though succinate has been identified as a consumed product for certain bacterial species, there is no typical organization for consumers and producers of succinate. However, it should be mentioned that the majority of this research is established on cells that are grown in cultures without cross-feeding interactions, such as the exchange of nutrients. Understanding the metabolic interactions between the various ecological niches within the community is crucial when it comes to gut microbiota. This knowledge can be leveraged to gain a deeper understanding of the function of microbial metabolites, like succinate, in physiology and diseases linked to dysbiosis.

11.6.3 Microbiota-Derived Succinate in Health and Disease

While the amounts of these metabolites in circulation are in the micromolar range, the caecal and colonic expressions of SCFAs range from 1 to 3 mM in a usually healthy state. Acetate (5–200 μM) is the most prevalent SCFA at the systemic level, monitored by butyrate and propionate (≤12–13 μM; Connors et al. 2018). In contrast, succinate is found in the gut lumen at comparatively low concentrations, which are probably caused by distinct gut bacteria cross-feeding one another and converting succinate to propionate (Cummings et al. 1987). Faecal succinate levels are nearly undetectable in germ-free mice compared to control animals, suggesting that gut microbiota is the primary source of luminal succinate (Nagao-Kitamoto et al. 2016; Kim et al. 2017). Numerous studies have demonstrated a clear correlation between disruptions in the gut microbiota and disease, such as antibiotic-induced dysbiosis (Tulstrup et al. 2015), motility issues (Ferreyra et al. 2014) and in particular IBD (Macias-Ceja et al. 2019). More particularly, a multitude of studies in humans and animals have shown that IBD results in an elevation of faecal succinate that has been linked to illness activity (Hallert et al. 2003; Osaka et al. 2017). Although the exact role of intestinal injury versus dysbiosis of the gut microbiota in this rise in succinate remains unclear, metagenomic analysis of the gut microbial flora of IBD patients revealed a marked reduction in the abundance of some bacterial strains that consume succinate (Morgan et al. 2012). In contrast, a chemically induced colitis model in mice has been shown to exhibit an increase in succinate-producing Bacteroides (Ariake et al. 2000). Surprisingly, in a dextran sulphate sodium-induced colitis mice model, colonization of germ-free animals with succinate-producing bacteria from IBD patients exacerbates intestine inflammation, which is linked to increased levels of faecal succinate (Setoyama et al. 2003). Therefore, the data that is now available points to a connection between inflammation, intestinal succinate buildup and dysbiosis. It is unclear, though, if this situation has a direct bearing on how diseases turn out. Like SCFAs, it is fair to assume that circulating succinate levels could be influenced by splenic extraction ratio, food and the make-up of the microbiota. This may be especially important in pathological situations where the function of the intestinal barrier is altered (leaky gut). Compared to healthy individuals, patients with Crohn's disease have been shown to have higher serum and intestinal succinate levels (Macias-Ceja et al. 2019). According to Meng et al. (2019), moms with IBD have been found to have higher levels of succinate in their breast milk when compared to their healthy counterparts. It is interesting to note that we recently discovered a high correlation between human circulating succinate and microbiological gut flora (Serena et al. 2018).

The researchers also showed that altering the gut microbial flora through dietary weight loss interventions, as well as microbiota that evolve on their own without regard to body weight, affects mingling succinate levels. These changes are closely related to a particular metabolic function and molecular entity of the microbiota that is connected to the metabolism of succinate (Serena et al. 2018). While a direct correlation between succinate levels and obesity-related diseases could not be established, the available data suggests that succinate is a novel participant in the pathophysiology of obesity-related metabolic disorders. Enteric pathogens are known to benefit from microbiota dysbiosis as a competitive advantage. Furthermore, there exists a clear correlation between the community structure and the metabolic activities of individual bacterial strains. In this case, the choice of succinate or propionate as a microbiological product appears to be determined mostly by carbon dioxide levels (Fischbach and Sonnenburg 2011). Nevertheless, only a small number of research studies have explicitly looked at how succinate affects the resilience and upkeep of gut microbiota, with the majority pointing to a negative impact on succinate. According to earlier publications, succinate is a virulence factor that may make enteric infections worse (Rotstein et al. 1988). As was previously established, the gut lumen of typical mice contains minimal amounts of succinate. However, the administration of antibiotics or chemically induced disruptions of intestinal motility create a brief rise in its concentration that may be effectively utilized by *Clostridium difficile*, the main source of antibiotic-linked diarrhoea, to multiply (Ferreyra et al.

2014). Similarly, enterohemorrhagic pathogens like *Citrobacter rodentium* and *Escherichia coli* sense surroundings rich in succinate and use that information to activate their virulence components (Curtis et al. 2014). Furthermore, as defined by Kim and colleagues recently, succinate may also benefit bacterial ecosystems in neonates. Specifically, in the situation of the neonatal gut, Bacteroides-derived succinate may promote colonization by anaerobes like *Clostridia* species, which inhibits the development of diarrhoea-causing bacteria like *S. typhimurium*. Moreover, the mucosal inflammatory reaction induced by *S. typhimurium* triggers a metabolic adaptation in the pathogen itself to use microbiota-derived succinate as a nutrient for colonization of the intestinal tract (Spiga et al. 2017). In summary, succinate is essential in commensal-pathogen interactions. Subsequent investigation is undoubtedly necessary to gain a more comprehensive understanding of succinate's function in the gut ecosystem. It is plausible that succinate serves as a main cross-feeding metabolite that is necessary for the preservation of a healthy resident gut microbiota; in this regard, the elevation of succinate in certain pathological situations would be indicative of dysbiosis.

One potential source of circulating succinate could be the intestinal efflux caused by a particular gut microbiota, which is influenced by a number of endogenous and exogenous variables. Like other metabolites, succinate may function in peripheral host tissues as a fuel or signalling metabolite to control energy metabolism. It has been documented that succinate activates SUCNR1 in Tuft cells, which in turn activates intestinal gluconeogenesis and type 2 immunity. One cannot rule out the possibility that succinate produced from gut bacteria may have additional immunological and metabolic impacts in peripheral tissues. In fact, succinate has been identified as the metabolite that controls the activation of brown adipose tissue. In other tissues, SUCNR1 signalling has been linked to the regulation of some metabolic processes (e.g. activator of the renal renin-angiotensin system, negative regulator of lipolysis in adipose tissue). Furthermore, the impact of succinate on the microbial community within the intestinal lumen may be particularly noteworthy. As a result, succinate has become known as a cross-feeding molecule that is crucial for both commensal pathogen interactions and the reconstruction of the gut bacterial community.

11.6.4 Role of Probiotic Interventions Focused on Modifying Gut-Derived Succinate to Be Used to Treat Obesity-Related Disorders

While most of the printed research associates clinical symptoms linked to dysbiosis with an environment rich in succinate, numerous recent research studies have reported the metabolic advantages of some bacteria through their succinate production, especially in obesity-associated metabolic disorders. Kovatcheva-Datchary et al. (2015) discovered that *P. copri*, a well-known producer of succinate, was more prevalent in the gut microbial flora of healthy individuals who demonstrated an enhancement in sugar digestion after consuming bread made from barley kernels.

Furthermore, in comparison to animals reconditioned with the microbial flora of non-responders to barley kernel-based bread eating, the microbiota from these patients enhanced sugar digestion in mice free from germs. Therefore, *P. copri* colonization enhanced caecal succinate and enhanced sugar tolerance in mice fed a regular diet (Kovatcheva-Datchary et al. 2015). Meanwhile, co-colonization with a *Bacteroides thetaiotaomicron* mutant incapable of changing succinate to propionate was unsuccessful in progressing the sugar reaction. The researchers established that the great amount of succinate seen in *P. copri*-treated mice was insufficient to explicate the beneficial effects of such a bacterial strain (Kovatcheva-Datchary et al. 2015). In a later investigation, a similar research group found that *P. copri* colonization reduces mice's ability to produce glucose in the liver (De Vadder et al. 2016). Additionally, it has been demonstrated that succinate is a crucial microbial product for the advantageous dietary fibre metabolic effects intake, as it raises the amount of succinate produced by Prevotella (Koeth et al. 2013). It is interesting to note that there is a systemic negative correlation between the abundance of *Akkermansia muciniphila* and metabolic problems connected to obesity. Succinate is one of the key metabolites produced by mucin breakdown (Chia et al. 2018). Growth in the plenty of *A. muciniphila* has been linked to bariatric surgery, antidiabetic medications like metformin and some prebiotic therapies that alleviate metabolic problems linked to obesity (Cani 2018). According to Wang et al. (2019a, b), colonization with further succinate makers, like *Parabacteroides distasonis*, has shown promise in treating metabolic dysfunctions linked to obesity. It is interesting to note that there is a negative correlation between the quantity of this commensal bacterium and IBD, which has been extensively shown to be associated with elevated levels of caecal succinate (Macias-Ceja et al. 2019). However, it has also been determined that secondary bile acids may have a role in the advantageous metabolic benefits of *P. distasonis* (Wang et al. 2019a, b). Probiotics' potential application in the treatment of metabolic illnesses, including T2D and obesity, is gaining attention in the scientific community. Succinate is suggested as a possible target microbial metabolite in the initial research on this topic (Kovatcheva-Datchary et al. 2015; De Vadder et al. 2016). However, additional study is necessary to elucidate the mechanisms of succinate biology, as these disorders are typically linked to elevated levels of succinate. Future studies should examine if this viewpoint holds for additional dysbiosis-related illnesses.

11.7 Conclusion

Succinate is involved in the TCA cycle and many other pathological processes, including proinflammatory macrophage mediation, ROS overproduction, interaction with GPR91 and energy metabolism. One metabolite that builds up in the gut during periods of inflammation and alteration of the microbiota, such as in inflammatory bowel disease (IBD), is succinate, which is produced by both host and microbiota cells. In response to metabolic stress or tissue damage, succinate can start vital defence mechanisms. However, when combined with other inflammatory stimuli, these reactions may become dysregulated or inappropriately expanded, which can lead to disease. The complex actions of succinate have a role in the development and advancement of several cardiovascular conditions, including atherosclerosis, hypertension, cardiac hypertrophy and impairment from reperfusion injury. Approaches that target succinate and the related metabolic pathway offer a unique and targeted treatment opportunity, given that succinate plays a part in signal transmission and the development of several cardiovascular illnesses. Many questions about succinate and the gut flora remain unresolved despite the growing body of research in this area. Changes in the relative abundance of commensal versus pathogenic bacteria and the amounts of anti-inflammatory SCFAs are also expected to occur concurrently with disruptions in the gut lumen that lead to the accumulation of succinate. To completely comprehend the effects of succinate on intestinal inflammation and its place in the larger scheme of host-microbiome interactions, more investigation is necessary. Research has indicated that the injection of succinate into mice can improve their tolerance to glucose and insulin, indicating a good effect on glycaemic control. Furthermore, the role of succinate in reducing intestinal inflammation has been demonstrated. To completely comprehend the chronic impacts of succinate supplementation on host metabolism and immunological function, more research is also required. It is important to recognize that animal models have been used in a large amount of research on succinate and the gut microbiota. Therefore, more investigation is required to confirm these results in human trials.

Acknowledgement The authors would like to acknowledge Dr. AP Pawar, Principal, Poona College of Pharmacy, Bharati Vidyapeeth (Deemed to be) University, Pune, for providing helpful article writing suggestions.

Author's Contribution All authors contributed to the manuscript. DM has collected the relevant articles. DM and SM were responsible for the literature search, drafting and critically revising of the manuscript. All the authors have read and approved the manuscript carefully.

Conflict of Interest Statement The authors declare that they have no conflict of interest.

Funding No funding was received for this research work.

References

Aguiar CJ, Rocha-Franco JA, Sousa PA, Santos AK, Ladeira M, Rocha-Resende C, Ladeira LO, Resende RR, Botoni FA, Barrouin Melo M, Lima CX (2014) Succinate causes pathological cardiomyocyte hypertrophy through GPR91 activation. Cell Commun Signaling 12:1–7

Akram M (2014) Citric acid cycle and role of its intermediates in metabolism. Cell Biochem Biophys 68(3):475–478

Alleyn M, Breitzig M, Lockey R, Kolliputi N (2018) The dawn of succinylation: a posttranslational modification. Am J Phys Cell Phys 314(2):C228–C232
Ariake K, Ohkusa T, Sakurazawa T, Kumagai J, Eishi Y, Hoshi S, Yajima T (2000) Roles of mucosal bacteria and succinic acid in colitis caused by dextran sulfate sodium in mice. J Med Dent Sci 47(4):233–241
Ariza AC, Deen PM, Robben JH (2012) The succinate receptor as a novel therapeutic target for oxidative and metabolic stress-related conditions. Front Endocrinol 3:22
Armstrong N, Storey CM, Noll SE, Margulis K, Soe MH, Xu H, Yeh B, Fishbein L, Kebebew E, Howitt BE, Zare RN (2022) SDHB knockout and succinate accumulation are insufficient for tumorigenesis but dual SDHB/NF1 loss yields SDHx-like pheochromocytomas. Cell Rep 38(9):110453
Avula S, Parikh S, Demarest S, Kurz J, Gropman A (2014) Treatment of mitochondrial disorders. Curr Treat Options Neurol 16:1–20
Beverly JK, Budoff MJ (2020) Atherosclerosis: pathophysiology of insulin resistance, hyperglycemia, hyperlipidemia, and inflammation. J Diabetes 12(2):102–104
Brereton MF, Rohm M, Ashcroft FM (2016) β-Cell dysfunction in diabetes: a crisis of identity? Diabetes Obes Metab 18:102–109
Budzinska M, Zimna A, Kurpisz M (2021) The role of mitochondria in Duchenne muscular dystrophy. J Physiol Pharmacol 72(2):157–166
Canfora EE, Meex RC, Venema K, Blaak EE (2019) Gut microbial metabolites in obesity, NAFLD and T2DM. Nat Rev Endocrinol 15(5):261–273
Cani PD (2018) Human gut microbiome: hopes, threats and promises. Gut 67(9):1716–1725
Carrico C, Meyer JG, He W, Gibson BW, Verdin E (2018) The mitochondrial acylome emerges: proteomics, regulation by sirtuins, and metabolic and disease implications. Cell Metab 27(3):497–512
Caspari D, Macy JM (1983) The role of carbon dioxide in glucose metabolism of *Bacteroides fragilis*. Arch Microbiol 135:16–24
Chen J, Xiang X, Nie L, Guo X, Zhang F, Wen C, Xia Y, Mao L (2023) The emerging role of Th1 cells in atherosclerosis and its implications for therapy. Front Immunol 13:1079668
Chia LW, Hornung BV, Aalvink S, Schaap PJ, de Vos WM, Knol J, Belzer C (2018) Deciphering the trophic interaction between Akkermansia muciniphila and the butyrogenic gut commensal Anaerostipes caccae using a metatranscriptomic approach. Antonie Van Leeuwenhoek 111(6):859–873
Chinopoulos C (2021) The mystery of extramitochondrial proteins lysine succinylation. International Journal of Molecular Sciences 22(11):6085
Chouchani ET, Pell VR, Gaude E, Aksentijević D, Sundier SY, Robb EL, Logan A, Nadtochiy SM, Ord EN, Smith AC, Eyassu F (2014) Ischaemic accumulation of succinate controls reperfusion injury through mitochondrial ROS. Nature 515(7527):431–435
Chu DT, Nguyet NT, Dinh TC, Lien NV, Nguyen KH, Ngoc VT, Tao Y, Le DH, Nga VB, Jurgoński A, Tran QH (2018) An update on physical health and economic consequences of overweight and obesity. Diabetes Metab Syndr Clin Res Rev 12(6):1095–1100
Connors J, Dawe N, Van Limbergen J (2018) The role of succinate in the regulation of intestinal inflammation. Nutrients 11(1):25
Cummings JH, Pomare EW, Branch WJ, Naylor CP, MacFarlane G (1987) Short chain fatty acids in human large intestine, portal, hepatic and venous blood. Gut 28(10):1221–1227
Curtis MM, Hu Z, Klimko C, Narayanan S, Deberardinis R, Sperandio V (2014) The gut commensal Bacteroides thetaiotaomicron exacerbates enteric infection through modification of the metabolic landscape. Cell Host Microbe 16(6):759–769
Dalla Pozza E, Dando I, Pacchiana R, Liboi E, Scupoli MT, Donadelli M, Palmieri M (2020) Regulation of succinate dehydrogenase and role of succinate in cancer. Semin Cell Dev Biol 98(4):4–14
de Castro FM, Aguiar CJ, da Rocha Franco JA, Gingold RN, Leite MF (2016) GPR91: expanding the frontiers of Krebs cycle intermediates. Cell Commun Signaling 14(1):1–9

De Vadder F, Kovatcheva-Datchary P, Zitoun C, Duchampt A, Bäckhed F, Mithieux G (2016) Microbiota-produced succinate improves glucose homeostasis via intestinal gluconeogenesis. Cell Metab 24(1):151–157

Dunn KA, Moore-Connors J, MacIntyre B, Stadnyk A, Thomas NA, Noble A, Mahdi G, Rashid M, Otley AR, Bielawski JP, Van Limbergen J (2016) The gut microbiome of pediatric Crohn's disease patients differs from healthy controls in genes that can influence the balance between a healthy and dysregulated immune response. Inflamm Bowel Dis 22(11):2607–2618

Edalat A, Schulte-Mecklenbeck P, Bauer C, Undank S, Krippeit-Drews P, Drews G, Düfer M (2015) Mitochondrial succinate dehydrogenase is involved in stimulus-secretion coupling and endogenous ROS formation in murine beta cells. Diabetologia 58:1532–1541

Ferreyra JA, Wu KJ, Hryckowian AJ, Bouley DM, Weimer BC, Sonnenburg JL (2014) Gut microbiota-produced succinate promotes C. difficile infection after antibiotic treatment or motility disturbance. Cell Host Microbe 16(6):770–777

Fischbach MA, Sonnenburg JL (2011) Eating for two: how metabolism establishes interspecies interactions in the gut. Cell Host Microbe 10(4):336–347

Flint HJ, Duncan SH, Scott KP, Louis P (2015) Links between diet, gut microbiota composition and gut metabolism. Proc Nutr Soc 74(1):13–22

Flores RE, Brown AK, Taus L, Khoury J, Glover F, Kami K, Sarangarajan R, Walshe TE, Narain NR, Kiebish MA, Shelton LM (2018) Mycoplasma infection and hypoxia initiate succinate accumulation and release in the VM-M3 cancer cells. Biochim Biophys Acta 1859(9):975–983

Foks AC, Bot I (2017) Preface: pathology and pharmacology of atherosclerosis. Eur J Pharmacol 816:1–2

Fried M, Yumuk V, Oppert JM, Scopinaro N, Torres AJ, Weiner R, Yashkov Y, Frühbeck G (2013) Interdisciplinary European guidelines on metabolic and bariatric surgery. Obes Facts 6(5):449–468

Gevers D, Kugathasan S, Denson LA, Vázquez-Baeza Y, Van Treuren W, Ren B, Schwager E, Knights D, Song SJ, Yassour M, Morgan XC (2014) The treatment-naive microbiome in new-onset Crohn's disease. Cell Host Microbe 15(3):382–392

Gilissen J, Jouret F, Pirotte B, Hanson J (2016) Insight into SUCNR1 (GPR91) structure and function. Pharmacol Ther 159:56–65

Gorman GS, Chinnery PF, DiMauro S, Hirano M, Koga Y, McFarland R, Suomalainen A, Thorburn DR, Zeviani M, Turnbull DM (2016) Mitochondrial diseases. Nat Rev Dis Primers 2(1):1–22

Goto H, Takemura N, Ogasawara T, Sasajima N, Watanabe J, Ito H, Morita T, Sonoyama K (2010) Effects of fructo-oligosaccharide on DSS-induced colitis differ in mice fed non-purified and purified diets. J Nutr 140(12):2121–2127

Grimolizzi F, Arranz L (2018) Multiple faces of succinate beyond metabolism in blood. Haematologica 103(10):1586–1592

Gut P, Matilainen S, Meyer JG, Pällijeff P, Richard J, Carroll CJ, Euro L, Jackson CB, Isohanni P, Minassian BA, Alkhater RA (2020) SUCLA2 mutations cause global protein succinylation contributing to the pathomechanism of a hereditary mitochondrial disease. Nat Commun 11(1):5927

Haffke M, Fehlmann D, Rummel G, Boivineau J, Duckely M, Gommermann N, Cotesta S, Sirockin F, Freuler F, Littlewood-Evans A, Kaupmann K (2019) Structural basis of species-selective antagonist binding to the succinate receptor. Nature 574(7779):581–585

Halban PA, Polonsky KS, Bowden DW, Hawkins MA, Ling C, Mather KJ, Powers AC, Rhodes CJ, Sussel L, Weir GC (2014) β-cell failure in type 2 diabetes: postulated mechanisms and prospects for prevention and treatment. J Clin Endocrinol Metabol 99(6):1983–1992

Hallert C, Björck I, Nyman M, Pousette A, Grännö C, Svensson H (2003) Increasing fecal butyrate in ulcerative colitis patients by diet: controlled pilot study. Inflamm Bowel Dis 9(2):116–121

Haythorne E, Rohm M, van de Bunt M, Brereton MF, Tarasov AI, Blacker TS, Sachse G, Silva dos Santos M, Terron Exposito R, Davis S, Baba O (2019) Diabetes causes marked inhibition of mitochondrial metabolism in pancreatic β-cells. Nat Commun 10(1):2474

He W, Miao FJ, Lin DC, Schwandner RT, Wang Z, Gao J, Chen JL, Tian H, Ling L (2004) Citric acid cycle intermediates as ligands for orphan G-protein-coupled receptors. Nature 429(6988):188–193

Hirschey MD, Zhao Y (2015) Metabolic regulation by lysine malonylation, succinylation, and glutarylation. Mol Cell Proteomics 14(9):2308–2315

Hoffman DJ, Powell TL, Barrett ES, Hardy DB (2021) Developmental origins of metabolic diseases. Physiol Rev 101(3):739–795

Honda K, Littman DR (2016) The microbiota in adaptive immune homeostasis and disease. Nature 535(7610):75–84

Husted AS, Trauelsen M, Rudenko O, Hjorth SA, Schwartz TW (2017) GPCR-mediated signaling of metabolites. Cell Metab 25(4):777–796

Jiang S, Yan W (2017) Succinate in the cancer–immune cycle. Cancer Lett 390:45–47

Kajiura T, Takeda T, Sakata S, Sakamoto M, Hashimoto M, Suzuki H, Suzuki M, Benno Y (2009) Change of intestinal microbiota with elemental diet and its impact on therapeutic effects in a murine model of chronic colitis. Dig Dis Sci 54:1892–1900

Kaplan GG (2015) The global burden of IBD: from 2015 to 2025. Nat Rev Gastroenterol Hepatol 12(12):720–727

Keiran N, Ceperuelo-Mallafré V, Calvo E, Hernández-Alvarez MI, Ejarque M, Núñez-Roa C, Horrillo D, Maymó-Masip E, Rodríguez MM, Fradera R, de la Rosa JV (2019) SUCNR1 controls an anti-inflammatory program in macrophages to regulate the metabolic response to obesity. Nat Immunol 20(5):581–592

Kim YG, Sakamoto K, Seo SU, Pickard JM, Gillilland MG III, Pudlo NA, Hoostal M, Li X, Wang TD, Feehley T, Stefka AT (2017) Neonatal acquisition of clostridia species protects against colonization by bacterial pathogens. Science 356(6335):315–319

Ko SH, Choi GE, Oh JY, Lee HJ, Kim JS, Chae CW, Choi D, Han HJ (2017) Succinate promotes stem cell migration through the GPR91-dependent regulation of DRP1-mediated mitochondrial fission. Sci Rep 7(1):12582

Koeth RA, Wang Z, Levison BS, Buffa JA, Org E, Sheehy BT, Britt EB, Fu X, Wu Y, Li L, Smith JD (2013) Intestinal microbiota metabolism of L-carnitine, a nutrient in red meat, promotes atherosclerosis. Nat Med 19(5):576–585

Koh A, De Vadder F, Kovatcheva-Datchary P, Bäckhed F (2016) From dietary fiber to host physiology: short-chain fatty acids as key bacterial metabolites. Cell 165(6):1332–1345

Komers R (2013) Renin inhibition in the treatment of diabetic kidney disease. Clin Sci 124(9):553–566

Kovatcheva-Datchary P, Nilsson A, Akrami R, Lee YS, De Vadder F, Arora T, Hallen A, Martens E, Björck I, Bäckhed F (2015) Dietary fiber-induced improvement in glucose metabolism is associated with increased abundance of Prevotella. Cell Metab 22(6):971–982

Krzak G, Willis CM, Smith JA, Pluchino S, Peruzzotti-Jametti L (2021) Succinate receptor 1: an emerging regulator of myeloid cell function in inflammation. Trends Immunol 42(1):45–58

Kurmi K, Hitosugi S, Wiese EK, Boakye-Agyeman F, Gonsalves WI, Lou Z, Karnitz LM, Goetz MP, Hitosugi T (2018) Carnitine palmitoyltransferase 1A has a lysine succinyl transferase activity. Cell Rep 22(6):1365–1373

Larsen N, Vogensen FK, Van Den Berg FW, Nielsen DS, Andreasen AS, Pedersen BK, Al-Soud WA, Sørensen SJ, Hansen LH, Jakobsen M (2010) Gut microbiota in human adults with type 2 diabetes differs from non-diabetic adults. PLoS One 5(2):e9085

Lee S, Xu H, Van Vleck A, Mawla AM, Li AM, Ye J, Huising MO, Annes JP (2022) β-Cell succinate dehydrogenase deficiency triggers metabolic dysfunction and insulinopenic diabetes. Diabetes 71(7):1439–1453

Lei W, Ren W, Ohmoto M, Urban JF Jr, Matsumoto I, Margolskee RF, Jiang P (2018) Activation of intestinal tuft cell-expressed Sucnr1 triggers type 2 immunity in the mouse small intestine. Proc Natl Acad Sci 115(21):5552–5557

Leitner DR, Frühbeck G, Yumuk V, Schindler K, Micic D, Woodward E, Toplak H (2017) Obesity and type 2 diabetes: two diseases with a need for combined treatment strategies-EASO can lead the way. Obes Facts 10(5):483–492

Li F, He X, Ye D, Lin Y, Yu H, Yao C, Huang L, Zhang J, Wang F, Xu S, Wu X (2015) NADP+-IDH mutations promote hypersuccinylation that impairs mitochondria respiration and induces apoptosis resistance. Mol Cell 60(4):661–675

Li L, Shi L, Yang S, Yan R, Zhang D, Yang J, He L, Li W, Yi X, Sun L, Liang J (2016) SIRT7 is a histone desuccinylase that functionally links to chromatin compaction and genome stability. Nat Commun 7(1):12235

Littlewood-Evans A, Sarret S, Apfel V, Loesle P, Dawson J, Zhang J, Muller A, Tigani B, Kneuer R, Patel S, Valeaux S (2016) GPR91 senses extracellular succinate released from inflammatory macrophages and exacerbates rheumatoid arthritis. J Exp Med 213(9):1655–1662

Liu P, Wang J, Wen W, Pan T, Chen H, Fu Y, Wang F, Huang JH, Xu S (2020) Cinnamaldehyde suppresses NLRP3 derived IL-1β via activating succinate/HIF-1 in rheumatoid arthritis rats. Int Immunopharmacol 84:106570

Louis P, Flint HJ (2017) Formation of propionate and butyrate by the human colonic microbiota. Environ Microbiol 19(1):29–41

Ma ZA, Zhao Z, Turk J (2012) Mitochondrial dysfunction and β-cell failure in type 2 diabetes mellitus. Exp Diabetes Res 2012:703538

Macfarlane GT, Gibson GR (1997) Carbohydrate fermentation, energy transduction and gas metabolism in the human large intestine. Gastrointest Microbiol 1:269–318

Macias-Ceja DC, Ortiz-Masiá D, Salvador P, Gisbert-Ferrándiz L, Hernández C, Hausmann M, Rogler G, Esplugues JV, Hinojosa J, Alós R, Navarro F (2019) Succinate receptor mediates intestinal inflammation and fibrosis. Mucosal Immunol 12(1):178–187

Malpartida AB, Williamson M, Narendra DP, Wade-Martins R, Ryan BJ (2021) Mitochondrial dysfunction and mitophagy in Parkinson's disease: from mechanism to therapy. Trends Biochem Sci 46(4):329–343

Manichanh C, Borruel N, Casellas F, Guarner F (2012) The gut microbiota in IBD. Nat Rev Gastroenterol Hepatol 9(10):599–608

Martínez-Reyes I, Chandel NS (2020) Mitochondrial TCA cycle metabolites control physiology and disease. Nat Commun 11(1):102

Matsumoto M, Suzuma K, Maki T, Kinoshita H, Tsuiki E, Fujikawa A, Kitaoka T (2012) Succinate increases in the vitreous fluid of patients with active proliferative diabetic retinopathy. Am J Ophthalmol 153(5):896–902

Maurya R, Bhattacharjee G, Gohil N, Khambhati K, Shakhreliya S, Bhatnagar A, Singh P, Ramakrishna S, Singh V (2023a) Low density lipoprotein receptor endocytosis in cardiovascular disease and the factors affecting LDL levels. Prog Mol Biol Transl Sci 194:333–345

Maurya R, Bhattacharjee G, Khambhati K, Gohil N, Singh P, Mani I, Chu DT, Ramakrishna S, Show PL, Singh V (2023b) Amyloid precursor protein in Alzheimer's disease. Prog Mol Biol Transl Sci 196:261–270

Meng X, Dunsmore G, Koleva P, Elloumi Y, Wu RY, Sutton RT, Ambrosio L, Hotte N, Nguyen V, Madsen KL, Dieleman LA (2019) The profile of human milk metabolome, cytokines, and antibodies in inflammatory bowel diseases versus healthy mothers, and potential impact on the newborn. J Crohn's Colitis 13(4):431–441

Mills E, O'Neill LA (2014) Succinate: a metabolic signal in inflammation. Trends Cell Biol 24(5):313–320

Mills EL, Pierce KA, Jedrychowski MP, Garrity R, Winther S, Vidoni S, Yoneshiro T, Spinelli JB, Lu GZ, Kazak L, Banks AS (2018) Accumulation of succinate controls activation of adipose tissue thermogenesis. Nature 560(7716):102–126

Mills EL, Harmon C, Jedrychowski MP, Xiao H, Garrity R, Tran NV, Bradshaw GA, Fu A, Szpyt J, Reddy A, Prendeville H (2021) UCP1 governs liver extracellular succinate and inflammatory pathogenesis. Nat Metab 3(5):604–617

Misawa T, Takahama M, Kozaki T, Lee H, Zou J, Saitoh T, Akira S (2013) Microtubule-driven spatial arrangement of mitochondria promotes activation of the NLRP3 inflammasome. Nature Immunology 14(5):454–460

Morgan XC, Tickle TL, Sokol H, Gevers D, Devaney KL, Ward DV, Reyes JA, Shah SA, LeLeiko N, Snapper SB, Bousvaros A (2012) Dysfunction of the intestinal microbiome in inflammatory bowel disease and treatment. Genome Biol 13(9):1–8

Murphy MP, Chouchani ET (2022) Why succinate? Physiological regulation by a mitochondrial coenzyme Q sentinel. Nat Chem Biol 18(5):461–469

Murphy MP, O'Neill LA (2018) Krebs cycle reimagined: the emerging roles of succinate and itaconate as signal transducers. Cell 174(4):780–784

Nadjsombati MS, McGinty JW, Lyons-Cohen MR, Jaffe JB, DiPeso L, Schneider C, Miller CN, Pollack JL, Gowda GN, Fontana MF, Erle DJ (2018) Detection of succinate by intestinal tuft cells triggers a type 2 innate immune circuit. Immunity 49(1):33–41

Nagao-Kitamoto H, Shreiner AB, Gillilland MG III, Kitamoto S, Ishii C, Hirayama A, Kuffa P, El-Zaatari M, Grasberger H, Seekatz AM, Higgins PD (2016) Functional characterization of inflammatory bowel disease–associated gut dysbiosis in gnotobiotic mice. Cell Mol Gastroenterol Hepatol 2(4):468–481

Ng AC, Delgado V, Borlaug BA, Bax JJ (2021) Diabesity: the combined burden of obesity and diabetes on heart disease and the role of imaging. Nat Rev Cardiol 18(4):291–304

Nguyen G, Park SY, Le CT, Park WS, Choi DH, Cho EH (2018) Metformin ameliorates activation of hepatic stellate cells and hepatic fibrosis by succinate and GPR91 inhibition. Biochem Biophys Res Commun 495(4):2649–2456

Ni J, Wu GD, Albenberg L, Tomov VT (2017) Gut microbiota and IBD: causation or correlation? Nat Rev Gastroenterol Hepatol 14(10):573–584

O'Neill LA, Pearce EJ (2016) Immunometabolism governs dendritic cell and macrophage function. J Exp Med 213(1):15–23

Ooi M, Nishiumi S, Yoshie T, Shiomi Y, Kohashi M, Fukunaga K, Nakamura S, Matsumoto T, Hatano N, Shinohara M, Irino Y (2011) GC/MS-based profiling of amino acids and TCA cycle-related molecules in ulcerative colitis. Inflamm Res 60:831–840

Osaka T, Moriyama E, Arai S, Date Y, Yagi J, Kikuchi J, Tsuneda S (2017) Meta-analysis of fecal microbiota and metabolites in experimental colitic mice during the inflammatory and healing phases. Nutrients 9(12):1329

Park J, Chen Y, Tishkoff DX, Peng C, Tan M, Dai L, Xie Z, Zhang Y, Zwaans BM, Skinner ME, Lombard DB (2013) SIRT5-mediated lysine desuccinylation impacts diverse metabolic pathways. Mol Cell 50(6):919–930

Patil NK, Bohannon JK, Hernandez A, Patil TK, Sherwood ER (2019) Regulation of leukocyte function by citric acid cycle intermediates. J Leukoc Biol 106(1):105–117

Pattarabanjird T, Li C, McNamara C (2021) B cells in atherosclerosis: mechanisms and potential clinical applications. Basic Transl Sci 6(6):546–563

Paturi G, Mandimika T, Butts CA, Zhu S, Roy NC, McNabb WC, Ansell J (2012) Influence of dietary blueberry and broccoli on cecal microbiota activity and colon morphology in mdr1a−/− mice, a model of inflammatory bowel diseases. Nutrition 28(3):324–330

Peace CG, O'Neill LA (2022) The role of itaconate in host defense and inflammation. J Clin Invest 132(2):e148548

Peruzzotti-Jametti L, Bernstock JD, Vicario N, Costa AS, Kwok CK, Leonardi T, Booty LM, Bicci I, Balzarotti B, Volpe G, Mallucci G (2018) Macrophage-derived extracellular succinate licenses neural stem cells to suppress chronic neuroinflammation. Cell Stem Cell 22(3):355–368

Peti-Peterdi J (2010) High glucose and renin release: the role of succinate and GPR91. Kidney Int 78(12):1214–1217

Peti-Peterdi J, Kang JJ, Toma I (2008) Activation of the renal renin–angiotensin system in diabetes—new concepts. Nephrol Dial Transplant 23(10):3047–3049

Quartararo CE, Hadi T, Cahill SM, Blanchard JS (2013) Solvent isotope-induced equilibrium perturbation for isocitrate lyase. Biochemistry 52(51):9286–9293

Rani V, Deep G, Singh RK, Palle K, Yadav UC (2016) Oxidative stress and metabolic disorders: pathogenesis and therapeutic strategies. Life Sci 148:183–193

Regard JB, Sato IT, Coughlin SR (2008) Anatomical profiling of G protein-coupled receptor expression. Cell 135(3):561–571

Reichardt N, Duncan SH, Young P, Belenguer A, McWilliam Leitch C, Scott KP, Flint HJ, Louis P (2014) Phylogenetic distribution of three pathways for propionate production within the human gut microbiota. ISME J 8(6):1323–1235
Ren W, Xia Y, Chen S, Wu G, Bazer FW, Zhou B, Tan B, Zhu G, Deng J, Yin Y (2019) Glutamine metabolism in macrophages: a novel target for obesity/type 2 diabetes. Adv Nutr 10(2):321–330
Rexen Ulven E, Trauelsen M, Brvar M, Lückmann M, Bielefeldt LØ, Jensen LK, Schwartz TW, Frimurer TM (2018) Structure-activity investigations and optimisations of non-metabolite agonists for the succinate receptor 1. Sci Rep 8(1):10010
Ristic B, Bhutia YD, Ganapathy V (2017) Cell-surface G-protein-coupled receptors for tumor-associated metabolites: a direct link to mitochondrial dysfunction in cancer. Biochim Biophys Acta Rev Cancer 1868(1):246–257
Rotstein OD, Nasmith PE, Grinstein S (1988) pH-dependent impairment of the neutrophil respiratory burst by the Bacteroides byproduct succinate. Clin Invest Med 11(4):259–265
Rubic T, Lametschwandtner G, Jost S, Hinteregger S, Kund J, Carballido-Perrig N, Schwärzler C, Junt T, Voshol H, Meingassner JG, Mao X (2008) Triggering the succinate receptor GPR91 on dendritic cells enhances immunity. Nat Immunol 9(11):1261–1269
Rubić-Schneider T, Carballido-Perrig N, Regairaz C, Raad L, Jost S, Rauld C, Christen B, Wieczorek G, Kreutzer R, Dawson J, Lametschwandner G (2017) GPR 91 deficiency exacerbates allergic contact dermatitis while reducing arthritic disease in mice. Allergy 72(3):444–452
Rutter J, Winge DR, Schiffman JD (2010) Succinate dehydrogenase – assembly, regulation and role in human disease. Mitochondrion 10:393–401
Ryan DG, Murphy MP, Frezza C, Prag HA, Chouchani ET, O'Neill LA, Mills EL (2019) Coupling Krebs cycle metabolites to signalling in immunity and cancer. Nat Metab 1(1):16–33
Sabari BR, Zhang D, Allis CD, Zhao Y (2017) Metabolic regulation of gene expression through histone acylations. Nat Rev Mol Cell Biol 18(2):90–101
Sadagopan N, Li W, Roberds SL, Major T, Preston GM, Yu Y, Tones MA (2007) Circulating succinate is elevated in rodent models of hypertension and metabolic disease. Am J Hypertens 20(11):1209–1215
Sage AP, Tsiantoulas D, Binder CJ, Mallat Z (2019) The role of B cells in atherosclerosis. Nature Reviews Cardiology 16(3):180–196
Schneider C, O'Leary CE, von Moltke J, Liang HE, Ang QY, Turnbaugh PJ, Radhakrishnan S, Pellizzon M, Ma A, Locksley RM (2018) A metabolite-triggered tuft cell-ILC2 circuit drives small intestinal remodeling. Cell 174(2):271–284
Schug TT, Xu Q, Gao H, Peres-da-Silva A, Draper DW, Fessler MB, Purushotham A, Li X (2010) Myeloid deletion of SIRT1 induces inflammatory signaling in response to environmental stress. Mol Cell Biol 30(19):4712–4721
Schweda F, Friis U, Wagner C, Skott O, Kurtz A (2007) Renin release. Physiology 22(5):310–319
Selak MA, Armour SM, MacKenzie ED, Boulahbel H, Watson DG, Mansfield KD, Pan Y, Simon MC, Thompson CB, Gottlieb E (2005) Succinate links TCA cycle dysfunction to oncogenesis by inhibiting HIF-α prolyl hydroxylase. Cancer Cell 7(1):77–85
Serena C, Ceperuelo-Mallafré V, Keiran N, Queipo-Ortuño MI, Bernal R, Gomez-Huelgas R, Urpi-Sarda M, Sabater M, Pérez-Brocal V, Andrés-Lacueva C, Moya A (2018) Elevated circulating levels of succinate in human obesity are linked to specific gut microbiota. ISME J 12(7):1642–1657
Setoyama H, Imaoka A, Ishikawa H, Umesaki Y (2003) Prevention of gut inflammation by Bifidobacterium in dextran sulfate-treated gnotobiotic mice associated with Bacteroides strains isolated from ulcerative colitis patients. Microbes Infect 5(2):115–122
Shen R, Ruan H, Lin S, Liu B, Song H, Li L, Ma T (2023) Lysine succinylation, the metabolic bridge between cancer and immunity. Genes Dis 10(6):2470–2478
Shio MT, Christian JG, Jung JY, Chang KP, Olivier M (2015) PKC/ROS-mediated NLRP3 inflammasome activation is attenuated by Leishmania zinc-metalloprotease during infection. PLoS Negl Trop Dis 9(6):e0003868
Sommer F, Bäckhed F (2013) The gut microbiota—masters of host development and physiology. Nat Rev Microbiol 11(4):227–238

Song Y, Wang J, Cheng Z, Gao P, Sun J, Chen X, Chen C, Wang Y, Wang Z (2017) Quantitative global proteome and lysine succinylome analyses provide insights into metabolic regulation and lymph node metastasis in gastric cancer. Sci Rep 7(1):42053
Spiga L, Winter MG, de Carvalho TF, Zhu W, Hughes ER, Gillis CC, Behrendt CL, Kim J, Chessa D, Andrews-Polymenis HL, Beiting DP (2017) An oxidative central metabolism enables Salmonella to utilize microbiota-derived succinate. Cell Host Microbe 22(3):291–301
Sreedhar A, Wiese EK, Hitosugi T (2020) Enzymatic and metabolic regulation of lysine succinylation. Genes Dis 7(2):166–171
Steckelings UM, Rompe F, Kaschina E, Unger T (2009) The evolving story of the RAAS in hypertension, diabetes and CV disease–moving from macrovascular to microvascular targets. Fundam Clin Pharmacol 23(6):693–703
Strobel HJ (1992) Vitamin B12-dependent propionate production by the ruminal bacterium Prevotella ruminicola 23. Appl Environ Microbiol 58(7):2331–2333
Sun F, Huo X, Zhai Y, Wang A, Xu J, Su D, Bartlam M, Rao Z (2005) Crystal structure of mitochondrial respiratory membrane protein complex II. Cell 121(7):1043–1057
Tabas I, Lichtman AH (2017) Monocyte-macrophages and T cells in atherosclerosis. Immunity 47(4):621–634
Tabas I, García-Cardeña G, Owens GK (2015) Recent insights into the cellular biology of atherosclerosis. J Cell Biol 209(1):13–22
Tang X, Fuchs D, Tan S, Trauelsen M, Schwartz TW, Wheelock CE, Li N, Haeggström JZ (2020) Activation of metabolite receptor GPR91 promotes platelet aggregation and transcellular biosynthesis of leukotriene C4. J Thromb Haemost 18(4):976–984
Tannahill GM, Curtis AM, Adamik J, Palsson-McDermott EM, McGettrick AF, Goel G, Frezza C, Bernard NJ, Kelly B, Foley NH, Zheng L (2013) Succinate is an inflammatory signal that induces IL-1β through HIF-1α. Nature 496(7444):238–242
Toma I, Kang JJ, Sipos A, Vargas S, Bansal E, Hanner F, Meer E, Peti-Peterdi J (2008) Succinate receptor GPR91 provides a direct link between high glucose levels and renin release in murine and rabbit kidney. J Clin Invest 118(7):2526–2534
Trauelsen M, Ulven ER, Hjorth SA, Brvar M, Monaco C, Frimurer TM, Schwartz TW (2017) Receptor structure-based discovery of non-metabolite agonists for the succinate receptor GPR91. Mol Metab 6(12):1585–1596
Tulstrup MV, Christensen EG, Carvalho V, Linninge C, Ahrné S, Højberg O, Licht TR, Bahl MI (2015) Antibiotic treatment affects intestinal permeability and gut microbial composition in Wistar rats dependent on antibiotic class. PLoS One 10(12):e0144854
van Diepen JA, Robben JH, Hooiveld GJ, Carmone C, Alsady M, Boutens L, Bekkenkamp-Grovenstein M, Hijmans A, Engelke UF, Wevers RA, Netea MG (2017) SUCNR1-mediated chemotaxis of macrophages aggravates obesity-induced inflammation and diabetes. Diabetologia 60:1304–1313
Van Vranken JG, Na U, Winge DR, Rutter J (2015) Protein-mediated assembly of succinate dehydrogenase and its cofactors. Crit Rev Biochem Mol Biol 50:168–180
Vargas SL, Toma I, Kang JJ, Meer EJ, Peti-Peterdi J (2009) Activation of the succinate receptor GPR91 in macula densa cells causes renin release. J Am Soc Nephrol 20(5):1002–1011
Wagner GR, Bhatt DP, O'Connell TM, Thompson JW, Dubois LG, Backos DS, Yang H, Mitchell GA, Ilkayeva OR, Stevens RD, Grimsrud PA (2017) A class of reactive acyl-CoA species reveals the non-enzymatic origins of protein acylation. Cell Metab 25(4):823–837
Wallace DC (1999) Mitochondrial diseases in man and mouse. Science 283(5407):1482–1488
Wang M, Lin H (2021) Understanding the function of mammalian sirtuins and protein lysine acylation. Annu Rev Biochem 90:245–285
Wang C, Zhang C, Li X, Shen J, Xu Y, Shi H, Mu X, Pan J, Zhao T, Li M, Geng B (2019a) CPT 1A-mediated succinylation of S100A10 increases human gastric cancer invasion. J Cell Mol Med 23(1):293–305
Wang K, Liao M, Zhou N, Bao L, Ma K, Zheng Z, Wang Y, Liu C, Wang W, Wang J, Liu SJ (2019b) Parabacteroides distasonis alleviates obesity and metabolic dysfunctions via production of succinate and secondary bile acids. Cell Rep 26(1):222–235

Wang W, Zhao F, Ma X, Perry G, Zhu X (2020) Mitochondria dysfunction in the pathogenesis of Alzheimer's disease: recent advances. Mol Neurodegener 15:1–22

Wang C, Wang Y, Shen L (2021) Mitochondrial proteins in heart failure: the role of deacetylation by SIRT3. Pharmacol Res 172:105802

Wang Z, Zhang X, Lu S, Zhang C, Ma Z, Su R, Li Y, Sun T, Li Y, Hong M, Deng X (2023) Pairing of single-cell RNA analysis and T cell antigen receptor profiling indicates breakdown of T cell tolerance checkpoints in atherosclerosis. Nat Cardiovasc Res 2(3):290–306

Watanabe Y, Nagai F, Morotomi M (2012) Characterization of *Phascolarctobacterium succinatutens* sp. nov., an asaccharolytic, succinate-utilizing bacterium isolated from human feces. Appl Environ Microbiol 78(2):511–518

Weber C, Noels H (2011) Atherosclerosis: current pathogenesis and therapeutic options. Nat Med 17(11):1410–1422

Wei YH, Ma X, Zhao JC, Wang XQ, Gao CQ (2023) Succinate metabolism and its regulation of host-microbe interactions. Gut Microbes 15(1):2190300

Weinert BT, Schölz C, Wagner SA, Iesmantavicius V, Su D, Daniel JA, Choudhary C (2013) Lysine succinylation is a frequently occurring modification in prokaryotes and eukaryotes and extensively overlaps with acetylation. Cell Rep 4(4):842–851

Wittenberger T, Schaller HC, Hellebrand S (2001) An expressed sequence tag (EST) data mining strategy succeeding in the discovery of new G-protein coupled receptors. J Mol Biol 307(3):799–813

Wojtovich AP, Smith CO, Haynes CM, Nehrke KW, Brookes PS (2013) Physiological consequences of complex II inhibition for aging, disease, and the mKATP channel. Biochim Biophys Acta 1827:598–611

Wollheim CB, Maechler P (2002) Beta-cell mitochondria and insulin secretion: messenger role of nucleotides and metabolites. Diabetes 51(Suppl. 1):S37–S42

Wong JM, De Souza R, Kendall CW, Emam A, Jenkins DJ (2006) Colonic health: fermentation and short chain fatty acids. J Clin Gastroenterol 40(3):235–243

World Health Organization (2018) Global Health Observatory (GHO) data: obesity. http://www.who.int/gho/ncd/risk_factors/obesity_text/en/. Access 07/ Mar 2023

Xiangyun Y, Xiaomin N, Yunhua X, Ziming L, Yongfeng Y, Zhiwei C, Shun L (2017) Desuccinylation of pyruvate kinase M2 by SIRT5 contributes to antioxidant response and tumor growth. Oncotarget 8(4):6984

Yang Y, Gibson GE (2019) Succinylation links metabolism to protein functions. Neurochem Res 44:2346–2359

Yang L, Ma X, He Y, Yuan C, Chen Q, Li G, Chen X (2017) Sirtuin 5: a review of structure, known inhibitors and clues for developing new inhibitors. Sci China Life Sci 60:249–256

Zhang W, Liu Z, Xu X (2021) Navigating immune cell immunometabolism after liver transplantation. Crit Rev Oncol Hematol 160:103227

Zhao G, Zhen J, Liu X, Guo J, Li D, Xie J, Xie L (2022) Protein post-translational modification by lysine succinylation: biochemistry, biological implications, and therapeutic opportunities. Genes Dis 10(4):1242–1262

Zhu W, Gregory JC, Org E, Buffa JA, Gupta N, Wang Z, Li L, Fu X, Wu Y, Mehrabian M, Sartor RB (2016) Gut microbial metabolite TMAO enhances platelet hyperreactivity and thrombosis risk. Cell 165(1):111–124

Retraction Note to: SDH and Its Impact on Neurodegenerative Diseases: Autophagy in Neuroinflammation Perspective

Sukesh Kumar Gupta, Archit Gupta, Dhruv Kansal, Devendra Kumar, and Vaishali Manikrao Patil

Retraction Note to:
Chapter 8 in: V. M. Patil et al. (eds.), *Deciphering the Role of Succinate Dehydrogenase in Drug Discovery*, https://doi.org/10.1007/978-981-96-6831-1_8

The Publisher has retracted this chapter due to presence of multiple non-standard terms such as "Alzheimer's infection", "mitochondrial harm" and "atomic DNA".

The Publisher is no longer confident in this chapter's contents.

The authors did not state whether they agree to this retraction.

The retracted version of this chapter can be found at https://doi.org/10.1007/978-981-96-6831-1_8

V. M. Patil et al. (eds.), *Deciphering The Role of Succinate Dehydrogenase in Drug Discovery*, https://doi.org/10.1007/978-981-96-6831-1_12

The manufacturer's authorised representative in the EU is Springer Nature Customer Service Centre GmbH, Europaplatz 3, 69115 Heidelberg, Germany. If you have any concerns regarding our products, please contact ProductSafety@springernature.com

Printed and bound by CPI Group (UK) Ltd, Croydon, CR0 4YY

13/10/2026

02232444-0002